THE DRAINING OF THE SOMERSET LEVELS

THE DRAINING OF THE SOMERSET LEVELS

MICHAEL WILLIAMS

Senior Lecturer in Geography, University of Adelaide

CAMBRIDGE

AT THE UNIVERSITY PRESS

1970

CAMBRIDGE UNIVERSITY PRESS
Cambridge, New York, Melbourne, Madrid, Cape Town, Singapore, São Paulo, Delhi

Cambridge University Press
The Edinburgh Building, Cambridge CB2 8RU, UK

Published in the United States of America by Cambridge University Press, New York

www.cambridge.org
Information on this title: www.cambridge.org/9780521106856

First published 1970
This digitally printed version 2009

A catalogue record for this publication is available from the British Library

Library of Congress Catalogue Card Number: 73–75830

ISBN 978-0-521-07486-5 hardback
ISBN 978-0-521-10685-6 paperback

CONTENTS

Contents

PLATES

(Between pages 112 and 113)

MAPS AND DIAGRAMS

Note on the construction of maps

The present-day coastline has been used as a basis of all maps, but it should be borne in mind that the coastline must have undergone changes from both natural and human activity during the last 900 years. In particular, Steart Island at the mouth of the Parrett estuary was not separated from Steart peninsula until 1798.

TABLES

PREFACE

The Somerset Levels are the largest marshland area on the west coast of Britain and lie astride the line of the familiar physical and cultural division between the highland and lowland zones of the country. These 250 square miles of lowland form a physically distinctive region, and steep slopes and rapid changes of relief on its edges leave no doubt about its individuality. The regional character is further enhanced and emphasized by the problem of flooding and its control, which dominates all aspects of life and activity. It is with this problem of flooding and the steps taken to solve it that this study is concerned, for the formation of the present landscape has evolved from the solving of this problem.

It is surprising that this distinctive and interesting region has escaped the attention of geographers and historians for so long. Apart from the palaeo-botanical research of Professor H. Godwin, and the archaeological work of A. Bulleid and H. St. George Gray on the Meare and Glastonbury lake villages, there has been no serious investigation into even a part of the formation, settlement, and draining of the Levels. In this neglect we can see something of the lack of appreciation of the great changes that have occurred in the region, something which is illustrated well in regional studies of the British Isles where the Levels are regarded as a convenient vacuum in which to end the 'Bristol area' and begin the 'West Country'. Consequently it is treated in neither.

The approach and framework to this study are set by the nature of the particular problem and by the character of the landscape investigated. We must reach back many centuries to gain some insight into the significance and distribution of the landscape features of today. Yet this is not a history in the usual sense of the word; many topics such as the history of the Court of Sewers, aspects of trade and finance, and the history of personalities and institutions in the region, are not elaborated or are totally excluded. Rather, this is an historical investigation into matters of geographical interest, and the landscape is the focus of attention. The marked pulsations of draining and reclamation set a ready framework for the study, high points of activity and change being the nebulous period from the early thirteenth century to the early fourteenth century, the sixty years between 1770 to about 1830, and the years since 1939. Between these periods lie the troughs of inactivity, never completely void of new works and changes, to be sure, but of relatively little significance compared with the periods before or after.

I have also tried to put this study into its national setting, for although the Levels lie apart on the western edge of the country, their draining was not accomplished in total isolation from the other marshland areas. In this way I should like to think that this book has some cohesion and balance, that the great mass of local details falls into patterns of significant landscape and economic change, and that the regional story can be viewed within the context of the historical geography of other marshland areas in particular and of the country as a whole.

This book began after a visit to the Levels in 1956. The fascination of that flat landscape and of its untold story proved to be the impetus to the work which was to be completed later as a doctoral dissertation at Swansea University College, University of Wales. The help and encouragement I received in those years came from many, but in particular from Dr Stewart Cousens of the Department of Geography, and Professor John Oliver who was my director of studies and whose incisive pen and rigorous criticism chastened that early work. To them I owe many thanks.

A space of five years in Australia separated the completion of that early work from the opportunity that occurred in 1966 for study at University College, London, in the course of which I was able to revisit the Levels and witness the great changes which had occurred in the brief interval. My thanks, therefore, are due, first to Professor Graham H. Lawton and the University of Adelaide for enabling me to spend this year in Britain; and then to Professor H. C. Darby for inviting me to his department where I was able to share in its stimulating atmosphere of ideas and research. I am grateful for the interest and hospitality with which he and, later, Professor W. R. Mead, and the entire staff welcomed me.

In the course of writing I have received much help from many people. Foremost amongst these is Mr E. L. Kelting, Chief-Engineer to the Somerset River Authority. His active enthusiasm in the subject of the history of the draining of the Levels has culminated in a store of records and plans that must be the envy of any River Authority, and his willingness to put these documents, together with the results of over twenty-five years of his own research work, at my disposal was as generous as it was undoubtedly helpful. I owe special thanks to Mr G. Thomson the Deputy-Engineer who carefully read an earlier draft of this work and to Mr D. R. May for his valuable criticism. The patience of the entire staff of the Authority in discussing the intricacies and significance of land drainage techniques and plans was far greater than any visitor could ever dare hope for.

Mr I. P. Collis and his staff at the Somerset Record Office facilitated my work in every way, but I owe particular thanks to Mr Mirams, whose untimely death has robbed future students of the county of a sympathetic ear and an encouraging hand in the task of their research.

I wish to thank Dr Joan Thirsk, Professor M. M. Beresford, Professor H. Godwin, Professor H. E. Hallam, Dr R. Glasscock, Mr L. Curtis, Mr Neil Stacey, and Mr R. H. Adams, Engineer to the Isle of Wight River Authority, for their criticism and help, and Mr W. A. Cowan, Lecturer in Classics at the University of Adelaide, and formerly Librarian of the Barr Smith Library of the University for his patience and thoroughness in reading the final draft of this book. I also wish to thank Professor H. C. Darby for reading the manuscript and for making valuable suggestions. My debt of gratitude to Mr M. Foale, who drew the maps, is very great.

Portions of maps in this book have already appeared elsewhere, and I should like to thank the following for their permission to reproduce these: Cambridge University Press for material from *The Domesday Geography of South-West England*, ed. H. C. Darby and R. Welldon Finn, for Figure 9; the publishers of the *Climatological Atlas of the British Isles* (1952) for Figure 3; C. H. Dobbie and Partners, for Figure 42B; the editors of *Geography* for Figures 39 and 40; the editor of *Geographical Publications Ltd* for Figure 42A; the editor of the *Journal of Ecology* for Figure 2; and the editor of the *Transactions of the Institute of British Geographers* for Figures 16, 17, 18, 26, 34 and 36.

Finally, I owe particular gratitude to Loré, who showed me the subject, read the results, and encouraged me to continue. I only wish this book was a better token to convey to her my appreciation of her patience and good-cheer throughout these years.

M.W.

ABBREVIATIONS

Acland	*The Farming of Somersetshire:* the combined essays of T. D. Acland and W. Sturge. Reprinted from the *Journal of the Royal Agricultural Society of England*, XI, 1851.
Billingsley	J. Billingsley, *A General View of the Agriculture of the County of Somerset*, 2nd ed. 1798.
Brit. Mus.	British Museum.
B.P.P.	*British Parliamentary Papers.*
Clark	J. A. Clark, 'On the Bridgwater and other Levels of Somersetshire', *Journal of the Bath and West of England Society*, II (1854), 99–128.
C.S.P.D.	*Calendar of State Papers, Domestic.*
E.A.	Enclosure Award: followed by Somerset Record Office numeration and the dates of the Act and of the Award.
G.C.	'The Great Chartulary of Glastonbury Abbey', ed. A. Watkin, 3 vols. *Somerset Record Society*, LIX (1944), LXIII (1948), and LXIV (1949/50).
Locke Survey	'A Manuscript Survey of Somerset' made *c.* 1798 by Richard Locke. S.R.O., *Som. Arch. Soc., Parochial MSS.* 86.
MSS.	Manuscripts.
P.R.O.	Public Record Office.
S.A.N.H.S.	*Proceedings of the Somerset Archaeological and Natural History Society.*
S.R.B.	Somerset River Board (Bridgwater).
,, /A.D.	Axe Documents.
,, /B.D.	Brue Documents.
,, /D.C.	Documents of the Somerset Drainage Commissioners (1881–1931).
,, */P.B.C.*	*Proceedings of the Brue Commissioners.*
,, */P.C.S.*	*Proceedings of the Commissioners of Sewers.*
,, /P.N.	Parrett Navigation Documents.
,, /S.W.	Wrington Sewers Papers.
S.R.O.	Somerset Record Office (Taunton).
S.R.S.	*Publications of the Somerset Record Society.*
V.C.H.	*Victoria County History of Somerset.*
W.M.	Wells Manuscripts: being *The Calendar of the Manuscripts of the Dean and Chapter of Wells.* Vol. I, ed. W. H. B. Bird (1907); vol. II, ed. W. P. Baildon (1914). The two volumes compose vol. XII of the *Historical Manuscripts Commission*; series XII.

I

DRAINING: THE GENERAL SETTING

The draining and reclamation of swamplands and marshes is a peculiarly fascinating topic that has long commanded the attention of geographers and historians alike. Part of this fascination undoubtedly lies in the rapid and revolutionary visual changes that occur as a result of draining; new landscapes are created, sometimes new lands appear, and a new society and economy are established. The whole interest is heightened by the knowledge of human endeavour and ingenuity and the co-operative enterprise of individuals, and by the unending nature of their task, which is displayed dramatically with every flood and tidal inundation.

While such statements give a general picture of the field of interest they do not define the particular points of study for the geographer. For him, draining can be viewed as one of the major 'resource-converting' techniques or processes whereby man changes the face of the earth, such as clearing the woodland, reclaiming the heaths, irrigating the desert or building towns. These have been termed 'vertical-themes' in historical geography,[1] and in the investigation of these themes in relation to the total geography of an area lies the key to much of what we see around us. But, more than that, draining is a particularly rewarding topic for the geographer because of the very close interplay which exists between the physical elements of the environment and man's efforts to control them, something which is emphasized, perhaps, by the essential need for the constant maintenance and upkeep of drainage works. No part of the earth's surface reverts more quickly to its natural state through lack of careful maintenance and attention than does a swamp, unless it be an irrigated desert area, or a clearing in an equatorial rain forest.

Draining is really no more than a conscious alteration and control of the natural hydrographical features of the landscape. Swamps

[1] H. C. Darby, 'On the relations of geography and history', *Trans. Instit. of British Geog.* XIX (1953), 8. Most of these themes and many others are dealt with in *Man's Role in Changing the Face of the Earth*, ed. William L. Thomas, Jr., with the collaboration of Carl O. Sauer, Marston Bates and Lewis Mumford (1956), yet it is surprising that draining is not dealt with as such and receives only incidental attention.

become dry land, water-tables are altered, old drainage channels are regraded and reorientated, and new ones dug; along the coast, tidal deposition is speeded up by building embankments and groynes. The effects of draining are more far-reaching than this, however, since the biotic features of the area change. One major vegetative cover is replaced by another, and the soil structure is disturbed and eventually altered, while land surface elevations are permanently altered with the irreversible shrinkage and wastage of organic soils. Indeed, at one end of the scale, because of its emphasis on the creation of new hydrological features, and the alteration and abandonment of old ones, together with the effort to control the natural processes of erosion and deposition in the river channels and along the coast, draining almost leaves the realms of human history and becomes a part of physical landscape interpretation. Concomitant with these changes in the physical and biotic landscape come the production of new crops, the founding of new settlements, both urban and rural, and the creation of new patterns of communication. In brief, there are new landscapes and new geographies.

This is a study of one such area, where the process of controlling the problem of flooding did change the landscape and create a new geography. Like other local studies, it seeks to achieve nothing more than thoroughness by investigating carefully the attempts to drain this particular region throughout the time of its human occupancy. But, in concentrating upon the particular example, one should not lose sight of the wider context of the study. While the Somerset Levels were being drained and reclaimed, so were vast stretches of wet lands throughout western and northern Europe, from the coasts of northern France, Belgium, Holland, Germany and Denmark to the interior swamps of the Urstromtäler in Poland and the Pripet marshes of Russia. Yet none of these efforts to drain the land was more spectacular than the activity of the Dutch in reclaiming Holland, their epic story being succinctly summarized in the saying that 'God made the world, but the Dutch made Holland'. The draining of the Somerset Levels, then, was a part of the fundamental process that occurred throughout the medieval and later centuries in Europe of reclaiming the waste.

To press analogies between these areas of draining without more knowledge of events would be wrong, for each was peculiar to itself and an outcome of the efforts of different peoples, working at different times and levels of technology, under different institutions. But if the draining of the Somerset Levels was another manifestation of a continental occurrence, it was more narrowly one example of seven or eight British regions so reclaimed. Along the eastern and southern coasts of Britain are the Fens, the largest of the British marshland areas, the Isle of Axholme, the Hull valley, the Vale of York and the Humberhead Levels, the Norfolk Broads, the Essex marshes, and Romney Marsh and the Pevensey Level in Kent. On the west coast of Britain, in the highland zone, there are only the Lancashire Mosses, and the Severn lowlands and the Vale of Berkeley, which merge into the Somerset Levels.[1] It is amongst these regions that there lies scope for fruitful comparative studies and some generalizations.

That all are basically flat expanses of land is obvious, although it is none the less important for that. This fact, coupled with their propensity to flooding and the consequent need to take remedial steps, gives these marsh areas a distinctive individuality that is reflected in the descriptive regional names of the Fens, the Carrs, the Levels, the Broads and the Mosses. Despite this general similarity, however, there are some differences in the physical setting and in the flood problems of the marsh regions. In the Fens, the Hull valley and the Levels, the presence of a coastal clay belt and of peat at a lower level behind gives rise to special drainage problems not encountered elsewhere. Even within these three marsh regions there are differences with respect to the rate and extent of peat shrinkage; it has been no real problem in the Levels but it has been serious in the east-coast

[1] It would be impossible to cite the almost endless references that exist about the draining of these areas, but the following are perhaps the most recent, accessible, and comprehensive accounts of landscape evolution and change in these areas: H. C. Darby, *The Medieval Fenland* (1940), and *The Draining of the Fens* (1940); H. E. Hallam, *Settlement and Society: A Study of the Early Agrarian History of South Lincolnshire* (1965); J. Thirsk, 'The Isle of Axholme before Vermuyden', *Agric. Hist. Rev.* I (1953), 16, and *English Peasant Farming* (1957); J. A. Sheppard, *The Draining of the Hull Valley*, and *The Draining of the Marshlands of South Holderness and the Vale of York*, East York. Local Hist. Soc. Publications, No. 8 (1958) and No. 20 (1966) respectively; J. M. Lambert, J. N. Jennings, C. T. Smith, Charles Green and J. N. Hutchings, *The Making of the Broads*, Roy. Geog. Soc. Research Series, No. 3 (1960); D. W. Gramolt, 'The Coastal Marshland of East Essex between the seventeenth and mid-nineteenth centuries', unpublished M.A. thesis, University of London (1961); R. A. L. Smith, 'Marsh embankment and sea defences in medieval Kent', *Econ. Hist. Rev.* x (1940), 29–37; D. B. Hardman, 'The Reclamation and Agricultural Development of North Cheshire and South Lancashire Mossland Areas', unpublished M.A. thesis, University of Manchester (1961); and W. Rollinson, 'Schemes for the reclamation of land from the sea in North Lancashire during the eighteenth century', *Trans. Hist. Soc. Lancs. and Cheshire*, CXV (1964), 107–47.

lowlands. Other contrasts in the physical setting of the marsh areas arise in connexion with their principal source of flooding; in the Essex marshes, Romney Marsh, and the Pevensey Level the danger is mainly tidal; in the Lancashire Mosses, the Vale of Pickering and the Isle of Axholme it is mainly from land floods; while in the Fens and Levels, in particular, there is a combination of both.

The settlement and the draining of these marsh regions also provide some common themes. The inaccessibility of some of them proved to be the very reason for their utilization because religious houses often sought defensive and secluded 'island' sites. The medieval houses in the Levels, the Fens, and the Hull valley grew into some of the largest and wealthiest properties in England, and were in the forefront of draining activity. That they also kept good estate records, which have been preserved, is a matter of no mean significance to our ability to perceive the distinctiveness of these regions from the twelfth to the fourteenth century. During the movement towards agricultural improvement in the seventeenth century, the marshlands were obvious places for new colonization, and this, coupled with the financial problems of the Stuart monarchy, produced an interplay of personalities and projects between the regions that had not occurred before. But it is curious that so little was done in the Somerset Levels at this time. It fell behind the other marshland areas in its improvement, a state which was clearly reflected in the land itself by the absence of windmills, which became such a distinctive feature of the lowland landscapes of eastern England. However, the agricultural and industrial revolutions, and the increased demand for food in the late eighteenth and early nineteenth centuries, brought about a concerted effort in all regions to improve the adverse conditions of the environment, an effort which was put on a new footing when wind and wood gave way to steam and steel, and which enabled the lowlands to be effectively pumped dry. The distinctive differences in land utilization which were occurring in areas of apparently similar type were revealed for the first time with the compilation of accurate agricultural statistics in the mid nineteenth century, e.g. the difference between the predominantly arable Fens and the predominantly pastoral Levels.

These are but a few of the themes of undoubted importance in the changing geography of the marshlands of Britain that have not been fully explored by geographers and which would repay closer study.

But while these broader implications form a framework into which the individual study can be fitted, it is not the purpose of this work to explore them in detail. Such general comparisons and contrasts can be based only on the local study, and where they are relevant to this work attention is drawn to them. Each reclamation story is distinctive, and is, as Professor H. C. Darby says, 'an epic in itself'.[1] It is essential that what happened in the draining of a particular area, at a particular time, be examined first.

[1] H. C. Darby, 'The changing English landscape', *Geog. Journ.* CXVII (1951), 381.

2

DRAINING: THE SETTING OF THE SOMERSET LEVELS

THE CAUSES OF FLOODING

Of the causes which give rise to flooding in the Levels perhaps the most important are the physique of the region, tidal behaviour and marine siltation along the coast, and the rainfall over the area. Although it is the interplay of these three factors that causes the worst floods to occur, they are best considered separately.

Physique

The main body of the Somerset Levels is flanked on the north by the Mendip Hills, and on the south-west by the Quantock Hills. To the south and south-east lie the Blackdown Hills and the Oolitic escarpment (Fig. 1). In the past, erosion of the Triassic and Jurassic rocks in the basin between these surrounding uplands produced an uneven surface. The remaining higher parts of the original surface-cover stood above the general level of the basin to become the prominent ridges of Curry Rivel, Stathe and the Polden Hills, and the 'islands' of Brent, Glastonbury, Puriton, Meare, Sowy, Burtle, and Wedmore, as well as others. To the north of the Mendip Hills, and skirting the coast, lies the smaller and in many ways different basin of the Northern Levels.[1]

Within these basins post-glacial deposition has given the Levels their present aspect. There have been three kinds of deposition. First, low moor (Fen) peats and raised moor or bog peats cover the greater part of the lowland. Secondly, the peat is overlaid on the landward side by freshwater alluvial deposits which occur alongside the river courses and other areas of habitual flooding. Thirdly, it is overlaid on the seaward side by a belt of marine clay (Figs 1 and 2).[2]

[1] For long-standing accounts of the geology of the area, see W. A. E. Ussher, *The Geology of the Quantock Hills and of Taunton and Bridgwater*, Memoirs of the Geological Survey (1908), and H. B. Woodward, *The Geology of the Somerset and Bristol Coal-Fields*, Memoirs of the Geological Survey (1876).

[2] For details of the origin and distribution of the peat and clay soils, see H. Godwin,

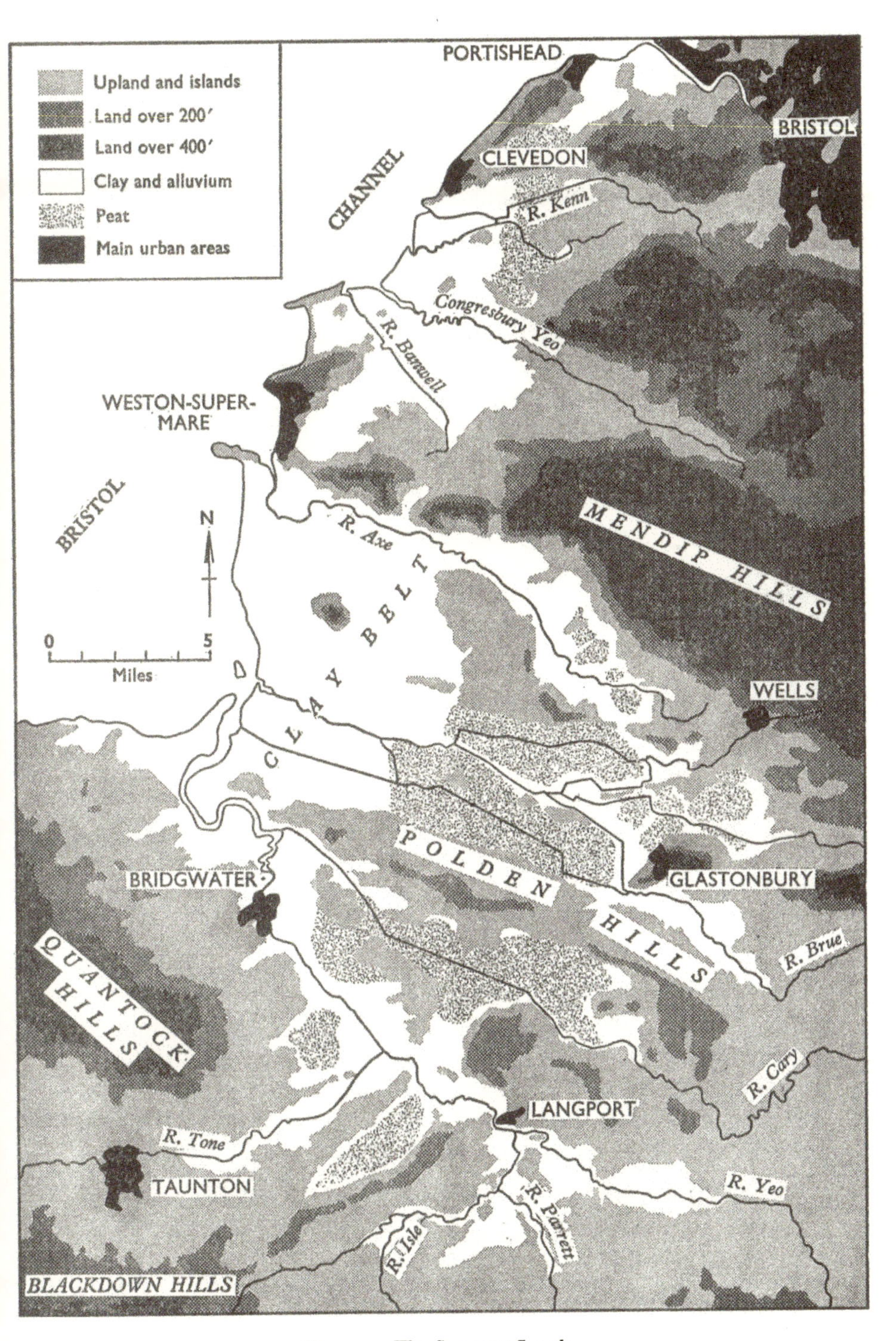

Figure 1. The Somerset Levels.

The peats are of varying thickness and composition, and range from less than a foot in parts of the Axe valley to 10 ft in King's Sedgemoor and up to 24 ft in the Brue valley. There are no statistics available with which to measure the shrinkage and wastage of the peat surface and thus its possibly increased susceptibility to flooding; but, with the exception of one region, all evidence points to the conclusion that shrinkage has been minimal. The frequent admixture of alluvium in the low moor peats has given the soil a stability against wastage by erosion or shrinkage. The one exception is in parts of King's Sedgemoor where better drainage and some dry summers have resulted in a little shrinkage in recent years.[1]

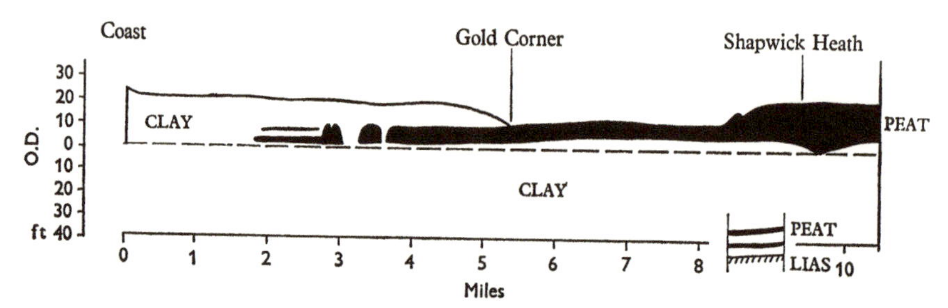

Figure 2. Schematic profile of the Levels from the coast near Huntspill, inland to the region of the relic raised bogs near Meare. Thick ombrogenous peat extends seawards below the coastal clay belt for about two miles. (Source: after H. Godwin, *Journ. of Ecology*, XXXI (1943).)

The low-lying tracts of peat and freshwater alluvium, or moors as they are known locally, penetrate inland alongside the courses of the rivers Tone, Parrett, Isle, Yeo, Cary, Brue, Axe, Kenn, Banwell and Congresbury Yeo. The general elevation of these moors is about 10–12 ft O.D. The rivers which flow across the moors have a very slight fall (e.g. the Parrett falls one foot per mile for 11½ miles between Langport and Bridgwater, the Brue falls 8·4 in. per mile for 8 miles above Highbridge Clyse, and the King's Sedgemoor Drain falls 6·5 in. per mile for 8 miles above Dunball Clyse) and in places the level of the river beds is nearly equal to that of the surrounding land. Repeated overflowing of the silt-laden water of the rivers has

'Studies in the post-glacial history of British vegetation: VI, Correlations in the Somerset Levels', *New Phytologist*, XL (1941), 108; B. W. Avery, *The Soils of the Glastonbury District of Somerset*, Memoirs of the Soil Survey of Great Britain (1955); and D. C. Findlay, *The Soils of the Mendip District of Somerset*, Memoirs of the Soil Survey of Great Britain (1965).

[1] Avery, *Soils of the Glastonbury District*, 63. See also pp. 255–6 below, where the shrinkage of the peat is discussed again.

resulted in the formation of natural levees which contain the water of the rivers during periods of natural flow but which are overtopped during periods of abnormal flow.

In contrast with the low ground of the inland moors there is the belt of marine clay which extends along the coast of the Levels; it has a general elevation of between 18 and 20 ft O.D. This clay belt remains free from flooding but acts as a barrier to the seaward flow of landward flood water, and thus causes flooding in the peat moors.[1]

Tides and siltation

The immunity of the coastal clay belt from landward floods is balanced, however, by the danger of flooding from the sea. The behaviour of the tides in the Bristol Channel is shown in Table 1.

Table 1. *Tidal levels along the Somerset coast (1959)*
(in feet, O.D. Newlyn)

	Bridgwater	Burnham	Weston-super-Mare	Avonmouth
Highest recorded level	25·0	26·0	?	29·0
Mean high water springs	21·0	20·3	19·7	22·1
Mean high water neaps	11·0	11·0	9·9	11·3
Mean low water springs	−6·0	−17·3	−17·4	−18·2
Mean low water neaps	−6·0	−9·6	−10·5	−9·5

Based upon information in the *Admiralty Tidal Handbook (European Waters)*, (1959 ed.), 214, and from tidal records at the Somerset River Board.

It can be seen that the mean high water spring tide is either equal to, or just above, the height of the coastal clay belt at all points along the coast, almost 10 ft above the present-day general level of the inland peat moors, and possibly 15 ft above the very lowest moors in the Brue valley. Adverse wind and pressure conditions in the Bristol Channel will quickly raise these tides to 6 ft or more above their predicted level; under these conditions the inundation of the coastal lands is prevented only by a complex series of sea walls. Yet the danger of flooding is limited because high water does not remain for long, except under the most adverse conditions.

A greater and more regular threat of flooding lies in the penetration

[1] H. Godwin, 'The botanical and geological history of the Somerset Levels', *Proc. of the British Association for the Advancement of Science*, XII, No. 47 (1955), 319. This is a concise summary of the complex history of the post-glacial phases and changes in the Levels.

of tide into the river estuaries, blocking the river outfalls and causing the accumulation and overflow of water behind the coastal clay belt. The inland penetration of tidal water can be prevented only by the erection of a tidal sluice, or clyse, as it is known locally. One was constructed on the Brue at Highbridge some time before 1485, and others on the King's Sedgemoor Drain in 1791, on the Axe at Batch in 1806, and on the Huntspill river in 1944. However, the prevention of tide-initiated floods by tidal doors may result in landward floods caused by the blocking of the outfall by the doors themselves. If the cessation of river discharge (tide-lock) is approximately 4½ hours at each tide, then the channel must have a storage capacity sufficient to accommodate 4·5 × the flow in cubic feet per second (cusec) × 3,600, that is, 16,200 times the normal river flow per second. Clearly these requirements cannot be met if the river is at a high level to begin with, and inevitably flooding will soon result.

Of all the Levels' rivers, only the Parrett and its tributaries are open to the influx of the tide. Mean high spring tides can rise to Hook Bridge on the Tone (16 miles inland), and to Oath Lock on the Parrett itself, even overtopping Langport Lock (19½ miles inland) on an exceptionally high spring tide. A river with an open outfall cannot continue discharging against a rising tide, and the disadvantage of an open channel to the Parrett is therefore obvious, particularly as its natural storage capacity is severely restricted by 'freshes', which continually fill its channel.

Associated with tidal conditions is the accumulation of silt at the river outfalls; it reduces the effectiveness and capability of the major channels to hold and ultimately to pass off freshwater floods, and it restricts the working of the tidal doors.

The gradually tapering waterway of the Severn Estuary causes the tidal wave to move up it with increasing momentum, culminating in the famous 'bore' of the Severn.[1] The turbulent currents associated with this movement keep a great amount of silt in suspension. Similar conditions exist in the Parrett estuary where a fairly moderate silt content of 1,450 grains of silt per cubic foot sample of water (high spring tide) in the lower estuary near Steart Island increased to 41,050 grains between Combwich and Dunball, and then fell to 26,020 grains near Bridgwater. The magnitude of deposition that

[1] For discussion on the tidal behaviour of the Severn, see R. Bassindale, 'Studies on the biology of the Bristol Channel, XI', *Journ. of Ecology*, XXXI (1943), 21.

occurs as a result of these conditions can be gauged from the fact that in one summer 7 ft of silt have been observed to accumulate on the seaward side of the Bleadon Clyse on the Axe, 8 ft at Highbridge Clyse and nearly 14 ft at the Dunball Clyse sill. These instances were not considered particularly abnormal.[1]

Fortunately, over a long period the rivers of the Levels would appear to achieve a general balance between tidal siltation and freshwater scouring, but this natural equilibrium is sometimes upset in favour of deposition during dry summers when the freshwater flow is low and the scouring power of the river is negligible.[2] Then a situation is created which worsens the flood problem in the subsequent autumn and winter.

Rainfall

Another basic cause of flooding is the relatively high rainfall on the upland periphery of the Levels, the average annual rainfall over Exmoor and the Mendip and Blackdown Hills being over 40 in. (Fig. 3). During the autumn months of September and October there is fairly consistent rainfall of moderate duration and intensity, which is associated with the prevailing south-westerly winds crossing the high ground of western and southern Somerset, and with frontal activity. This rainfall most readily affects the catchment areas to the west and south of the Levels. These weather conditions normally produce no flooding because the saturation potential of the river basins is sufficient to prevent too rapid a run-off, but towards the end of October the catchments are becoming increasingly waterlogged, and only moderate falls are needed in November and December to cause a flood.

The heaviness of rainfall in the upper catchments of the southern and western rivers of the Levels can be gauged from the following table which shows the frequency of monthly falls over 6 in. for a 35-year period from 1924 to 1958, for a random selection of stations, Crewkerne, Glanvilles-Wootton, and Wiveliscombe, in the upper

[1] W. Lunn, *Report on the River Parrett Floods Prevention* (1898), paras. 30–1. Up to 12 ft of silt was deposited annually on the slime-batches near Bridgwater (see p. 216 below), and 2 ft of mud had been observed to accumulate near Burrow Bridge in one week of a neap tide.

[2] This was amply demonstrated in A. H. Gibson, *Severn Barrage Committee Report* (1933); also in C. G. Du Cane, *River Parrett Estuary Scheme* (1941) which was a report on the working of the Parrett estuary tidal model. An earlier view, opposite to the above, is expressed in W. G. Sollas, 'The estuaries of the Severn and its tributaries', *Quart. Journ. Geol. Soc.* XXXIX (1883), 611–25.

Parrett and Tone catchments. These stations are compared with three stations selected from the Levels and its borders.

Although the distribution of rainfall in any one month will affect its ability to produce a flood, monthly totals of over 6 in., and definitely totals of over 7 in., are wet by any standards. In most west

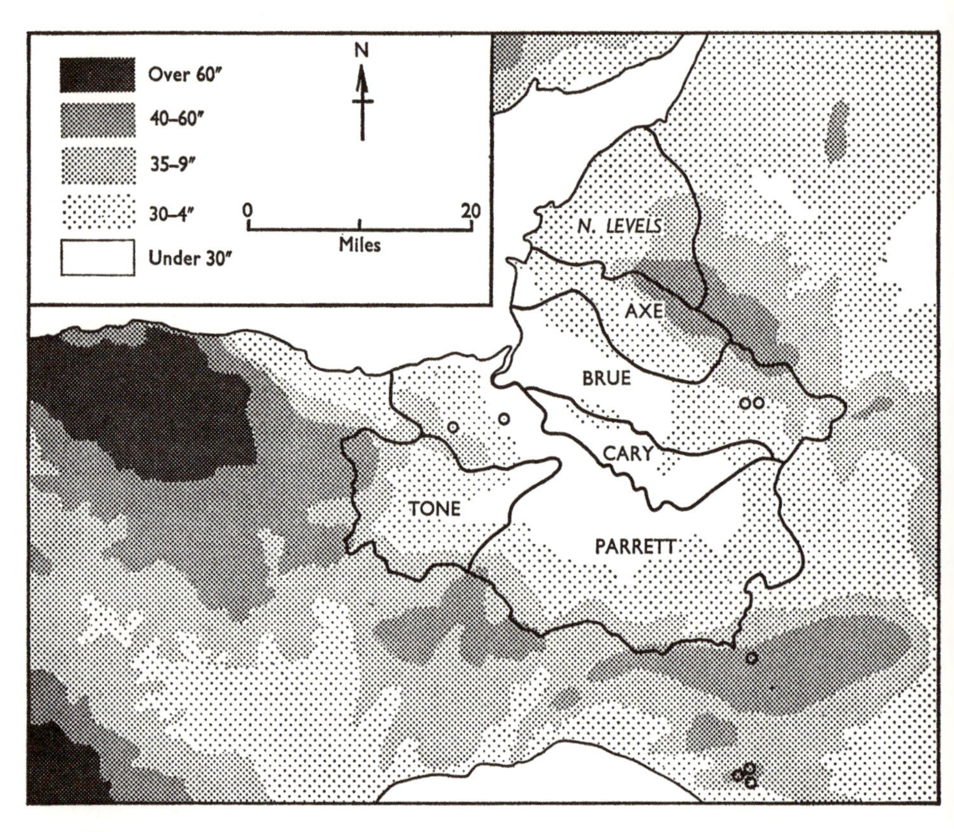

Figure 3. Average annual rainfall and river catchment areas in the Levels. (Source: *Climat. Atlas of the British Isles* (1952).) The circular symbols are located over eight of the nine rainfall stations which have had the highest recorded falls in twenty-four hours in the British Isles.

coast situations in the British Isles such falls are no problem for the natural drainage system, but the relief and natural drainage system of the Levels are ill-adapted to the rapid evacuation of falls of this magnitude.

Whilst the average falls of rain produce a continual risk of flooding, it is the falls of more than average intensity and duration which have

Table II. *Frequency of monthly falls in six stations in Somerset*

	Falls of					Over 35-yr period	
	6–7 in.	7–8 in.	8–9 in.	9–10 in.	over 10 in.	Falls over 6 in. as % of total no. of falls	Falls over 7 in. as % of total no. of falls
Crewkerne	17	8	4	1	3	7·8	3·8
Glanvilles-Wootton	14	7	3	2	4	7·1	3·8
Wiveliscombe	19	12	6	4	3	10·0	6·4
Cheddar	11	10	1	1	—	5·4	2·8
Street	6	1	—	—	—	4·0	0·16
Somerton	6	—	—	—	—	4·0	—

resulted in serious flood conditions in recent years. For example, in October 1960, there was an average fall of 9·72 in., or 24% above the normal for the month, over the whole of the Levels' river catchments, followed in November by a fall of 4·23 in. over the same area. In the Tone catchment alone the average rainfall in October was 11·33 in., which resulted in eighty-six nearly consecutive flood days through October to January; it was perhaps the wettest year in this century, and it produced some of the worst floods. Other comparable falls were 9·35 in. over the catchment of the Parrett and its tributaries in November 1951, which gave nearly a month of flooding; in December 1934, 7·16 in. fell, giving extensive flooding; and in 1929, 9·95 in. fell in November, followed by 7·94 in. in December, causing one of the highest floods ever known in the Parrett area. The only other reasonably well-documented extensive flood was in the winter of 1872–3, when the average rainfall over the Levels from October 1872 to January 1873 was 24·74 in. compared with an average of 16·71 in. for the same period for the preceding three years. Over 107 square miles of the Levels were under water from October to March, but this was obviously not unusual, since such intense rainfall had been noted before, and, as was said at the time, 'the floods though excessive were not altogether exceptional'.[1] Thus the probability of such falls seems fairly high.

Finally, while not attempting to argue too strong a case for exceptional conditions, the south-west peninsula of England in general,

[1] R. B. Grantham, *Report on the Floods in Somersetshire in 1872–73* (1873), 16.

and south Somerset and north Dorset in particular, appear to be areas of exceptional rainfall. Of the nine highest recorded falls in twenty-four hours in the British Isles, from 1865 to 1956, four have been in Somerset, and all nine have been on the southern and western edges of the Levels' catchment area (Fig. 3). They are as follows:[1]

Table III. *Highest recorded falls in twenty-four hours in the British Isles*

County	Station	Amount (in.)	Date
Dorset	Martinstown	11·00	18 July 1955
Somerset	Bruton	9·56	28 June 1917
Dorset	Upwey (1)	9·50	18 July 1955
Somerset	Cannington	9·40	18 August 1924
Dorset	Upwey (2)	9·00	18 July 1955
Devon	Langstone Barrow	9·00	9 August 1952
Somerset	Bruton	8·48	28 June 1917
Somerset	Aisholt	8·39	28 June 1917
Dorset	Upwey (3)	8·31	18 July 1955

None of these falls, with the exception of those at Cannington and Bruton, produced a sizable flood or damage in the Levels, but the potential danger they represented if they had been centred over the Levels is obvious.

THE OCCURRENCE AND EXTENT OF FLOODING

The combined effect of the flat relief, the fairly heavy rainfall and the high tides is to produce a flood problem in the Levels. Approximately 800 upland square miles[2] drain into 233 lowland square miles through rivers which have scarcely any fall, and whose outfalls may be blocked by high tides. But the flood problem varies both in its timing and in its extent; these two features of flooding are of the utmost significance to an appreciation of the utilization of the moors and the attempts to drain them.

The pattern of flood occurrence is indicated by Table IV which shows the days on which the rivers Brue, Parrett and Tone reached 'flood-stage', that is to say, when flooding of the banks was imminent or occurring, during the period April 1952 to March 1965. 'Flood-stage'

[1] *British Rainfall, 1956* (1957), 39.

[2] It is to be understood throughout this work that all land above the height of any known flood is termed 'upland'. Thus, both land on the edge of Meare Island which is 25 ft O.D. and land over 1,200 ft in the Quantock Hills are designated upland for the purposes of this study.

is an arbitrarily defined level which varies from river to river but it is between 18 in. and 2 ft below the level at which leakage through the banks becomes serious and at which patrolling of the banks becomes necessary. The attainment of 'flood-stage' for any length of time usually signifies overflowing of the banks; indeed, flood water often remains on the moors for up to ten days after the river has ceased to overflow and be recorded as at 'flood-stage'. Therefore, Table IV tends to underestimate rather than overestimate modern flood conditions.

Table IV. *The number of days on which the rivers Brue, Parrett and Tone reached 'flood-stage' from April 1952 to March 1965*

	J	F	M	A	M	J	J	A	S	O	N	D	Total
Brue	49	21	10	9	1	5	—	6	2	20	38	60	221
Parrett	58	48	24	8	2	—	—	—	—	24	53	77	294
Tone	86	55	30	9	1	—	—	—	—	28	55	95	359
Total	193	124	64	26	4	5	—	6	2	72	146	232	874

The most significant fact that emerges from this table is the concentration of flood days from November to February, and then their almost total absence from March to October. Analysis shows that 43·2 % of all such days occurred in November and December, 36·2 % in January and February, and of the remaining 20·6 % most occurred in March; however, the 1960–1 floods were unusual in that they augmented considerably the October proportion, adding sixty-two days to bring the total to seventy-two for a usually flood-free month. The few flood days in the spring months (10·7 %) are of greater consequence than at first appears because of the effect of floods in inhibiting the growth of young grass; at this time of the year 2 in. of flood water can have as devastating an effect as 2 ft. The summer months had only thirteen flood days, all of which occurred during thunderstorms in the Brue catchment.

Table IV does not include all known flood occurrences from 1952 to 1965. There were other times when rainfall in the upper part of the main river catchments produced levels which did not reach 'flood-stage', and were therefore not recorded, but which were still high enough to stop the gravitational discharge from adjacent moorland catchment areas where only moderate falls had occurred. This situation arises in the moorland basins alongside the rivers Parrett and Tone, where each basin has its own distinct catchment area and is ultimately dependent on the Parrett for the evacuation of its surplus water. The same thing happens in the moors to the north and south of the River Brue. It is not likely that this flooding would destroy the pattern in Table IV; rather it would intensify it because such floods occur most commonly just before or after 'flood-stage'.

Of the total number of days on which rivers reached 'flood-stage', 41·0 % were for the Tone, 33·8 % were for the Parrett, and 25·2 % were for the Brue, which reinforces the impression that there is a decreasing propensity for rivers to overflow as one moves north in Somerset. There is no information on the flood behaviour of the other rivers of the Levels; nevertheless, it appears that the River Cary has a regime that is intermediate between those of the Brue and Parrett rivers. The flow of the Axe is complicated because much of the catchment area is on the Carboniferous Limestone of the Mendip Hills from which the Axe and its main tributary the Cheddar Yeo emerge from underground caverns as already large streams. The outflow is so great as to lead the authorities to believe that a large volume of the water originates from an area outside that which has normally been considered its surface catchment area. The seasonal flow of the Axe is roughly in accord with that of the other rivers of the Levels, with peaks occurring from November to January, and with associated flooding in the lowest sections of the valley, but there is less variation in the flow, resulting from the uniform percolation of rain through the Carboniferous Limestone.

It is probable that these features of flood occurrence are indicative of conditions in the past, particularly the definite and regular onset of flooding in November. Recent drainage works have been directed mainly at the elimination of the damaging intermittent floods of late February, March and the early spring months; these may now be less frequent than they were in the past.

The extent of flooding varies both seasonally and annually, the variation depending upon combinations of rainfall, tides, and wind and pressure conditions, and the effectiveness of the rain in producing a flood depending upon evaporation and the absorptive qualities of the soil. Some years are dry and little flooding occurs; other years are extremely wet. Because of these variations it is possible to show only either an average flood situation or a situation in a particularly bad year when flooding was extensive and therefore mapped. Figures

18, 37A and 42A show the extent of flooding in 1794, 1853 and 1936 respectively and they depict what was regarded as an almost annual situation. Figures 37B and 42B on the other hand, show the extent of two particularly bad floods during the winter of 1872–3, and in October and November of 1960, respectively. It is noticeable that flooding is concentrated in the low peat moors. In this connexion, compare Figures 18, 37 and 42 with Figure 1.

THE PRE-DRAINAGE LEVELS

From the evidence available it seems certain that from the prehistoric centuries of the Iron Age until the early thirteenth century, the Levels were regarded as an unfavourable area of marsh and swamp that was to be avoided and therefore no draining took place.

The first well-attested occupancy of the Levels came with the Iron Age settlements of the Meare and Glastonbury villages which were situated on the edge of a vast lake which must have covered the area of Meare Pool and the lowland to the east, up to Glastonbury. But these lake villages represented a remarkable adaptation of life to an adverse environment and not an attempt to change the environment by draining. Each village consisted of about sixty to seventy huts, built on earth and timber mounds in the centre of a lake that must have covered the greater part of the eastern end of the Brue valley. Archaeological evidence indicates that the inhabitants relied on the abundant fish and fowl of the lake for their food and did not cultivate land. Theirs was a semi-aquatic existence, which did not require the reclamation and settlement of the marshes.[1]

With the deterioration of climatic conditions during the change from the Sub-Boreal to Sub-Atlantic periods, rainfall increased and the raised bogs of the western end of the Brue valley, hitherto fairly dry and traversable, were flooded by calcareous waters, and rose and became covered with *Cladium* sedge fen. This alteration in local land levels, and the increased wetness, caused the level of the lake to rise. The changes were reflected in the abandonment of the lake villages and in the construction of a complex pattern of wooden trackways (often of timber taken from the lake villages) over the increasingly unstable and rising peat bogs of the Brue valley to the surrounding upland. Ultimately, these were of no avail in the face of worsening

[1] A. Bulleid and H. St. George Gray, *The Glastonbury Lake Village* (1911), and *The Meare Lake Village*, I (1948) and II (1953).

conditions and both lake villages and trackways were submerged and covered by about 2 ft of the *Cladium* sedge fen.[1] The growth of the peat bogs continued until arrested by severe flooding of brackish and base-rich water, and also by the great marine transgression of about A.D. 250, which deposited the coastal clay belt over the western margins of the peat and filled the Axe and Parrett valleys.[2]

Because of these adverse conditions and changing sea levels, Roman settlement avoided the Levels and used existing early British occupation sites on the surrounding upland.[3] Settlement was scattered, but with definite concentrations near Somerton and Bath, where it was associated with the route of Fosse Way which ran along the eastern upland edge of the Levels.[4] It is possible that two other routes stretched westwards from Fosse Way towards the Levels, one across the Mendip Hills and the other along the Polden Hills; the former was in use in A.D. 44 when lead was being worked in the Mendip Hills, but the evidence for the Polden Hills road is less certain. It is possible that a third road extended into Wedmore 'island'.[5] But, in all cases the moors of the Levels were avoided and only the flood-free uplands were utilized.

Because of the western situation of Somerset in the British Isles, the activity of the early Celtic Christian church was an important phase in the cultural evolution of the region, and one of the utmost significance for the future. In their search for solitude and tranquillity, the early monks found the 'islands' of the Levels, with their surrounding stretches of impenetrable marsh, ideal situations for monasteries and a contemplative life. A shroud of uncertainty and

[1] H. Godwin and A. R. Clapham, 'Studies of the post-glacial history of British vegetation: IX, Prehistoric trackways in the Somerset Levels', *Phil. Trans. Roy. Soc. London*, B, CCXXXIII (1948), 249. See also Part X, 'Correlations between climate, forest composition, prehistoric agriculture and peat stratigraphy in the Sub-Boreal and Sub-Atlantic peats of the Somerset Levels', *Ibid.* 275.

[2] See pages 6 and 7 above.

[3] There is evidence of some exceptions; for example, the Roman villa at Wemberham; the buildings in Stoke Moor in the Axe valley, which air photographs would suggest are far larger and more complex than is commonly supposed; and the discovery of Roman artifacts 12 ft below the coastal clay belt at Bason Bridge, and 7 ft below at Highbridge. See W. Phelps, 'On the formation of peat bogs and turbaries...', *S.A.N.H.S.* IV (1853), 103, where these finds are noted, and H. Godwin, 'The coastal peat beds of the British Isles and the North Sea', *Journ. Ecology*, XXXI (1943), 199, where the relationship of these finds with the clay is discussed.

[4] See H. J. Haverfield's chapter on Romano-British Somerset in *V.C.H.* I (1906), 267, and especially the map of distribution of settlements and finds between pp. 206 and 207.

[5] I. D. Margary, *Roman Roads in Britain: South of Fosse Way—Bristol Channel* (1955), I, 95, 114, 115.

myth obscures the details of the origin of Glastonbury, but it seems certain that it was established by the beginning of the fourth century, if not earlier, and it had gathered together quite a colony of holy men by the following century. It was restored and rebuilt with a more permanent and lasting fabric between 705 and 709.[1] With its claim to be the first Christian church in Britain, it fostered a tradition of monastic and ecclesiastical life in the region that was soon an example to others. Thus in 878, Athelney Abbey was founded 'in the fastness of the fens'. William of Malmesbury described it as 'not an island of the sea', but 'so inaccessible on account of bogs and inundation of lakes that it cannot be approached but by a boat';[2] little wonder that Alfred had used the 'island' as a refuge against the Danes. Muchelney Abbey, on Muchelney 'island', was founded in 933, but early charters show that there had been a small monastic settlement in existence from about the middle of the eighth century.[3] Smaller cells were established later on Burtle 'island' some time before 1199,[4] and on Aller 'island' in 1328.[5] Contemporary with this monastic development was the growth throughout the eighth and ninth centuries of the great See of Bath and Wells, the large estates of which bordered upon the Levels.[6] The Bishops of Winchester and Coutance also had holdings in the Northern Levels.

With the pious munificence of kings and barons the estates of Glastonbury, Athelney, Muchelney and Wells, in particular, grew until nearly all the Levels and much of the surrounding uplands were in their possession (Fig. 4).[7] As the sole owners of these vast and wealthy estates, the ecclesiastical houses were in the forefront of all drainage activity that occurred during subsequent centuries,

[1] *Willelimi Malmesbiriensis Monachi: De Gestis Regum Anglorum*, ed W. Stubbs, Rolls Series No. 90 (1887), I, 36–7. The best modern account of Glastonbury's history is T. Scott-Holmes, 'Glastonbury Abbey' in *V.C.H.* II (1911), 81.

[2] Based on Asser's contemporary account in his *De Aelfredi Rebus Gestis*, Sect. 53, as printed in H. Petrie and J. Sharpe, *Monumenta Historica Britannica* (1848), I, 493. It reads 'in loco qui Aethelingaig quod per maxima gronnia paludisissime et intransmeabilia et aquis indique circumcingitur...'. For similar descriptions in other early chronicles, see J. W. E. Conybeare, *Alfred in the Chroniclers* (1900).

[3] W. Dugdale, *Monasticon Anglicanum* (1817–30 ed.), II, 335.

[4] G.C. I, No. 162.

[5] 'The Cartularies of Athelney and Muchelney Abbeys', ed E. H. Bates *S.R.S.* XIV (1899), The Muchelney Cartulary, No. 6.

[6] H. Wharton, *Anglia Sacra*, I (1691), 556.

[7] Based on a map of inferred Domesday estates in Somerset compiled by Bishop Hobhouse in *S.A.N.H.S.* XXXV (1889), p. x, with additional material from the Wells Manuscripts.

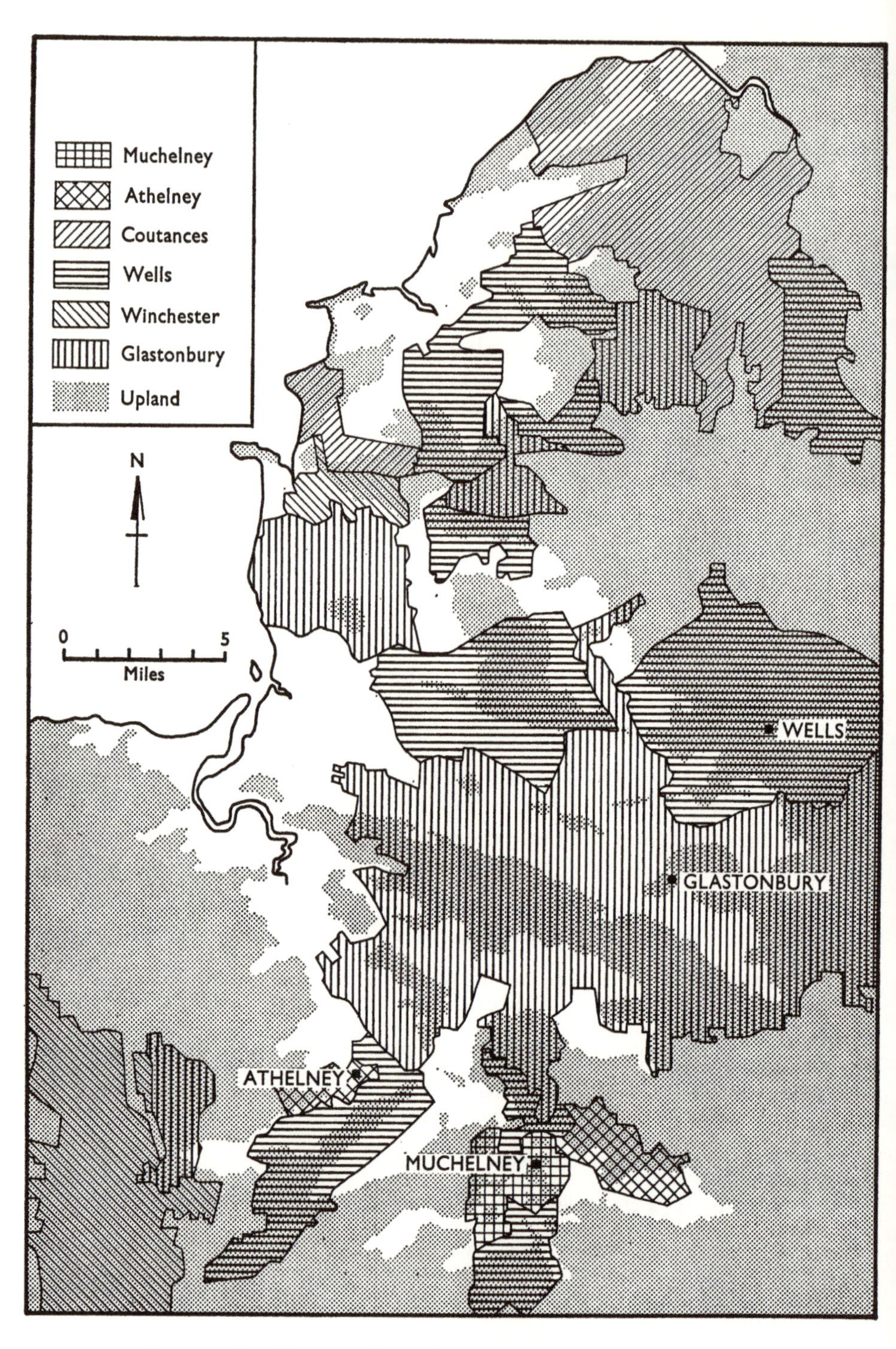

Figure 4. Possible extent of ecclesiastical estates in the medieval Levels.

so that it is not surprising to find that the Abbots of Glastonbury were regular members of the Commissions of Sewers.[1]

With the foundation and the growth of the great ecclesiastical estates went the Saxon settlement of the upland surrounding the Levels, and of the 'islands' within them. It is suggested that the Saxon penetration was along the flood-free ridges which plunge deeply into the heart of the Levels; this hypothesis is geographically reasonable.[2] Place-names testify to the extensive agricultural colonization of the period, especially on the insular sites; Athelney, Muchelney, Othery, Godney, Martinsey, Beckery, Andersey, Ardensey, Chedzoy, Middlezoy, Westonzoyland, Midelney, Pitney, Thorney, Horsey, Bradney, and others, all derive their ending from the Old English suffix 'ey' or 'iey', meaning 'island'.[3] An indication of the extent of Anglo-Saxon colonization may best be had from Figure 5, which shows the distribution of identified settlements in 1086. The location of settlements reflects the difference between floodable and flood-free land, which is the major topographical feature of the Levels, and which was to be the framework for the economy of the region for all subsequent time. Settlement was absent from the peat areas because of the constant flooding and because the soil provided no stable foundation on which to build. The settlements were on the upland edges, the 'islands', and on the coastal clay belt, where the marine clay provided more solid land and afforded the prospect for a well-organized and diversified economy of cultivation and animal husbandry.

Whether or not Saxon settlement resulted in any utilization of the Levels is uncertain. There is no substantial body of evidence from which one can attempt to reconstruct a picture of the economy, but an examination of an area for which charters from the seventh to ninth century are available reveals that Meare, Athelney, Long Sutton, Middlezoy, Huish Episcopi, Brent and Shapwick (which included Chilton, Stawell and Sutton Mallet) all contained portions of what must have been land liable to annual flooding.[4] This suggests that the peripheral moors were used for pasture at the height of the

[1] See *Calendar of Patent Rolls* (*Cal. Pat. Rolls*), *1348–1361*, 418.

[2] G. B. Grundy, 'The ancient highways of Somerset', *Arch. Journ.* XCVI (1930), 226.

[3] See E. Ekwall, *The Concise Oxford Dictionary of English Place-Names* (1936), under appropriate headings.

[4] G. B. Grundy, 'The Saxon charters and field names of Somerset', issued as a supplement to *S.A.N.H.S.* LXXIII (1927), and LXXX (1934). See also G.C. II, Nos. 644 and 648.

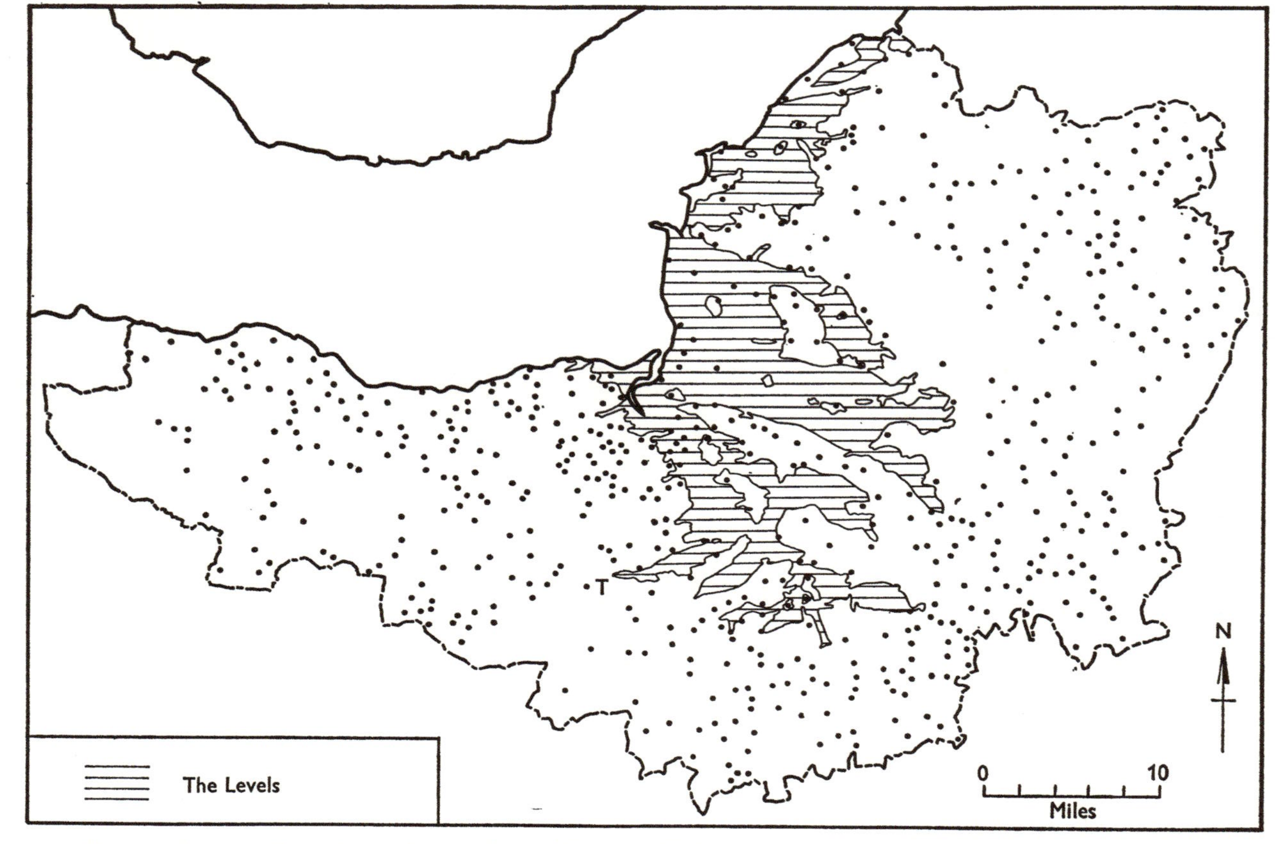

Figure 5. Somerset Domesday settlement. Around Taunton (T) lay a number of settlements that cannot be identified. They are not shown.

summer when they were at their driest, but no evidence of draining activity during this period has yet been detected.

The Domesday evidence for Somerset contributes surprisingly little to our knowledge of the moors and the draining of the Levels. Moors [*morae*] are mentioned for only eleven places, all but one of them, Milborne Port, situated not far from the Levels. They were Adsborough, Fiddington, Huntworth, North Newton, Milborne Port, Seavington, Tuxwell, Weston-in-Gordano, Wells, Yatton and Wedmore,[1] the moors at Wedmore being described as valueless.[2]

Equally surprising, considering the multiplicity of watercourses in the Levels, was the infrequent mention of fisheries in the marsh areas. At Cheddar there were three fisheries rendering 10*s*., at Meare there were ten fishermen and three fisheries rendering 20*s*., one at Martock rendering 5*s*., two at Somerton rendering 10*s*., two at Wedmore rendering 10*s*., and two at Muchelney, Midelney and Thorney which returned a total of 6,000 eels annually. Other fisheries are mentioned at Creech St. Michael and North Curry. In addition to these specific references to fisheries, there belonged to the manor of Staple Fitzpaine a manor which returned 60 eels as rent for a garden in Langport.[3]

On the other hand, information about meadow land in and near the Levels is more plentiful. Some of the individual returns for the Levels settlements were large; for example, Burnham, 300 acres; Huntspill, 120 acres; Wookey, 150 acres; Brandon and Martock, 100 acres each; and Wrington, 280 acres. But some others were surprisingly small, like the 70 acres at Wedmore, 62 acres at Glastonbury, 25 acres at Brent, and 30 acres at Sowy. While much of these meadows may well have been in periodically flooded land, there is no proof that they were reclaimed from the moorland edges. As a whole, the amount and density of meadow land in the Levels at this time was well below that of the Liassic, Oolitic and Oxford Clay upland regions of the northern and eastern parts of the county.[4]

[1] A transcript of the Somerset Domesday with additional information from the Exon Domesday, compiled by E. H. Bates, is to be found in *V.C.H.* I (1906). The material concerning the moors has been analysed by Welldon Finn and P. Wheatley, 'The Domesday Geography of Somerset', p. 187 of *The Domesday Geography of South-West England*, ed. H. C. Darby and R. Welldon Finn (1967).

[2] 'Praeter haec sunt ibi morae quae nichil reddunt', *V.C.H.* I, 458.

[3] See *Domesday Geography of South-West England*, ed. Darby and Finn, 187–8.

[4] F. W. Morgan, 'The Domesday geography of Somerset', *S.A.N.H.S.* LXXXIV (1938), 149.

Taken together, the evidence suggests that the Levels were largely regarded as a negative region which deterred occupation. Attempts were made to adapt life to the unfavourable environment, but none were made to control or radically modify the environment. No definite evidence has yet come to light which suggests any draining activity before the late twelfth century.

3

MEDIEVAL RECLAMATION

THE TRADITIONAL OCCUPATIONS

The condition of the large wastes of moor and marsh about which the Domesday Book is so silent becomes increasingly clear as we look at the evidence of the cartularies, rent rolls, and court rolls of the late twelfth to early fourteenth century. These abound with information about the landscape and livelihood of the Levels, about the traditional occupations such as fishing, fuel gathering and the pasturing of cattle, which led to the primary division of interests in the moors, and about the continuous piecemeal reclamation of meadows, and even of arable land, on the edges of the moors. It is against this background of traditional occupations and activities that the reclamation and draining of the Levels during the Middle Ages are in the first place best viewed.

The commonly-held view that the Levels at this time 'must have been either a gloomy waste of waters, or still more hideous expanse of reeds and other aquatic plants, impassable by human foot, and involved in an atmosphere pregnant with pestilence and death'[1] was true only in part, mainly for the hard core of badly drained peat lands. But, whilst technically part of the waste of the medieval manor, the Levels were far from being the unproductive, desolate and dangerous morasses they have sometimes been supposed to be. There was almost a 'hierarchy' of usefulness amongst the moors that varied with the state of their drainage. First there were the pools, the water-covered moors, and the natural watercourses, which were abundant in fish, fowl, rushes and reeds; secondly, there were the periodically water-covered lands where the turbaries and pastures were found, which with improved drainage became meadows that were flooded occasionally; these in turn merged into the flood-free arable land of the 'islands' and uplands on which had developed a more usual medieval arable economy. This distinction between ground of varying utility and value is constantly referred to in inquisitions post-mortem and in

[1] R. Warner, *A History of the Abbey of Glaston and of the Town of Glastonbury* (1826), 241.

similar inventories of property. For example, at North Petherton there was 'a carucate of land, 40 acres of meadow, 77 acres of pasture, 8 acres of moor, and 16 acres of marsh'; at Knappe, near the River Tone, '10 acres of land, 10 acres of meadow, 80 acres of wood, and 200 acres of marsh'; and at Aller, '12 acres of land, 60 acres of meadow and 100 acres of moor';[1] and these examples could be multiplied tediously.

Fisheries

Even in their worst state the moors were not a liability but a valuable addition to a settlement because the flooded land and standing water contained many fish. The livelihood obtained from the rhynes, rivers and meres is mentioned in the late twelfth-century survey of the Glastonbury estates, *Liber Henrici de Soliaco*; for example:

(1) The cellarer has a fishery at Middlezoy of which the lord abbot owns one part and the abbey two parts. At Andredesey [Nyland], another fishery whence he receives 2,000 eels. At Clewer, a fishery which ought to pay 7,000 eels. At Martinsey [Marchey Farm on the Axe, half a mile north of Bleadney], a fishery which ought to pay 7,000 eels. From Northload, 30 salmon at the Assumption of the Holy Mary. From Houndstreet, 10 small salmon at the Nativity.
(2) At Meare are two fisheries paying 30 pence.
(3) Wlgar [at Othery] pays for one fishery 12*d.*, and for the other 3,000 eels and 17 pence.[2]

In addition to these, and at about the same time, the Abbey of Athelney was granted 'three fisheries in the water which is called Tian [Tone] which are called by these names, Estwere, Merewere, and the third which Ianswine held with all their belongings on the land and in the water. And in addition thirty sticks of eels from the fishery which is called Hengestwere.' It is possible that Estwere can be identified with Morcock Estwere which in 1180 or 1181 was said to yield 5,000 eels annually; another fishery at Stathe on the Parrett yielded 1,000 eels, the combined catch of 6,000 eels being worth 20*s.*[3]

Surprisingly, information on fisheries became less in later years, particularly regarding their rents, either rents in kind or rents which

[1] 'Pedes Finium, 1307–1346', ed. E. Green, *S.R.S.* XII (1898), 177 and 196; and 'Pedes Finium, 1196–1307', ed. E. Green, *S.R.S.* VI (1892), 349.

[2] *Liber Henrici de Soliaco Abbatis Glaston. An Inquisition of the Manor of Glastonbury, 1189*, ed. J. E. Jackson (Roxburghe Club, 1882), pp. 9, 29, and 50 respectively.

[3] 'The Athelney Cartulary', Nos. 129 and 134, *S.R.S.* XIV. Eels were measured in sticks, the stick piercing the gills, and probably consisting of twenty-five eels.

may have been commuted to money payments; it is therefore difficult to gauge the importance of fisheries in the economy of the Levels compared with other activities. What we do know, however, is that there were numerous fisheries in all the main watercourses, and that there were, in particular, artificial weirs or gurgites, which bayed back the freshwater floods, caused flooding in the surrounding moors, and became objects of much debate. The debate and complaints were, perhaps, indicative of the growing realization of the value of the moors as pasture grounds, and of the potential and actual value of reclaimed meadows alongside the rhynes and river courses. Such was the case with Pouwere, Wozwere, Nywere, Wichok, Tappingwere, Buriwere, and another unnamed weir in the lower Parrett channel between Langport and Bridgwater, which caused severe flooding of the nearby land.[1] At Mark, in the western end of the Brue valley, the Bishop of Bath and Wells had built Newere and Northwere, which so stopped up the water that it was said 'that the lands of neighbours there are flooded every year', and similar flooding occurred at Mark, Huntspill, Burnham, and Huish by Highbridge, at Rodwere and Scaddleweres in the Pilrow cut, and at Hakewere and two weirs called Poundweres.[2] There were other weirs in the Brue valley; the bishop and his servants were accused of making weirs 'whenever they place "holies" [nets] and other instruments to catch fish there, by stopping up the water on the common ground of Thorlemore and More'.[3] On the other hand, the fisheries of the Brue, near Street bridge, seemed to have caused no trouble,[4] and the great expanse of Meare Pool, the largest and most productive fishery of all, with its fish-house, was of a different nature from the gurgites, and therefore did not cause flooding.

Sometimes complaint erupted into action, and £40 worth of

[1] G.C. II, No. 969. Stathewere, another weir in the Parrett, is mentioned in about 1270; see G.C. II, No. 956.

[2] *W.M.* I, 285 mentions repairs done to Rodwere fish-house in 1378, the presence of which suggests that it was a productive and valuable fishery.

[3] Coram Rege Rolls, Easter 32, Ed. III. Mem. 5 and 23, quoted in C. T. Flower, *Public Works in Medieval Law*, (Selden Society, 1923), XL, 131–4. Similar obstruction seems to be implied in a case heard in 1258 when John de Balun raised a weir in Saltmoor in Huntspill in the land of William de Marisco and also diverted a watercourse that used to run to William's fishery where he was entitled to half the fish, as far as Lodespill. The new fishery was ordered to be 'thrown down...and to be made as it was wont and ought to be': 'Somersetshire Pleas, Thirteenth Century', ed. C. E. H. Chadwyck-Healey, *S.R.S.* XI (1897), No. 1452. For a similar case in Weare, see *ibid.* No. 394.

[4] G.C. III, No. 1301, and G.C. II, Nos. 413 and 431.

damage was done at Glastonbury gurgites and other property by John de Ivethorne in 1336; and 'certain persons' had also destroyed weirs in Mark in 1313.[1] More spectacular than these minor outbursts was the action of the Bishop of Bath and Wells, who was accused of destroying seven fisheries, one each at Midelwere, Rugwere, Sugwere, Godenswere and Clewer and two at Martinsey, belonging to Glastonbury Abbey, and of depriving the abbey of the fishery at Meare. Then the Abbot of Glastonbury broke three fisheries of the bishop with his boats [*fregit tres piscari*] on the Axe between Glastonbury and Radcliffe (Rackley), 'so that he made them wider at the entrance outfall than they were wont to be before';[2] these were but two of many similar cases recorded in the prolonged and almost perpetual disputes between Wells and Glastonbury over the ownership and use of the peat moors and their natural resources on the frontier between their territories in the Brue valley. Whether such destruction was another manifestation of the general enmity and jealousy between them, whether it was in response to flooding caused by the weirs, or whether it was because of the actual value of the fisheries themselves, is difficult to say. But in subsequent years, the abbot carefully established his claim to repair the weirs of Hachwere, Bordenwere and Pariswere, all on the Dean of Wells' property in the Brue below Meare Pool, and also the sole right to fish in Ferlingmere (Meare Pool); however, the dean's tenants had a right to cross the pool daily to reach their pastures, but only between sunrise and sunset, perhaps to stop illicit fishing; all of this suggests that the fisheries were valuable. Later still, the boundary between both parties in Queen's Sedgemoor was defined along the Hartlake rhyne, and each agreed to 'raise a weir for taking fish'. Similar disputes occurred in the boundary lands between the estates of the Dean and the Abbot of Athelney in North Curry and in the course of the lower Tone, and in 1249 the abbot's nets in Burwere were seized by the moorward of the dean.[3]

[1] G.C., II, No. 592; *Cal. Pat. Rolls, 1313–1317*, 406. In response to similar trespass and destruction in the Tone river, the Bishop of Bath and Wells instructed his vicars 'with all solemnities to publish sentence of greater excommunication...upon persons unknown who have fished in the river and waters of the Tone'. *W.M.* I (1378), 287.

[2] *Adami de Domerham Historia de rebus gestis Glastoniensibus*, ed.T. Hearne (1727), II, 452, and 'Somersetshire Pleas', No. 787 (1242–3), ed. Chadwyck-Healey, *S.R.S.* XI.

[3] *W.M.* I, 226–8, 324 and 86–7 respectively. For the careful regulation of fisheries at Hamwere at Weare on the Axe, see 'Cartulary of St. Mark's, Bristol', ed. C. D. Ross, *Bristol Rec. Soc.* XXI (1959), No. 267 (1247).

Whatever the motives may have been for the destruction of and the disputes over the gurgites and fisheries, one thing is clear; there were a large number of fisheries in existence, and they were valuable enough to be protected by agreement and custom.

Turbaries and timber

The moors had two other natural resources which were exploited, turbaries and timber. In the waterlogged peat moors, particularly in the Brue valley, turfs were cut and dried for fuel, and on the periodically flooded edges of the moors there were extensive thickets of wood, mainly alder, but some ash and, of course, willow, which were used for constructional purposes as well as for fuel.

Both the right to dig turf and the right to collect brushwood were closely allied to the right to pasture animals, and a complicated body of custom had grown to control these operations; for example, the agreement between Wells and Glastonbury in 1327, to apportion pasture rights on the peat moors on the north side of the Brue valley, included rights of turbary. This combination of rights is something which documents for much later periods show quite clearly.[1] Strangely enough, there is little mention of turbaries in the otherwise abundant documents of this period. We do learn, however, that during the early thirteenth century the Dean of Wells had been taking peat illegally from Heath Moor near Meare, and that Abbot Michael, in order to re-establish the abbey's claim to the peat lands, 'stripped off on many occasions all the peat of that great moor so that for ten years or more, every hearth in Glastonbury was plentifully provided';[2] and also that shortly before 1262, Nicholas FitzRoger, lord of Tickenham in the Northern Levels, granted to the Priory of St. Mark's, Bristol, the right to dig turfs and collect rushes and sedge. This liberty was carefully regulated, and the priory could have only three men working there daily 'in suitable and accustomed places at times when others were accustomed to dig and cut'; and furthermore, the diggers were to take care not to work on land which he or his heirs managed to reclaim and convert into meadow or cultivate; this says much for the value of the turbary compared with other resources in the moors. Passing reference is also made to turbaries at Chilton, and Edington, in the Brue valley; at Weston-in-Gordano in the Northern

[1] *W.M.* I, 228 and G.C., II, No. 647.
[2] *Adami de Domerham*, ed. Hearne, II, 509.

Levels in 1270; and at Compton-juxta-Axbridge, Overwere, Burnham and South Brent, but that is all.[1] The recorded occurrence of turbary is so slight, and the amount dug so uncertain, that it is impossible to assess its importance in the medieval Levels; there is certainly no evidence of any systematic digging and marketing in the accounts of the Glastonbury estates in the fourteenth century.[2] Probably it was not of great importance because there was plenty of wood available on and near the moors, as well as on the upland ridges which penetrated deeply into the centre of the Levels.

The earliest reference to gathering wood fuel is in 1189, probably near Glastonbury.[3] Later, information is more explicit; for example there were extensive natural alder beds in the moors around Athelney which were obviously a resource that was eagerly sought after and jealously guarded. In 1213, the Abbot of Athelney exchanged his rights in North Moor with John de Erlega for 100 acres near the abbey, but still retained his rights of common pasture and the collection of brushwood and firewood [*claustura et fualia*] throughout the moor. Some years later, the abbey received a grant of 40 acres of moor in Lyng from Henry de Erlega 'for the acquirement of fuel [*fuallia*] of which the monks are often in want'. Rights of common and fuel had also been acquired by the abbey throughout most of Stan Moor; in Salt Moor the abbey had rights of common of pasture, but there was no fuel to be had there. The abbey's hold over Stan Moor was strengthened in 1250 by obtaining the crop of 80 acres of alder instead of taking one boat-load of brushwood daily from Michaelmas to Holeday.[4] Later, in 1337, we hear of the Dean and Chapter of Wells instituting proceedings against Robert Gyan of North Curry who cut down 'a great number of alder in a moor called Stathemoor' (Stan Moor) to the value of ten shillings, and terrorized the local and rightful commoners. After Robert had submitted to a physically brutal penance, the dean granted him six boat-loads of brushwood annually, to be taken between the

[1] 'Cartulary of St. Mark's, Bristol', ed. Ross, *Bristol Rec. Soc.* xxi, No. 440; *Cal. Inquisit. Post Mort., Henry VII*, iii, 310–11; 'Pedes Finium', Nos. 6 (1330–1), and 17 (1342–3), *S.R.S.* xii; and 'Pedes Finium', No. 144, *S.R.S.* vi.

[2] I. J. E. Keil, 'The Estates of the Abbey of Glastonbury in the Later Middle Ages'. Unpublished Ph.D. thesis, University of Bristol (1964).

[3] *Liber Henrici de Soliaco*, ed. Jackson, 11–12; 'In more buschia debet quod sufficiat ei sine wasto', for example.

[4] 'Athelney Cartulary', Nos. 99, 102 and 212, *S.R.S.* xiv; *W.M.* i, 317–18. The wording of this document suggests that it might have been a deliberately planted crop.

Exaltation of the Holy Cross and Michaelmas, under the view of the bailiff.[1]

Across the River Parrett, in the Sowy Moors, there were other extensive alder beds, which the Abbey of Glastonbury was careful to preserve against illegal depredations and enclosure, and which, similarly, individual lords were determined to preserve from abbey encroachments.[2] Other productive groves of alder lay on the moors in the Brue valley alongside the Polden Hills, particularly in Walton, Street and South (Alder) Moors,[3] and other groves were situated on Hearty Moor in eastern Queen's Sedgemoor.[4]

Although rights of gathering firewood (husbote and heybote) often went with rights of pasture, there seems to have been no similar right to take wood for constructional purposes, for this meant the total destruction of the trees. Thus in 1241, forty-one commoners of Godney Moor 'cut down about four score trees and carried them off', and some little while later took a further 'one hundred and thirty beams' [*cheverons*] to repair bridges, which, they argued, was their right because they had rights of pasture, fuel and sedge in the moor, but which was subsequently disallowed. Whether the same need to repair structures lay behind similar actions in uprooting and carrying off trees at Meare, a Glastonbury property, in 1299, and in Mudgley, a Wells property, in 1327, is not clear, but as they were accompanied by other acts of violence and destruction to houses and stock, and by firing of the peat moors, it seems that they were yet another example of the feuds between Wells and Glastonbury over the ownership of the Brue valley.[5] From what little evidence is available it seems that

[1] *W.M.* I, 415–16 (1337), and 71 (1338). This case is a little strange because Robert's father, William, had been given sufficient common of pasture for husbote and heybote some time before, and Robert probably had some claim to it. See *W.M.* I, 71–2.

[2] G.C. II, No. 966 (*c.* 1280); Walter de Eyr resigned rights in 20 acres of moor and alder in Sowy; and G.C. II, No. 962 (1330) which deals with an agreement between Sir William de Montacute and the abbey concerning 60 acres of alder near Westonzoyland.

[3] In the Abbot Bere's *Terrier* of the early sixteenth century, Alder Moor was described as being so thickly covered with alders that it was impossible to hunt there. Printed in *Joannis Confratis et monachi Glastoniensis chronica, sive Historia de rebus Glastoniensibus*, ed. T. Hearne (1726), II, 311.

[4] G.C., II, No. 461 (1305). William Pasturel was given twenty 'reasonable cartloads of alder wood for his fire each year' from Hearty Moor by the abbey. See G.C., II, Nos. 995 and 997 for the particulars of a similar grant to Sir Ralph de Counterville de Mariscus in 1267, and G.C., II, No. 779 (1281), where William de Yadwich was given four acres of moor and alder in Hearty Moor to enclose.

[5] 'Somersetshire Pleas', No. 572, ed. Chadwyck-Healey, *S.R.S.*, XI; *Cal. Pat. Rolls, 1292–1301*, 472; and *Cal. Pat. Rolls, 1327–1333*, 88.

the right to obtain large timber was given in specific grants to individual persons. Thus in 1267, John Channel, parson of Walton, was able to take saplings and spars from Walton and Street Moors for roofing and repairing houses; Sir Boniface de Falliaco, rector of the churches of Butleigh, Street and Walton, took wood from the same alder groves for firewood, and for 'house and hedge repairs', and when the moors were flooded he was allowed to take timber from 'Hundewood and Laghefrygh' on the Polden Hills. Similarly, Thomas and Lucy de Street were granted the manor of East Street near Glastonbury, and allowed to go into the moors and cut alder wood.[1]

Pasture and intercommoning

However interesting and peculiar these activities of fishing, wood-gathering and turf-digging were, they should not be over-emphasized, for the Levels were, above all else, a region of common pasturing. The pasture rights were an important part of the manorial economy of the Levels, perhaps more so than on the surrounding uplands, and these rights were highly valued. Villages, or groups of villages, on the 'island' and in peripheral situations, assumed or acquired the right to intercommon their cattle and other beasts on the moors which they adjoined, but the importance and the obvious value of the intercommoning arrangements neither lessened their vagueness nor rectified the basically unsatisfactory character of common pasturing. The topographical division of the Levels into distinct basins made the demarcation of some interests fairly clear, e.g. those of West Moor, West Sedgemoor, and Queen's Sedgemoor; but other commoning interests, like those in King's Sedgemoor and the Brue valley generally, were very complex and open to abuse by illegal pasturing.[2] In addition to this problem, the grazing was seasonal because flooding made the utilization of the moors for stock impossible except in summer and early autumn, and there was also the constant threat of a spring or summer flood which could cause great harm by checking the growth of the grass, and, if the water lay on the land long enough, by killing the herbage completely.

Another drawback to common grazing was the right of all

[1] G.C. II, Nos. 760 and 610.

[2] See p. 89 below and Figure 13 for an indication of these intercommoning arrangements in the seventeenth century.

freeholders to pasture unlimited numbers of their beasts throughout the year.[1] Under this generous custom there were great numbers of cattle on the moors during the limited dry season, the position being aggravated by the right of some commoners to take in cattle from areas outside the Levels for a fee, a practice which increased enormously during later centuries;[2] added to this, there must have been much illegal commoning by adjacent tenements and holdings without rights. Overstocking worked to the detriment of all commoners, and limitations were placed on the number of animals that could be kept, as in Street in 1291, Baltonsborough in 1244, and Walton and Ashcott in 1242–3, where Geoffrey de Langelegh was summoned to answer Abbot Michael of Glastonbury why he had in his pasture 'one hundred and fifty goats and twenty oxen and cows beyond the number which he and his ancestors were wont always to have, to wit, sixteen oxen only'.[3]

In order that the specified number of beasts could be checked periodically, and so that the cattle of outsiders could be detected, the 'drove' or 'drift' was held at least once annually; the cattle were rounded up, impounded and only released on payment of a fine. For example, one such attempt to regulate the grazing of the commons occurred to the north-west of Sowy where Abbot Michael of Glastonbury and William de Monte Acuto 'fought a battle of arms', in 1235, to determine the ownership of the moors. The ownership fell to William but it was agreed mutually that the abbot and his tenants of Chedzoy should have common in the marsh

> for all their animals and flocks and also, if any beasts other than those of the Abbot and his men or his own beasts or those of his men of the Manor of Chedzoy enter the said marsh for grazing, the poundage shall be collected by a servant of the Abbot, and that money shall be divided equally, and a pound for the animals shall be erected in the middle of the marsh and sustained at the cost of both parties.[4]

[1] *Inter alia*, 'Somersetshire Pleas, 1280', ed. L. Landon *S.R.S.* XLIV (1929), No. 763 (Crandon in 1280); 'Muchelney Cartulary', No. 73, *S.R.S.* XIV (Knappe and Cleve in 1238); 'Somersetshire Pleas, 1255–1272', ed. L. Landon, *S.R.S.* XXXVI (1921), 132 (Godney Moor in 1270); 'Pedes Finium', No. 139 (1268–9) *S.R.S.* VI; 'Athelney Cartulary', No. 120 *S.R.S.* XIV (Salt Moor in 1282) and 'Somersetshire Pleas, Thirteenth Century', No. 1509 *S.R.S.* XI (Beckery in 1254–5).

[2] *W.M.* I, 316–17, for example.

[3] G.C., II, No. 757; G.C. II, No. 770; and 'Somersetshire Pleas', No. 730 *S.R.S.* XI. For other restrictions see G.C. II, No. 953 (Westonzoyland Moor in 1303); 'Somersetshire Pleas', *S.R.S.* XI (Cossington, 1257); and G.C. I, No. 64 (Weston Moor).

[4] *W.M.* I, 326; G.C. I, No. 355; and 'Pedes Finium', *S.R.S.* VI, 80–1.

Other examples of driving and impounding exist for Curry Moor, in 1365, North Curry, and Stathe Moor, in 1311 and 1327.[1] Depositions in the sixteenth and seventeenth centuries concerning Mark Moor in the western part of the Brue valley, and King's Sedgemoor, suggest that the drives were continued throughout the Middle Ages.[2]

But the drive was not the only way in which stock on the moors was controlled; divisions were being made in the moors, and general rights in common with many others in large areas were being exchanged for specific rights in smaller areas. This primary organization of the land of the Levels was also accompanied by the increasing concern of the religious houses about their boundaries in the Levels, and there are some interesting examples which show how the division of the Levels took place. In west Hay Moor, in the western portion of the upper Tone valley, the Prior of Muchelney and his tenants at Hamme relinquished their rights in the moor to the Dean and Chapter of Wells in exchange for common of pasture in a specified portion of the North Curry moors, 'from Ladlake to La Mulelake for all their own beasts and hired beasts if necessary according to the quantity of their tenements, but only such as shall be required for tilling their land'. The dean undertook to 'renew and make visible' these bounds and, when making a drift, return the prior's and his tenants' cattle to Hamme.[3]

It is in the Brue and Axe valleys, however, that the material about division and dispute is fullest, and the disagreements rivalled in bitterness those that took place around Crowland Abbey in the Fens.[4] The Bishop of Bath and Wells, the Dean and Chapter of Wells, and the Abbot of Glastonbury were constantly quarrelling amongst themselves over their interests in the pastures, turbaries and fisheries in the northern part of the Brue valley and in the southern part of the Axe valley. (For many of these names, see Fig. 8.) Ownership of these broad open expanses of the Levels, with no clearly-defined topographical demarcation, was open to heated and quite often violent dispute, though it seems certain that they belonged to Glastonbury, but 'through negligence' the abbots 'had for long never set foot upon

[1] *W.M.* 269, 316 and 317; *W.M.* II, 585–6; and *W.M.* I, 416.

[2] For example, P.R.O., E134/6 Jas.1/Mich.12.

[3] *W.M.* I, 316–17 (1303). Such marking out of the boundaries was actively obstructed at times. See *Adami de Domerham*, ed. Hearne, 486–7, where Sir Ralph de Alre bribed Hamo, the Seneschal of Glastonbury, with 'a splendid cockerel' so that he might not mark out the boundary in Aller Moor.

[4] Darby, *Medieval Fenland*, 86–92.

the northern moors of Meare...' so that the abbey's right had almost ceased to exist.[1] Trouble started in 1278 when the abbot's men destroyed a piggery of the bishop's in Godney Moor, but the abbot agreed to make restitution and allow it to be rebuilt.[2] By 1283, however, he established his claim to Godney Moor, in exchange for relinquishing his interests in the moors (either Crannel or North Moors) east of the watercourse leading from Bleadney to Lineacre, which was Nineacre at Godney. A further rationalization of rights and territories came about with the gift by the abbey to the dean, in 1310, of 600 acres of moorland between Cocklake and Northlode, abutting the dean's moor at Wedmore, with the retention, however, of the rights of the abbot's tenants in Panborough, Bleadney and Clewer, in the moors.[3]

But despite this seeming peace there were still causes for dispute, particularly over the swinecots in Godney Moor, which were demolished again in 1315, together with a wall and some dykes in Blackford on the west side of Wedmore 'island', and over sluices at Mark, which, the bishop alleged, caused flooding, and his corn and herbage to be submerged. In order to avoid further conflict, Bishop Drokensford and Abbot Adam of Domerham appointed four persons, in March 1326, to discuss and settle the differences and transgressions which had arisen over the boundaries of the moorland pastures.[4] But four months later, in the middle of the dry season, someone set fire to the peat moor on the whole of the south side of the Brue valley, from Burtle Priory towards Glastonbury, so 'that the abbey might be burnt'. Dean Godelee was accused by the bishop, but this seemingly harsh act was probably a friendly one allowing the dean publicly to purge himself of the implication that might have been drawn by the abbey and its supporters. Already the bishop had published a ban of excommunication on all 'who infringe liberties of the Church', which was obviously aimed at the abbot. The battle was on. Further destruction of buildings and theft of property, allegedly by two servants of the abbot and two monks, occurred in the bishop's moor of Thealemoor in November, and the bishop proclaimed sentence of excommunication on the abbot and the four wreckers, later visiting the abbey and passing similar sentences on all

[1] *Adami de Domerham*, II, 509.

[2] G.C. I, No. 158.

[3] G.C. II, Nos. 643 and 597.

[4] *Cal. Pat. Rolls, 1313–1317*, 406, 411–13; and G.C. I, No. 160.

those who made 'illicit oaths of secrecy to defeat correction'.[1] Further acts of destruction were said to have taken place early in the next year, and the abbey was once more accused, this time of burning timber and grass in Mudgley, and later of taking 12 horses, 60 oxen, 50 cows, 100 bullocks and 10 swine worth £200 from the bishop's property at Meare.[2]

Such contention could not continue and had to end in a compromise,[3] which was the prelude to some important decisions to divide the moors. In the northern half of the Brue valley it was agreed to divide the moors into two equal parts, equal in value but not in acreage, the line of division running south from Consailleswalle, south of Mudgley on Wedmore 'island', to 'the rhyne below Coubrugg [Westhay Bridge] directly opposite the east corner of a close called Parismede below the hamlet of Westhay',[4] and upon this line 'a new dyke shall be cut and raised, upon which shall be erected four stone crosses, two at the charge of either party'. The present parish boundary between Meare and Wedmore runs along this line, and the rhyne alongside is called Bounds Ditch. Glastonbury had the Manor of Meare and Godney Moor to the east, and Wells had Tadham and Tealham Moors to the west. The rights of the tenants to common in the nearest moor were protected, although certain of the dean's tenants in Mudgley, Mark and Wedmore still had unlimited rights of pasture in Godney Moor, and the tenants of Moor and Biddesham had rights of unlimited pasture in Oxenmoor. The compromise agreement was extended to include permission for the dean and his tenants to cross Meare Pool and the Brue to reach their pastures, for Glastonbury to repair fisheries on the dean's land, and for both to have turbary, alders and piggeries in each other's half of the moors.[5]

[1] 'Bishop Drokensford's Register, 1309–1327', ed. Bishop Hobhouse, *S.R.S.* I (1887), 264, 271, 277, 279 and 153.

[2] *Cal. Pat. Rolls, 1327–1330*, 88.

[3] Strangely enough, despite the ill-feeling and violence, some advance had been made already in the division of the moors. The boundary between the abbey and Wells in Queen's Sedgemoor, between Launcherley near North Wootton and Bacchyngwere, east of Meare Pool, was discussed, but probably with no immediate result as a final agreement was not made until 1352. See G.C. I, No. 157 for the 1327 discussions.

[4] S.R.O. Som. Arch. Soc. Maps C212. Map of the Brue valley in 1779, before Parliamentary enclosure, in which many of these places can be identified.

[5] *W.M.* I, 228 and 226–7, and a copy with some differences and additions in G.C. II, No. 647. The foundations of the stone crosses were still standing in 1558. S.R.O., Som. Arch. Soc. Parochial MSS. 462: 'Perambulation and Customs of the Manor of Muddesley [Mudgley], Somerset'.

No sooner were these agreements over the Brue valley solemnly made than fresh disputes arose over the southern Axe valley, northern Godney Moor and Queen's Sedgemoor, where previous agreements by both parties to repair Hartlake bridge and rhyne were violated by Glastonbury, and 'a wall of pales and earth' for watering meadows and stock in Wookey was broken down. The 1327 partition agreement had contained provisions for settling new disputes by the appointment of three knights by each party. They were now summoned, as well as forty-eight 'good and loyal men of Somerset', from whom the knights obtained material for a plea between the two parties to be heard at the Court of Canterbury.[1]

What happened is not known; in 1327, however, a lengthy agreement was made by which the bishop was to have common rights for himself and his tenants at Easton, Burcote, Polsham and Coxley, for all their beasts, in the abbot's moor at Godney Moor. The abbot and his tenants at Meare, Bleadney, Panborough and Martinsey were to have similar rights in the bishop's moor of Wookey, and each agreed that if the beasts of the other strayed out of their respective moors they would not be impounded but driven back. Some division was made between these interests by straightening and making 12 ft wide the watercourse through northern Godney Moor between Monkenmede (Hurn) and Bleadney, which each party was to maintain equally, and over which the bishop and his tenants were to build two bridges for stock. An 'efficient ditch' was also to be dug to keep beasts on Wookey Moor from feeding on the cultivated land and meadows in Wookey, and this ditch is probably represented by Old Closes Ditch that skirts the inner margin of the moor today. The agreement over Wookey and Godney Moors was followed by one over Queen's Sedgemoor in 1352, whereby the land north of the line of Hartlake rhyne to Meare Pool was to be the bishop's, and land to the south to be the abbot's.[2]

These disputes between Glastonbury and Wells are full of details, but it is not hard to believe that there were other disputes which have passed out of memory but which may have been equally revealing about the medieval Levels. Nevertheless, the fact remains that the delimitation of intercommoning rights and the division of the moors occurred during the late thirteenth and early fourteenth centuries,

[1] G.C. II, No. 645; *Cal. Pat. Rolls, 1324–27*, 349; and G.C. I, No. 159.
[2] G.C. I, No. 161; and *W.M.* I, 324 and *W.M.* II, 617.

and the bitterness of the disagreements underlines the vital position occupied by pasture rights in the economy of the Levels. But the division had an even greater significance than that, for, taken one step further, the division of the moors inevitably led to the enclosure and reclamation of the Levels, and to a transformation of the landscape.

EVENTS LEADING TO RECLAMATION

Taken as a whole, the evidence suggests that the practice of inter-commoning on the open moors was unsatisfactory; the unlimited rights and stocking arrangements were ill-defined and open to abuse, and both the length of the grazing season and the value of the pasture were uncertain because of flooding. Therefore, in an effort to ensure pasture against both human and natural depredation, reclamations were made in the commons, a trend which was encouraged by the general division of the moors occurring at this time, and by the growing concern to obtain specific rights in more limited areas. In the creation of new meadows and in the extension of the grazing season lay the best means of enlarging the medieval economy of the Levels. The physical environment restricted arable farming to insular and upland locations, but the moors were capable of sustaining more intensive pastoral activity.

There can be little doubt that reclamation was encouraged by the knowledge that pasture land converted into meadow could equal or excel in worth a comparable area of arable land. The question of winter fodder was crucial to medieval livestock farming and needs little elaboration; insufficient fodder and a severe winter would result in the slaughter of stock and impoverishment of a settlement.[1] Thus, at Baltonsborough, an acre of pasture was worth 9*d.*, an acre of arable land worth just under 5½*d.*, and an acre of meadow, 2*s.*[2] In Glastonbury, pasture land ranged from 2*d.* to 6*d.* an acre, arable land from 4*d.* to 10*d.* an acre, but meadow was worth as much as 12*d.* and 13*d.* per acre. New meadow in Meare was worth only 2*d.* an acre, probably being little better than pasture land for it was not yet entirely free from regular flooding,[3] but reclamations in the Sowy moors

[1] See F. W. Maitland, *The Domesday Book and Beyond* (1897), 443; and other comments in *An Historical Geography of England before A.D. 1800*, ed. H. C. Darby (1936), 198.

[2] 'Rentalia et Custumaria: Michaelis de Amesbury, 1235–52, et Rogeri de Ford 1252–61', ed. C. J. Elton, *S.R.S.* v (1891), 196.

[3] *Ibid.* 196, 180–2, and 206.

showed that new meadows could often be more valuable than arable land.[1]

During the twelfth and thirteenth centuries the general laxity of manorial control led to many illegal encroachments upon the manorial waste. The great ecclesiastical estates of the Levels participated to the full in the manorial disintegration which characterized this period, particularly during the latter part of the twelfth century.[2] Demesne lands were let out on lease, labour services commuted, and assarts and enclosures made in the commons and wastes.[3] In addition, the years 1178 to 1219 were disastrous for the Abbey of Glastonbury and the See of Bath and Wells; Bishop Savaric's attempt to unite them under one rule led to protracted feuds which enfeebled them financially and administratively, and which must have contributed to their losing control of some of their estates. As a consequence of these events, vigorous efforts were made by the religious houses during the thirteenth century to stem the tide of commutation and leasehold, and to recover, and sometimes drain, their alienated lands. This was particularly true of the Glastonbury estates where the first impetus came during the reigns of Abbots Michael of Amesbury (1235–53) and Roger Ford (1252–61), both of whom sedulously concentrated on the cultivation of demesne lands, and pursued an energetic policy to recover the illegal enclosures on the moors and to recover some of the property lost to the Bishopric of Bath and Wells.[4] They were followed by Robert of Petherton (1261–74) and John of Taunton (1274–90), who carefully consolidated the progress of their predecessors by attempting to extend the demesne lands of the abbey, increase its income and prestige, and begin new reclamations.[5] These endeavours in Glastonbury, as well as those in the other large estates of

[1] See p. 50 below, and Figure 11.

[2] M. M. Postan, 'The chronology of labour services', *Trans. Roy. Hist. Soc.* Fourth Series, xx (1937), 189.

[3] For an important discussion on the organization and decline of the Glastonbury estates during the twelfth century, see M. M. Postan, 'The Glastonbury estates in the twelfth century', *Econ. Hist. Rev.* v (1953), 358, and *idem*, 'The Glastonbury estates in the twelfth century: A reply', *Econ. Hist. Rev.* ix (1956), 106. For some other views, see R. Lennard, 'The demesnes of Glastonbury Abbey in the eleventh and twelfth centuries', *Econ. Hist. Rev.* viii (1955), 355–63.

[4] See *Adami de Domerham*, ed. Hearne, 504–5, and 509–10, During the reigns of Abbots Michael, Roger and Robert, i.e. 1235–74, the abbey spent 9,878 marks on litigation in the Roman Courts in order to establish lost privileges and rights. G.C. i, No. 161.

[5] *Adami de Domerham*, 525–96. For an excellent summary of the agrarian policy of the Glastonbury estates during this period, see R. A. L. Smith, *Collected Papers* (1947), 103–16, 'The Benedictine Contribution to Medieval English Agriculture'.

Athelney and of Wells, must have been greatly facilitated by the passing of the Statute of Merton in 1235, which recognized the lord's right to occupy and enclose the commons or 'waste', provided he left sufficient pasture for his free tenants.[1]

It is from the details of the endeavours to recover land and re-establish lost privileges in the moors that we learn much of the reclamation activity of the Middle Ages. Reclamation and enclosure were much more common in the thirteenth century than was formerly thought to be the case, and there is a growing body of evidence to suggest that the experience of the Levels was but one manifestation of a general trend throughout the country.[2] Enclosure and reclamation lessened with the general downswing in economic conditions during the fourteenth century, which was accelerated by the Black Death and the subsequent decline in population, scarcity of labour and reduced demand for land and its produce.[3] But this is looking too far ahead; the point is that the energy and enterprise of the religious houses in the thirteenth and early fourteenth centuries in recovering and reclaiming land were potent factors in the draining of the Levels and in the creation of a new geography in the region.

THE NATURE AND LOCATION OF RECLAMATION

Generally speaking, the majority of medieval reclamations were piecemeal in character, and usually took the form of upgrading the common grazing pastures of more favourable areas into meadow, in order to produce an occasional hay crop, and, in some cases, of draining the land so thoroughly that cultivation could be undertaken.

[1] *The Statutes of the Realm*, I, 2 (Record Commission).

[2] For example, Joan Thirsk, *Tudor Enclosure* (Historical Association, G. 41), and R. H. Hilton, 'A study of the pre-history of English enclosure in the fifteenth century', in *Studi in Onore di Armando Sapori* (1957), I, 673 f. For specific examples in marshland areas in Britain, see Hallam, *Settlement and Society*, for the Lincolnshire Fenland, and R. A. L. Smith, *Christchurch Cathedral Priory* (1943) and 'Marsh embankment and sea defences in medieval Kent', *Econ. Hist. Rev.* x, 29–37 for the Kent marshes. In Somerset it is recorded that 845 acres were reclaimed from low-lying moors and upland areas between 1296 and 1328 by assarting and by the formal enclosure of waste ground over which commoning had been practised, originally illegally, for a number of years. (G.C. I, No. 290).

[3] For the best general account of this controversial subject of economic change and decline in the early fourteenth century, see M. M. Postan and others, 'L'Economie européenne des deux derniers siècles du Moyen-âge', *Reports of the 10th International Congress of Historical Science*, vol. 3: *Storia del Medioevo* (1955), 657; and also Barbara F. Harvey, 'The Population Trend in England between 1300 and 1348', *Trans. Roy. Hist. Soc.*, 5th ser., XVI (1966), 23–42.

There is very little detailed evidence of what was done to drain the land, but we do know that in some cases rhynes were dug across the surface of the new grounds to facilitate the more rapid evacuation of water and so ensure a longer growing or grazing season. In any case, the boundary ditches of each new enclosure quickly assumed a drainage function. Low walls were sometimes built to keep out the less severe floods, although, generally speaking, there could have been little hope of permanently excluding the great autumn and winter floods which swept over the area with such regularity and severity. In other places, stronger and more elaborate walls were constructed with more lasting effects.

The usual flooding could be prevented only by the overall control of water within the drainage system, and the achievement of this ambitious aim was beyond the knowledge or technical ability of the medieval drainer, even if it was ever contemplated. Instead, reclamation was of a local nature and was determined by the inherent ease or difficulty of draining and utilizing particular soils. These soil differences arose from local variations in the elevation, structure and composition of the clay and peat which predominate throughout the Levels. These differences are significant for an understanding of the distribution of medieval drainage activity.

The location of the areas in which upgrading, reclaiming and draining of the moors occurred is depicted in Figure 6. The location of many references can be plotted exactly, but, generally speaking, the nature of the evidence does not allow us to be too exact, and general areas of reclamation only are depicted. The map covers a long period of time and the distribution is necessarily incomplete in detail, but in spite of its obvious limitations, it does provide a basis of assured fact from which one can examine the medieval contribution to the draining of the Levels.

There are three features of the distribution which are striking, namely the lack of evidence for reclamation on the coastal clay belt and similar areas in the Northern Levels; the avoidance of peat soils; and the concentration of reclamation in distinct and definite areas.

The evidence for the clay belt presents many difficulties. The absence of indications of reclamation may be due to the fact that barely one-third of it was included within ecclesiastical estates, and that few records other than ecclesiastical ones have survived the Middle Ages. Except for the Glastonbury estates of Brent, Berrow and

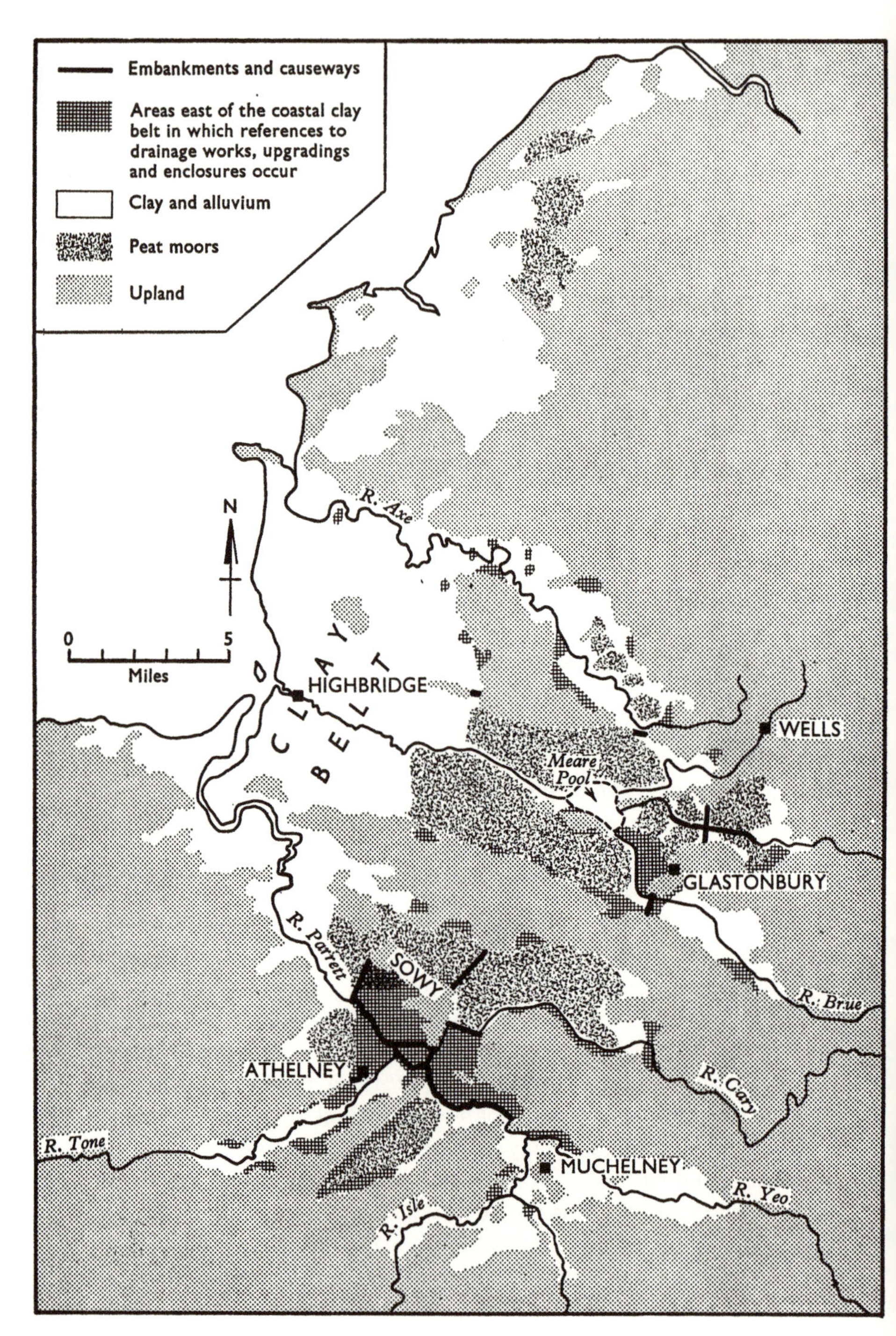

Figure 6. Medieval reclamation.

Lympsham, and the Winchester estate of Bleadon, there is scarcely any information. Another reason for the absence of evidence for reclamation at this time may be that the clay belt was largely flood-free. The Domesday survey shows that it was an area of early settlement, and by the time reclamation was taking place in the inland moors in the thirteenth century it was already utilized and integrated into what would appear to be a well-ordered and balanced medieval economy. Indeed, the Domesday evidence shows that the clay belt had a density of plough-teams as great as, and sometimes greater than, many an upland area (Figs. 5 and 9). The large amounts of meadow are equally indicative of its long-founded prosperity.

Some indication of the earliness of colonization on the clay belt is given in the *Liber Henrici de Soliaco* of 1189 where, as well as the evidence for cultivation and farming, there are references to the maintenance of 'gulets', possibly sluices or drainage channels, and to the occupation of salt marshes. For example, it mentions one ferdeller who paid 2*s*. 'et custodit unam thecam j Gulet cum opere unius hominis', a holder of 5 acres who paid 1*s*. and kept one gulet with one man, and 6 smallholders, each of whom 'vadit ad gulet maris'. There was also a ferdeller who paid 2*s*. for his tenure or 'una pecia est de Wicha et altera de saleland', which may mean salt land, i.e. reclaimed salt marsh or 'warth'; an entry for Lympsham also tells us that Alured held 2 acres 'de bocland in excambium terre sue quam mare occupavit', which must have been an all too common setback along the coast and which is more than substantiated by examples from later centuries.[1] Away from the coast in Brent there is a suggestion of reclamation in the entry that 12 tenants did 20*s*. worth of work *in mora* and only a small service at harvest time, the lightness of the service suggesting that they were fully occupied in turning the moor into meadow.[2]

Although the clay belt was flat and required artificial channels, or at least the scouring of the natural ones, to carry away the surface water, the danger of inundation was not so much from land floods but rather from sea floods, which could only be stopped by large sea

[1] *Liber Henrici de Soliaco*, ed. Jackson, 65, 67 and 79. For later evidence of flooding in Huntspill where land was only 'occasionally arable on account of inundation', see *Cal. Inquisit. Post Mortem, Edward I*, 321, and for flooding at Stockland Bristol, see 'Cartulary of St. Mark's Bristol', ed. Ross, *Bristol Rec. Soc.*, XXI, No. 199 (1316) and No. 244 (1326). See also p. 83 below.

[2] *Liber Henrici de Soliaco*, 67.

walls. Sea walls are mentioned alongside the Axe as early as 1129,[1] and in the survey of Brentmarsh in the *Inquisicio magistri Reginaldi de Fontibus* of 1201, 'Wreseldi' are mentioned at Berrow, 2 at East Brent, and 6 at South Brent.[2] The word is Saxon in origin and probably means settlers on or in the 'Wre', but the meaning of 'Wre' is unknown. They might be the forerunners of the Wickarii or Wickmen (dwellers on salt marshes) who figure in the *Custumaria* of Michael of Amesbury in the early thirteenth century and whose sole duty it was to maintain the sea walls. There were 4 in South Brent, 3 in Berrow, and 4 in Lympsham.[3] Other wall duties are mentioned in the *Custumaria*; for example, 5 tenants in East Brent, holding 3 acres each, owed a service described as 'custodire wallam'; at Lympsham, Henry, son of William, was obliged to make up his section of the sea wall for his customary day's labour between Michaelmas and hoeing time, as were 14 other tenants holding half virgates, and 10 ferdellers.[4] Contemporaneous with this evidence for the Glastonbury estates is a custumal of Bleadon, in the great looping meander of the lower Axe river, which was an estate of the See of Winchester. Here three great 'Hammes' and pastures had been made, called the 'innynge', and the surrounding sea wall was maintained by 10 virgaters, and another one who paid 8*s*. in commutation for his services, together with 12 half virgaters.[5] It seems that by the beginning of the thirteenth century a line of sea walls was in existence along the lower estuary of the Axe river and probably in gaps in the natural defences of the dunes along the coasts in the Glastonbury estates.[6] Indeed, sea walls must also have extended further south

[1] *Adami de Domerham*, ed. Hearne, 307–8, 'offendi terram hinc circulari aggere vallatam, illinc profunditate fluminis circumseptam'.

[2] Trinity MSS. (Cambridge), R5, 33*b* 112*v*. I am indebted to Mr Neil Stacy of Magdalen College, Oxford, for drawing my attention to this information.

[3] 'Rentalia et Custumaria', *S.R.S.* v, 39, 48 and 51. For confirmatory evidence see Adam Bere's *Terrar* of 1517, in *Joannis confratis et monachi Glastoniensis*, ed. Hearne, II, 333–4, where the Wickarii are called Moor-men and Wykemen. There were eighteen of them and they were charged with the care of the walls and sluices as other tenants were with ditching duties.

[4] 'Rentalia et Custumaria', *S.R.S.* v, 43 and 48–50. For example, 'debet facere wallam suam pro handanis suis inter Festum Sancti Michaelis et tempus serclandi'.

[5] 'Notice of the Custumal of Bleadon, Somersetshire, and of the Agricultural Terriers of the 13th century', *Proc. Arch. Soc.* Salisbury (1849), 182–210. For example, 'Si fossare debeat pro dayua sua, tunc messor vel prepositus debet ponere per perticam dayuam suam'.

[6] *Adami de Domerham*, ed. Hearne, 507, when speaking of this period from 1234–52, says 'Maris accessum ibidem quantum ad sortem suam pertinuit multi multis sumptibus potenter abstraxit'. See also G.C. II, No. 1,002 which mentions a ditch with a wall called Newewall in Berrow Village. The wall and ditch were 80 perches long and 4½ perches wide.

towards Huntspill and the Parrett estuary because a Commission of Sewers was issued in 1327 for the view and repair of sea walls in a strip of coastal land between Steart and Buctle (Burtle?) in length, and between Highbridge and Woolavington in width.[1]

In the landward margin of the clay belt, the evidence of the *Liber Henrici de Soliaco* concerning ditching duties can be supplemented by looking at the *Custumaria* of Abbot Michael where 'fossare' was a common villein service, particularly in the 'Mordich' which figures so conspicuously in all the manors that one wonders if this is a reference to the Pilrow cut (Fig. 8).[2] Other ditching duties in rivers throughout 'Brentmarsh' are itemized in an early thirteenth-century document in the Glastonbury Cartulary but the river courses are very difficult to identify. In a Custumal of Neifs of 1332 for Biddesham one reads of John Benshef, who, amongst other things, had to 'make drains towards the sea and fetch timber (value 2*d.*)' and do 'half a day's ditching with two men (value 1*d.*)'.[3] Examples of these and similar customary ditching duties could be multiplied endlessly.

Altogether, the evidence for the clay belt reclamation is scanty and incomplete, but from what is available one must conclude that a very large part of it was settled and utilized. Evidence in the analogous area of the Lincolnshire siltlands shows how complete their settlement was by the end of the thirteenth century, and the same must have been true of the Somerset coastal clay belt.[4]

The second feature of the distribution is that reclamation activity did not occur in peat areas. The peat soils of the raised bogs in the central Brue valley, and to a lesser extent in the 'Sedgemoor' peats in the rest of the Levels, were avoided because they were acid, sterile, and deficient in plant nutrients. Moreover both peat areas were equally subject to periods of prolonged inundation; flood waters were trapped in the peat moors for months on end because their general level was between six and ten feet below that of the coastal clay belt, and below the main rivers and their natural levees. Even if the water did not remain on the surface the peat continued to be so

[1] W. Dugdale, *History of Imbanking and Drayning* (1662), 2nd ed. by C. N. Cole (1772), 105, using Patent Rolls, 14 Edward III, pt. I, m. 28*d.* For evidence of smaller walls around individual enclosures, see 'Somersetshire Pleas', No. 1452 *S.R.S.* XI.

[2] 'Rentalia et Custumaria', *S.R.S.* V. For example in East Brent (p. 40) 'fossare in Mordich per j diem et in cursu aque x perchatas'. About 100 persons were involved in the scouring-out of Mordich. See Figure 8 for its location.

[3] G.C. II, No. 1015, and *W.M.* I, 347.

[4] Hallam, *Settlement and Society*.

waterlogged as to be of little use. In this connexion it is relevant to note that the present outlet of King's Sedgemoor, and possibly that of the Brue valley, were not then in existence, and the evacuation of water from the moors behind the clay belt barrier would therefore have been more difficult.

Another important reason for the avoidance of the peat soils in the Brue valley was their liability to swell with prolonged wet weather; seasonal variations of up to six feet in the peat bog centres were said to have been observed as late as the eighteenth century.[1] Thus the foundations of the larger drainage structures such as sluices and walls would have tended to be unstable, and the flow in drainage ditches possibly interrupted at the very time it was needed most. The Axe valley was also a negative area from the point of view of reclamation. The valley was floored by a gray silty estuarine clay on which some very poorly drained soils had developed. Much of this clay was situated below the level of the natural levees of the rivers and was often flooded and waterlogged. In places the surface was covered with one to two feet of peaty clay, which in many places gave way to a true peat.[2] Consequently, the valley was more akin to the peat lands than to the clay lands in its unsuitability for reclamation.

The absence of evidence for reclamation in the peat moors does not arise from a lack of documents, as may be the case in the clay belt. The whole of the Brue valley, most of the Axe valley, all King's Sedgemoor, West Sedgemoor, Queen's Sedgemoor, in fact all the peat lands, were under the control of either Wells, Glastonbury, Athelney or Muchelney (Fig. 4); yet, despite the comprehensive and numerous surveys and cartularies from the twelfth to the fourteenth century covering all these ecclesiastical estates, there are barely any references to reclamation or draining in the peat lands. The few examples of reclamation that do exist in King's Sedgemoor and the Axe valley can probably be explained by the presence of either flood-derived or down-wash deposits, which had ameliorated the condition of the peat.

Thirdly, the areas of draining activity were mainly on alluvial soils. In contrast to the peat lands, the alluvium and clay were more favourable for reclamation; the very nature of their origin as flood

[1] For example, Billingsley (1798), 169; W. Buckland and W. D. Conybeare, 'Observations on the South-western coal district of England', *Trans. Geol. Soc.* I (1824), 309; and Phelps, 'On the formation of peat bogs and turbaries...', *S.A.N.H.S.* IV (1853), 101.

[2] See Findlay, *Soils of the Mendip District*.

deposits meant a slight, but nevertheless significant, slope away from the river, and this facilitated drainage. Therefore, even if they were the most frequently flooded lands of the Levels, they were often the most rapidly cleared of water. In addition, the ground was more tenacious than the peat, and the deposits left by the flood waters constantly restored the fertility of the soil and permitted the growth of more nutritive strains of grass.

Each of the areas of activity can now be examined in turn, with regard to its physical setting and historical development.

THE REGIONS OF RECLAMATION

The basin of the lower River Parrett

One important area in which reclamation took place was in the lower portion of the Parrett, in the vicinity of the junction of the River Tone and of the River Cary, which at this time flowed into the Parrett. Here was a wide expanse of alluvial soils. The nucleus of activity was south and west of Sowy 'island', and adjacent to this centre were two other areas of reclamation activity, one following the line of the Tone valley, the other in Aller Moor (Fig. 7*b*).

Sowy. It seems certain that the great concentration of drainage works south of Sowy was due in part to the prosperity of the 'island' itself. The freely drained sandy loams of the 'island' were some of the most easily worked and fertile soils that were favourable to early cultivation in Somerset. The 'island' contained about 1,850 acres of flood-free land, of which 1,275 acres were already under cultivation in 1235 in the large open fields which surrounded the three main settlements of Westonzoyland, Middlezoy and Othery.[1] But when one adds the area of Thorngrove Wood and of the settlements and roads to the area of the open fields, it is obvious that there could have been little waste land for the expansion of either agricultural or pastoral activity on the 'island'. The only area available was in the surrounding moors.

Already by the time of Abbot Michael's Survey in 1234, 722 acres of meadow, in sizes varying between half an acre and 24 acres, had been reclaimed from the low grounds surrounding the 'island', and

[1] 'Rentalia et Custumaria', ed. Elton, *S.R.S.* v, 25–33.

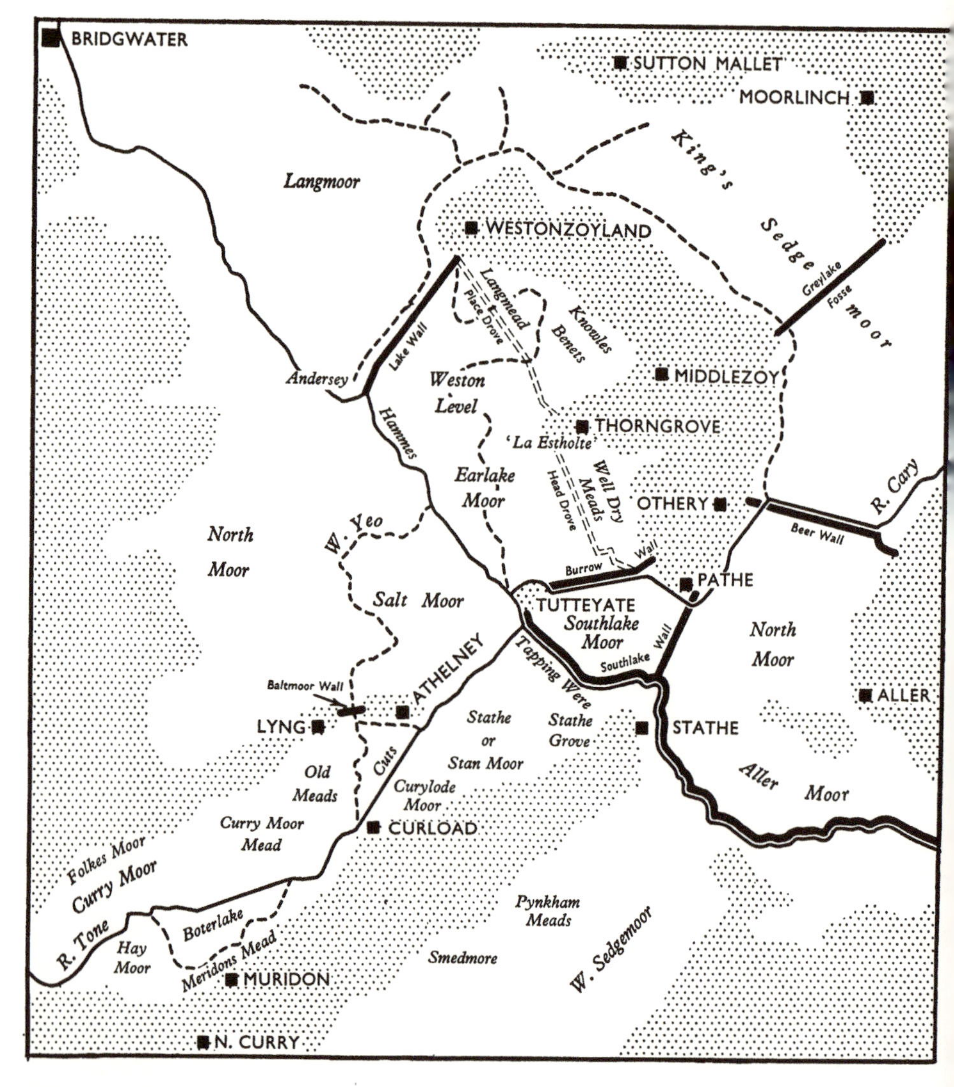

Figure 7*a*. The 'island' of Sowy and the lower Tone valley in the Middle Ages. (Key as in Fig. 7*b*.)

forty houses had recently been built on them.[1] The new meadows lay scattered throughout the moors, some near Tutteyate (Burrow Bridge), some alongside the Parrett at Hammes, and others to the north-west in Langmoor, between Chedzoy 'island' and Sowy. Most were in Knowles, Benets and Langmead, which were situated

[1] 'Rentalia et Custumaria', ed. Elton, *S.R.S.* v, 175, 'Memorandum quod tempore M. Abbatis erecta sunt xl[a] mansa in moris de Sowy.'

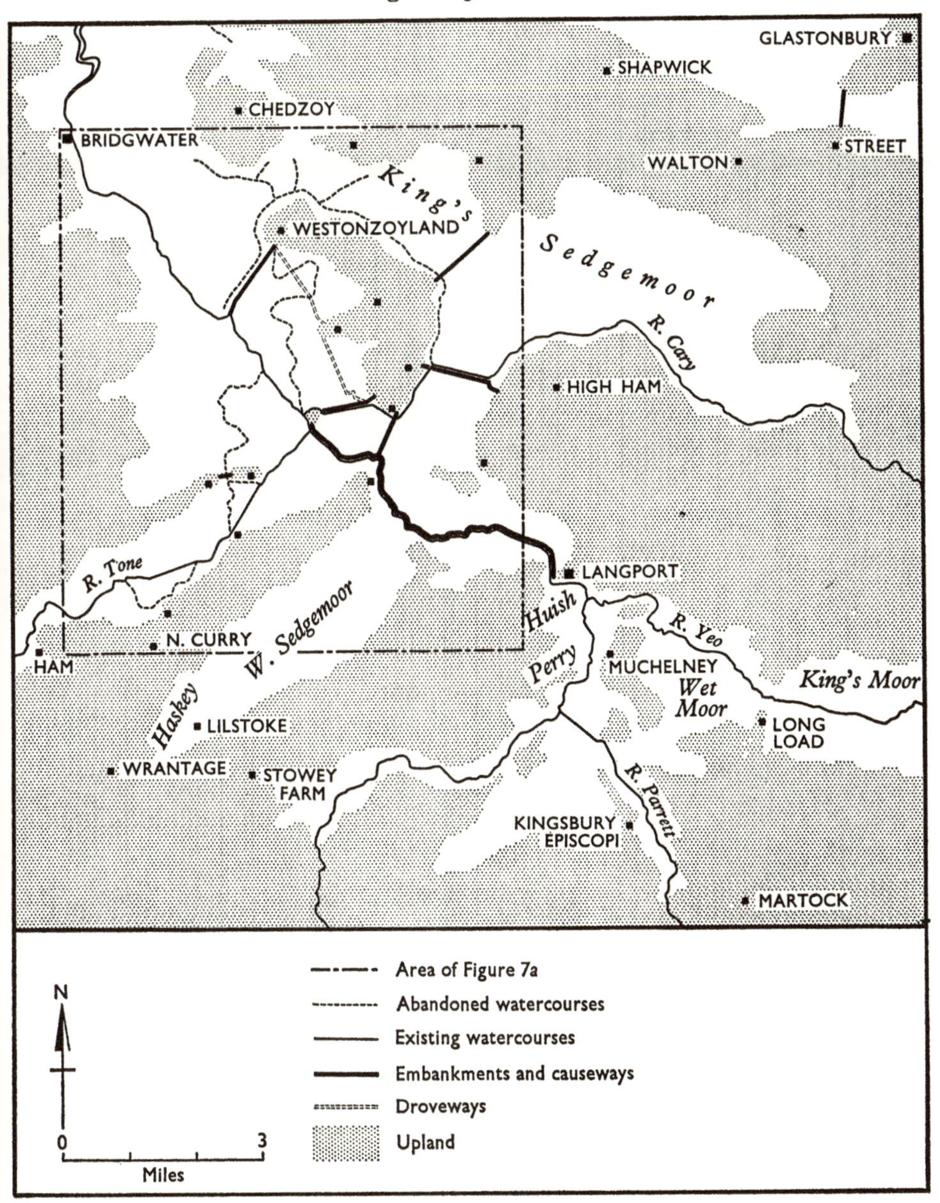

Figure 7*b*. The Southern Levels and King's Sedgemoor in the Middle Ages.

between Westonzoyland and Thorngrove Peninsula, and behind Place Drove, many others were in Well Dry Meads behind Head Drove between Thorngrove and Pathe. These meadows paid 'morgabulum' or moor-penny, which was a money rent paid for an enclosed and reclaimed piece of land in the moors on which a house had also probably been built.[1] The morgabulum amounted to £4. 8*s*. 10½*d*. for Othery, £3. 11*s*. 5*d*. for Westonzoyland, and £4. 8*s*. for Middlezoy, a total of £12. 8*s*. 3½*d*., which was over one half of the total rent roll of the manor of Sowy. Other lands were being reclaimed early in the thirteenth century, for under the heading 'Hoc est incrementum de toto manoria de Sowy de mora tradita pratum faciendum tercio anno M. Abbatti' (1238)[2] appears a list of 250½ acres of meadow, 231 of which were valued as follows:

	1 acre worth 10*d*. per acre
	8 acres worth 9*d*. per acre
	5 acres worth 7*d*. per acre
	16 acres worth 6*d*. per acre
	151 acres worth 4*d*. per acre
	38 acres worth 2*d*. per acre
	12 acres worth 1*d*. per acre
Total	231 at an average of about 4*d*. per acre.

This average value for new meadow was exactly the same as the average per acre for the 1,275 acres of cultivated land on the 'island', and it is a measure of the obvious desirability of, and success in, reclaiming the moors.

Therefore, by the end of the fourth decade of the thirteenth century a total of 972½ acres is known to have been reclaimed from the moors surrounding Sowy. Some general confirmation of this reclamation is given in a document of 1247, in which Abbot Michael granted St. John's Hospital in Glastonbury £13. 16*s*. 8*d*. 'from approvements in the moors at Sowy manor', £5 each from Westonzoyland and Othery, and £3. 16*s*. 8*d* from Middlezoy, these rents to be

[1] P. J. Helm, in his 'The Somerset Levels in the Middle Ages', *Journ. Brit. Arch. Ass.* 3rd ser., XII (1949), 37, suggests that morgabulum was an annual payment for the privilege of intercommoning, but the entries in the survey clearly relate the payment to individual, small closes of meadow, and in the Westonzoyland entries it is stated specifically that 'holdings of meadow in the moors' contributed morgabulum. See also N. Neilson, 'Customary Rents', in *Oxford Studies in Social and Legal History* (Oxford, 1910), II, 47.

[2] 'Rentalia et Custumaria', ed. Elton, *S.R.S.* v, 33–5. This was a period of very dry weather in south and west England. See J. Titow, 'Evidence of weather in the Account Rolls of the Bishopric of Winchester, 1209–1350', *Econ. Hist. Rev.* XII (1960), 360.

paid by the manorial bailiff and reeves to the almoner each Christmas.[1]

The way in which some of these reclamations came about is also known. For example, in 1250, Nicholas, son of Godfrey de Sowy, was given 12 acres of waste land in the northern part of the Weston Level by Abbot Michael of Glastonbury in exchange for various illegal enclosures that he had made in the common pastures at some previous time. At about the same time, Hugh and John le Rous received a new enclosure of 8 acres, once held by William de Tonhgrave, to the south of Thorngrove, in 'la Estholte', in exchange for the illegal improvements in 'Zoymoor' in which Abbot Michael reserved his right to upgrade and extend if he so wished. Later, in 1255, they also gave up common rights in Burrow Moor (Southlake Moor) and in the moors on the south side of Thorngrove 'peninsula'.[2]

By the time of the survey of Abbot Roger Ford in 1260, the total area of meadow in the manor had increased to 1036 acres, and it is tempting to suggest that the difference between this figure and that of 1238, which amounts to 63½ acres, represents a true increase in reclaimed meadow, but this may have been affected by other factors.[3] The number of acres of meadow in 1260 was fairly evenly distributed between the three settlements on the 'island', 361½ acres in Othery, 297½ acres in Middlezoy, and 377 acres in Westonzoyland, but the morgabulum had risen markedly from the figure of 1238; it was £7. 8*s*. 0*d*. in Othery, £5. 12*s*. 2*d*. in Middlezoy and £7. 7*s*. 8*d*. in Westonzoyland, or a total of £20. 7*s*. 10*d*., which represented an increase in value of more than 25% over a span of approximately twenty-two years. This increase must be accounted for largely by the improving value of the meadows as their drainage was made more effective.

There is ample evidence to suggest the continuation of reclamation in the moors in later years. In 1281, Walter de Knolton was allowed to retain 30 acres of illegally made meadows in Southlake Moor for a rent of 7*s*. 6*d*., in return for renouncing all common rights he possessed in land then being enclosed by the abbot in the same moor.[4] In the same year a similar agreement was made with Richard and Nichola Pyke who forfeited their claim to common in the Weston

[1] G.C. III, No. 1,318.

[2] G.C. II, Nos. 927; 922 (confirmed in 1272, No. 914); and 912.

[3] 'Rentalia et Custumaria', *S.R.S.* v, 170–5.

[4] G.C. II. No. 905.

Level, where the abbot was about to make a new enclosure, and they were given in return 10 enclosed acres of 'moor and pasture' in 'Estholt' for 5*s*. per annum, and 5 other unspecified closes.[1] Similarly, Nichola de Sowy gave up 20 acres of moor and alder in Southlake Moor adjacent to Burrow Mound (Tutteyate) and Tappingwere. Eleven years later, in 1292, the same Richard Pyke was allowed to hold a piece of pasture in the east end of Greylake, fringing the north-east corner of the 'island', for a rent of 6*d*. a year, once more exchanging it for enclosures he had made in Westonzoyland moors and in King's Sedgemoor. By 1302, the abbey had reclaimed a further 100 acres of pasture and meadow in various places along the eastern edge of the 'island' between Greylake and Othery, and 60 acres in the Westonzoyland Level, whilst other persons enclosed 5 acres in Middlezoy, 20 acres in 'Estwer', and 3 acres in 'Westwer'.[2]

For the years that follow, references to reclamations and 'approvements' are again frequent. Among the more important agreements was one in 1307 whereby Joan, widow of Walter de Knolton, gave to Abbot Geoffrey de Fromund all pasture rights in enclosures made in the Sowy moors some time before, retaining 8 acres of meadow in 'Campingwere' (?), which was enclosed sometime between 1234 and 1352, and 30 acres of marsh in Southlake Moor, enclosed between 1274 and 1290. She also held 20 acres of meadow and pasture in Southlake Moor for 10*s*. a year, but her ownership of this meadow was illegal and was going to be investigated by the abbot. In the same year, Godfrey de Sowy resigned to Glastonbury all his rights in the 'moors and marshes' of Sowy.[3]

These reclamations did nothing to alleviate the problem of conditions in the main river, on which the successful draining of the moors ultimately depended. On the contrary, the Parrett must have overtopped its banks nearly every autumn and winter and so rendered the new meadows and improved pastures useless. The flooding of the Parrett had a twofold origin, arising, as a pre-drainage account put it,

[1] G.C. II, No. 921. This agreement is endorsed in No. 922, which adds that Richard renounced purprestures made in Zoy, Shapwick, Walton, Street and Beckery, and in Sedgemoor, and in the moor near Beckery.

[2] G.C. II, Nos. 926; 923 and 924; and 384.

[3] G.C. II, Nos. 906 and 940. See also No. 939 (Sept. 1316) where Godfrey renounced all rights in 5 acres of meadow and pasture in Sowy Manor. Other examples and indications of improvement are contained in G.C. II, No. 957 (*c*. 1250), No. 954 (n.d.), No. 955 (1262), No. 959 (1302), No. 939 (1316) and Nos. 947–52 (1325–38) which concern the involved transactions whereby William, son of Godfrey de Sowy, who was in debt, gave up all land in Sowy, which included many meadows, to Glastonbury.

from 'the inundation of the tides flowing in by the said river as of the fresh water descending by that and other channels', this obviously being a reference to the rivers Cary and Tone The effect of the rising tide on a full river was particularly feared; Sowy inhabitants had long complained that 'the sea defences [were] not sufficient to prevent the sea water flowing into and scouring the main river'. The answer to this problem lay in building a clyse downstream near the outfall, but this was hampered by the fact that the coastal clay lands in which the outfalls were situated were not owned by the same people who owned the peat moors; consequently there was no identity of interest. The lack of co-operation between the inhabitants of the relatively flood-free coastal lands and those of the peat moors was to be a recurring theme in later years. Nor was the danger of flooding wholly of a natural origin; there were many 'gurgites', or 'fishing weirs', in the Parrett, which raised the water-level to their own height. There were six weirs between Sowy and Langport alone, which included 'Pouwere which blocks the water course and makes the river flow back among the moors...a weir called Wozwere which does great damage...a weir called Nywere which does harm in the same way...a weir called Wichok...a weir called Tapping Were which is a great obstruction...[and] a weir called Burwere'.[1]

The only solution to flooding in this portion of the Parrett near Sowy lay in building embankments to surround the moors and exclude the flood waters. Therefore, during the thirteenth century an impressive system of embankments or walls was constructed, but whether the walls were built before the individual reclamations of moorland took place or whether the opposite was true will probably never be known. Southlake Wall was built from near Pathe to Tutteyate, which is probably the strange mound beside the Parrett known as Burrow Mump. The wall all but encircled the moor and effectively sealed the eastern and south-eastern corners of Southlake Moor, and the Sowy moors in general, from the Parrett (Figs. 6 and 7*b*). The course of the Parrett was also straightened where the new wall abutted upon the river.[2]

Southlake Moor was still subject to flooding in the north from the

[1] Dugdale, *History of Imbanking and Drayning*, 107, using Patent Rolls, 8 Rich. II, pt. 2, m. 36; G.C. II, No. 969 (early fourteenth century); and G.C. II, No. 969.

[2] For evidence of this wall, see P.R.O. C7/77/2/39; *Cal. Pat. Rolls 1313–1317*, 411–12; and Dugdale, *History of Imbanking and Drayning*, 106, using Patent Rolls, 8 Rich. II, pt. 2, m. 26 (1385).

River Cary; therefore, in order to obviate this constant threat, 'Burwall', or Burrow Wall as it is now known, was built between Tutteyate and Sowy.[1] From the evidence of air photographs it seems possible that the Cary previously meandered through Earlake Moor and was canalized to run alongside the new wall.[2] Thus, Southlake Moor was entirely surrounded by protective embankments, and, later, rhynes were dug across its surface.[3] To the north-west was Lake Wall which joined the high ground of Westonzoyland with the bank of the Parrett. The wall protected the Sowy moors from the flood waters that accumulated in King's Sedgemoor and which at this time had their exit through the gap between Chedzoy and Sowy 'islands', and met the Parrett at Andersey.

There remains the problem of how the Parrett was excluded from the western side of Earlake Moor and the Weston Level. Air photographs reveal the presence of a huge former channel of the Parrett cutting into Langmead; yet, as these meadows in Langmead were in existence by 1235, the channel must have been defunct by that time. Either natural siltation of the levees of the Parrett raised them and so cut off the channel, or the banks were artificially raised to exclude the Parrett water from the moors, which is likely when one recalls the other impressive works undertaken in the vicinity.

Two other walls were built from Sowy in the late thirteenth century, but these were causeways rather than drainage works. Beer Wall linked Sowy with High Ham across North Moor, and Greylake Fosse united the 'island' with the rest of the Glastonbury estates by joining it to the Polden Hills, across King's Sedgemoor (Fig. 7*b*). This latter causeway impeded the flow of water from east to west in King's Sedgemoor and was said to have produced serious flooding.[4]

A final indication of the extent of reclamation in the Sowy moors

[1] Dugdale, *History of Imbanking and Drayning*, 106.

[2] For an early nineteenth-century description of this wall see Brit. Mus. Add. MSS. 33,691, ff. 121–2 '... this bank was evidently risen in remote times not only as a pass across the country but to prevent the floods from injuring the lower level of the Marsh... [It] measures thirty feet in width by twelve in height. A small stone wall is carried along the top inasmuch to prevent the wearing of the water when it rises over the mound in winter. Some idea may be formed of the number of years the boundary has stood from the circumstances of the ground having risen four or five feet from the deposit of the water on the upper level...'

[3] Dugdale, *History of Imbanking and Drayning*, 106–7.

[4] G.C. II, No. 384. It reads, 'quod abbas Glaston levavit unum fossatum per medium more versus manerium suum de Sowy et super fossatum fecit calcetum de petris et ita obstupavit aquam ad grave dampnum patrie...' The date of the document is 1302.

in the Middle Ages can be ascertained from the fact that a breach made in Southlake Wall in 1311 by Sir Matthew de Clyvedon, who owned the adjoining Aller Moor, led to the inundation of 1,000 acres of barley, beans, peas and oats, 50 acres of meadow land and 300 acres of pasture land. This total of 1,350 acres is approximately 70 % of all the lowland that lies between Southlake Wall in the south, Lake Wall in the north, and the River Parrett and Sowy 'island' in the west and east. The reference to crops in the reclaimed land is also significant (although probably exaggerated), because these crops relied on a thorough and efficient drainage system and not on a seasonal one, which says much for the effectiveness of the walls in excluding floods.[1]

The Tone valley. From Figure 6 it is evident that a subsidiary concentration of reclamation was centred in the Tone valley, to the south-west of the Sowy nucleus. Reclamation in this locally complex area of alluvial flats and peat moors, which were separated by prominent ridges, arose largely through the activities of the Abbot of Athelney but more especially through those of the Dean and Chapter of Wells, who acquired at the end of the twelfth century the whole of the Stathe ridge, parts of West Sedgemoor and the moors of the Tone valley. (For the holdings, see Fig. 4.) The narrow alluvially-covered valley of the Tone, containing Hay, Boterlake and Curry Moors, widened out to the east into Stan or Stathe Moor. North of the River Tone and the Isle of Athelney, and west of the River Parrett, the alluvium of Salt Moor overlapped the peat of the wide expanse of North Moor, further to the west.

Draining activity followed much the same pattern as in the Sowy area, with minor reclamations and upgradings preceding more ambitious schemes. Much of the reclaiming activity was also part of a concerted effort by the dean to consolidate and define his new estates in the area, particularly those south of the Lyng ridge and the River Tone. Thus, in 1233, John and Agnes de Alre were accused by the dean of 'setting up a dyke' in Salt Moor and enclosing Stathe Grove. The inevitable counter-claims of wrongful enclosures followed, but whatever the rights of these conflicting claims, John and Agnes gave the dean, amongst other things, 'leave to convert into

[1] P.R.O. C7/77/2/39 and G.C. II, Nos. 967 and 968. There is also a suggestion of cultivation in the moors in G.C. II, No. 958.

meadow 60 acres of Stathemoor' and in recompense they received six new enclosures in the common, permission to enclose Stathe Grove, 'and as much as they will of their land and meadow adjoining...and such part of Salt Moor as they think fit and convert it into meadow'. The Abbot of Athelney was the other landowner with a major interest in the moors, therefore the dean met him in the middle of Stathe Moor and compounded an agreement with him, in 1250, by which the abbey renounced all rights in the moor except for firewood, the pasture of Prior's Moor and the moor belonging to their tenant, Hugh de Curylade. The abbot further conceded to the dean 'all the purprestures made in Curymoor and of the whole island of Haygrove [?] and others new and old in the manor of Northcury'. The agreement also mentions that a new rhyne, Willaumsdich, had been excavated, but this cannot be identified.[1]

During the next few years the dean consolidated his West Sedgemoor holdings and established his rights against a number of smaller landholders to purprestures made 'from the Rhine called Peret to the town of Wrentichford [Wrantage?] belonging to the manor of Northcury', that is to say, throughout the whole length of West Sedgemoor; and he also established his right to enclose the 'lands and meadows' of Lutstoke [Lilstoke] near Cathanger and Stawey [Stowey Farm] which belonged to Geoffrey de Scoland and Richard le Walys, Richard and Geoffrey retaining their common of pasture over the rest of the adjacent moor and the right to enclose their lands 'which lie everywhere towards the sun facing the said moor'.[2]

When compared with what was to follow in 1263, these agreements were of limited significance, and a mere prelude to a series of more complicated settlements over rights of enclosure and reclamation throughout the whole of the Tone valley. Both of the two major remaining landholders, Sir John de Alre and the Abbot of Athelney, recognized the right of the dean to enclose the meadows and pastures of Corymormede, Leverymede Yetta, Meridonesmede, Richemannesmede, Almenshey [Alouereshay] Nelesmede, Henrysmede, Edwynmede, Hughsmede, Slomede, Babbemede, Curymede, Haymoor, Babbecraft Stadmormede, Wythry (many of which were in the area known today as Oldmeads) and 40 acres of newly-enclosed moor in

[1] *W.M.* I, 10–11, and 317–18.

[2] *W.M.* I, 319 (1256–7), I, 32 (1259–60) and 'Pedes Finium', No. 205 (1262), *S.R.S.* VI; see also *W.M.* I, 319 (n.d.) for 2 acres of purpresture in 'Haskemore'.

Curry Moor, all in the Tone valley; and Haskeymoor, Pynkham, Smedmor, and Jutteshullesmore in West Sedgemoor. In return, the abbot was given a meadow near Lyng called 'la Hocmede', which was enclosed by a dyke, and Sir John was given 40 acres of moor in Stathe Moor on the Stathesmede, 'with power to enclose the same, saving the tithe thereof if converted into meadow or arable land [*cultura*]'.[1]

Despite this settlement, the continual disagreements and disputes between the dean and the abbot and other landholders over the use and apportionment of the moors throughout the Tone valley finally culminated in an even more extensive and comprehensive agreement in 1311. The various parties agreed to surrender 886 acres of moor land in 'Curymor, Boterlak, and Stathmor...which the Dean and Chapter are about to enclose and hold in severalty by consent of the commoners'. John de Murdene was recompensed for his loss of common with 7 acres in Curry Moor, 'with liberty to enclose and hold the same in severalty at a yearly rent of 3*s*. 6*d*. (namely 6*d*. an acre)'; he also agreed to allow the dean to enclose a further 20 acres in Stathe Moor. Under similar arrangements Walter le Freye was given 4 acres in Stathe Moor; Robert le Hyreys, Simon de Domeram, and John Dreu, 4½ acres in Curry Moor; Richard Broun, 2 acres in Curry Moor; William Hughet, 4½ acres in Curry Moor, and Robert de Ocham, 1 acre in Curry Moor; William and Agnes Bussell, 8 acres in Stathe Moor; and the Abbot and Convent of Athelney, 41 acres in Stathe Moor. Athelney also received another 100 acres of moor in an unspecified close in northern Curry Moor, which was an additional recompense for its loss of the moors.[2]

Obviously, there was a large measure of acquiescence over the division of these moors; Bishop Drokensford of Wells wrote to the dean, 'Glad to hear that the Partitioners in the common pasture of their moor at North Curry (without whose consent the ownership of the soil would be profitless) agree to apportionment and separate use to their great advantage'. Little was done immediately, but by 1316

[1] *W.M.* I, 317 (1263) for agreement with Athelney. See 'Athelney Cartulary', No. 135, *S.R.S.* XIV, for confirmatory details, and *W.M.* I, 322 (1263) for the agreement with Sir John de Alre. See also *W.M.* I, 319–20 (1280) for later confirmation and expansion of these agreements and details of a small enclosure of 2 acres in Stathe Moor, worth 12*d*. yearly, near 'Westere trench'.

[2] *W.M.* I, 408; *W.M.* II, 585–6, Nos. 175, 176 and 177; *W.M.* I, 423; *W.M.* II, 585, No. 174; 'Athelney Cartulary', Nos. 136 and 137, *S.R.S.* XIV; *W.M.* I, 313–14; see also *Cal. Pat. Rolls, 1307–1313*, 265.

two people were appointed 'to cause the moors to be measured and assign portions to the freeholders and villeins'. The dean's freeholders in Muridon were to have 'one acre of the moor newly enclosed beyond the portion formerly assigned to them', together with 30 acres of common in Curry Moor, and 'a wide road' to drive their cattle through Boterlake Moor to Folkesmoor. The cottars were to get 'one acre of moor in severalty attached to their cottages... [and] give 6*d.* yearly for every acre of moor assigned to them'.[1] These documents contain one of the clearest indications we have of the reclamation of the commons and of their conversion into individual closes.

In the ensuing years the dean handed over various enclosures in Stathe Moor, totalling 89½ acres, to various persons to stop their claiming common in his moors in the Tone valley, charging a rent of 3*d.* per acre, which suggests that these enclosures were probably not very well drained pieces of land.[2] An indication of the value of draining meadow in the Tone valley comes a little later when fourteen closes of pasture of 5 acres each, in Hay Moor, were valued at 7*d.* per acre, but when fully enclosed and upgraded were to be let at 1*s.* 4*d.*, or more than double their original value.[3]

To the north of the Lyng ridge and the 'island' of Athelney, the Abbey of Athelney had had almost undisputed sway over the alluvially-covered area of Salt Moor and the peat waste of North Moor, but by the beginning of the thirteenth century claims on these moors were reduced to a holding of 100 acres, and rights of common and firewood; other inroads were being made into its possessions in Salt Moor in 1241, and small concessions of 24 acres were made to John and Agnes de Alre on the north side of the moor near 'Ferling' and in the land alongside the Parrett towards 'Laberegh' (?). What is particularly significant about this last agreement is the evidence of cultivation in the moors, for it stipulated that John was not to 'break up or cultivate the said moor, more than was broken up and cultivated on the day this concord was made'. Athelney recovered some land in North Moor through gifts during the middle of the century, but on the whole the abbey would not appear to have been either as acquisitive or as aggressive in its claims on these moors to the north of

[1] 'Calendar of the Register of John de Drokensford, Bishop of Bath and Wells, 1309–1327', ed. Hobhouse, *S.R.S.* I, 35; *W.M.* I, 186–7.

[2] *W.M.* I, 416 (1337). See also *W.M.* I, 417, where a *cultura* is mentioned in Stathesmoremede, which was in the Levels.

[3] *W.M.* I, 411.

the island as the Dean of Wells was, at its expense, in the moors to the south.[1]

All these piecemeal reclamations contributed nothing, however, to a final solution to the flood problem of the Tone valley. Flooding in the lower portion of the valley was particularly serious, and it was caused by the overflow of the River Parrett, and more especially by the inadequate nature of the Tone outfall. At this time the Tone flowed towards the Athelney-Lyng gap and then bifurcated; one channel passed through the gap and meandered across North Moor; the other channel, and probably minor course, branched eastwards to run south of Athelney 'island' and re-entered the Parrett near the present Tone-Parrett confluence (Fig. 7*a*). These meandering channels had little chance of coping successfully with the sudden and devastating 'freshes' which swept down the Tone from the Western Hills; therefore, in an attempt to obviate the flooding, in 1374–5 the Abbot of Athelney and other landowners diverted the River Tone into a new embanked channel, three-quarters of a mile long, whereby, it was said, 'the flow of water was kept from its ancient course'. This channel ran to the south of Salt Moor, and any flooding from it was therefore more likely to affect Stathe Moor, which did not belong to the abbey. Further documentary evidence of this river diversion is supplied twenty years afterwards when the Abbot of Athelney admitted that he had illegally expropriated 'a water course where the river Tone used to flow, which belongs wholly to the Dean and Chapter until they diverted the said river and levelled the ancient course with the adjacent meadow...'.[2] The final protection of Salt Moor from any breakthrough of the Tone flood waters via the Athelney-Lyng gap was accomplished by the building of Balt Moor Wall, which must also have served as a causeway.

The evidence of air photographs, together with the evidence in the field, confirms the testimony of these documents. The abandoned course of the Tone towards the Athelney-Lyng gap is marked by unmistakable evidence of a large channel, which is now followed by the sinuous 'Crooked Drove'. To the north of the gap the course is

[1] 'Athelney Cartulary', Nos. 28, 99, 106, 107 and 118, *S.R.S.* XIV; and 'Pedes Finium', No. 118, *S.R.S.* VI.

[2] *W.M.* I, 425–8 and 324–5. This latter document also gives the boundaries of this triangular piece of land enclosed between the old river and the new cut, and these boundaries can be identified. It is also relevant to note that this piece of land is still known as 'Cuts'.

continued by the present West Yeo, a name which is significant in itself (Fig. 7*a*).[1] The other original branch of the river south of Athelney 'island' is also marked by a deep depression throughout its course.[2]

It is probably no coincidence that after the diversion of the River Tone a renewed interest was taken in the Salt Moor, near its mouth. Salt Moor was then estimated to contain 1,000 acres, and Athelney had exclusive rights of pasture over it, but the moor was also part of the Manor of Stathe, and three of its principal landholders, Sir John de Beauchamp, Matthew de Clyvedon and Elias Spelly successfully laid claim to the moor in 1382. Athelney was given about 220 acres of it to hold in severalty; Sir John took half of the remainder, and Matthew and Elias a quarter each. The Dean of Wells was not lax in putting forward his claim because the moor was part of the parish of North Curry and within his ecclesiastical jurisdiction. Eventually pasture for 80 beasts was granted to him, but, in lieu of taking his common of pasture, he settled for an annual payment of 4 marks from John, Matthew and Elias.[3] Whether the individual pasture allotments were then subdivided and drained is not known.

It is relevant here to note the probable results of the diversion of the Tone on the drainage of adjacent portions of the Levels. First, Stathe Moor was more likely to be flooded because the channel of the Tone passed nearer to it, and the western arm of the river was now blocked off by Balt Moor Wall. Secondly, the uniting of the two streams of the Tone into one channel, and the diversion of the whole of the flow of the river into a much smaller stream without widening the outfall, meant that there was great pressure on the river banks at flood time.[4] Thirdly, the Tone now entered the Parrett further upstream than formerly, at a point where the Parrett was already too narrow to deal with its own flood discharge and that of the River Cary which flowed into it here. Although it is difficult to assess the effect of this on the condition of the Parrett and its propensity to flood the Sowy moors and Aller Moor at this time, it seems certain that the burden of water was increased at a point where it was least

[1] 'Yeo' is a common local name for 'river', and is derived from the Old English 'ēa'. See Ekwall, *Concise Oxford Dictionary of English Place-Names*, 519.

[2] There is no documentary evidence for the other large diversion of the River Tone in Hay Moor.

[3] 'Athelney Cartulary', Nos. 120 and 119, *S.R.S.* xiv, and *W.M.* i, 322–3, and 326.

[4] There is no evidence that widening took place in the Tone channel below Athelney.

wanted. Therefore, the diversion of the Tone probably created many more problems than it solved.

Aller Moor. To the south-east of Sowy 'island', and alongside the Parrett, are Aller Moor and North Moor; these two moors constitute another secondary area of drainage activity around the Sowy nucleus (Fig. 6). The distribution of river-deposited clays, together with medieval documentary evidence, indicates that these two moors formed probably one of the most frequently flooded areas of the Levels. This is confirmed by present experience. The cause of this flooding was the constriction and winding of the course of the River Parrett, which formed the southern boundary of the moors.[1] Once the moors were flooded, their natural drainage was difficult because they sloped away from the Parrett and towards King's Sedgemoor, and evacuation through the King's Sedgemoor was not always assured because of the partial blockage of Beer Wall and the possibility of an already high flood-water level in the moor, which was ultimately dependent on the level of the River Parrett.[2] In addition to the danger from the Parrett, the flood liability of Aller and North Moors was increased by the lengthy course of the River Cary, which ran alongside and through them (Fig. 7*b*).

These adverse physical conditions may account for the absence of any preliminary reclamation in the area, and not until the very late thirteenth century was any known work begun. Even then the evidence is not abundant but it is clear. Matthew de Clyvedon and John de Acton had 'newly enclosed the moor and made rhynes "et plauntiz a graunt nombre" so that the water cannot follow its old course in that moor'. North Moor was similarly 'newly enclosed... with rhynes', and a large rhyne between Aller and Southlake had already been dug by 1280, if not earlier, to facilitate the drainage of both moors.[3]

To ensure the exclusion of the Parrett flood water a wall must have been built along its banks. There is no documentary evidence of its construction, except an indication that something was happening on

[1] See G.C. II, No. 969 for an account of the obstructions in the Parrett from 'Langport to Zoy' (Sowy).

[2] P.R.O. C7/77/2/39. In 1311 Matthew de Clyvedon cut the Southlake Wall in order to relieve Aller Moor of its flood water, which would not disperse in any other way. (See p. 55 above.)

[3] G.C. II, No. 696.

this portion of the river bank in 1280, when the burgesses of Bridgwater complained that Acton (who owned land only alongside the Parrett in Aller) had stopped them 'towing their boats on the waterway of Peret, along the moorlands and meadows between Brugewat' and Langport'.[1]

The Glastonbury area

The grouping of references to reclamation and drainage in the Glastonbury district is geographically distinct. Operations were mainly concentrated at the foot of the Polden Hills and extended northwards in a narrow belt, past Glastonbury itself, and towards the eastern edges of Meare Pool. Two lesser centres are evident to the east; one in the upper end of the Hartlake valley, the other in the Brue valley, near Butleigh (Fig. 8). Once more the very close relationship of reclamation to alluvial soils is evident, despite the fact that the local variations between alluvium and peat are more complex here than elsewhere in the Levels.

The individual reclamations, upgradings and enclosures of moorland followed much the same pattern as those in Sowy and the Tone valley, and for that reason it is not proposed to enumerate them all in detail, but the relevant references are noted below (see Fig. 6).[2] Some of the larger reclamations, however, are worthy of note; they are that of Lower New Close, west of Glastonbury, some time before 1294, and by implication that of the Higher New Close, the enclosure of 100 acres of moorland in the eastern end of Butleigh Moor, another 100 acres in Wootton Moor in the Hartlake valley, similarly enclosed and divided in 1269 and then let out for rent, and an unspecified, but probably large, 'embanked enclosure' in Godney Moor in 1276.[3] Certainly, by the time of the Survey of Abbot Roger Ford in

[1] 'Somersetshire Pleas', ed. Landon, No. 763, *S.R.S.* XLIV. See Bodleian Library, Rawlinson MSS. D. 706, 'The Survey and Perambulation of Allermore' (1577), for closer details of the drainage system in the moor, and of its maintenance.

[2] G.C. II, Nos. 398 (*c.* 1180), Edgarley; 741 (1263) and 742 (1267), either side of Street causeway; 816 (1269), Wootton Moor; 486 (1265), 487 (1265), 485 (1291), 482 (1295), 481 (1290), 483 (1305), 484 (1308), all concerned with the exchange of illegal purprestures in the moors for unspecified meadows and other grounds; 752 (1273), Prestmoor and east of Street causeway; 779 (1281), Hearty Moor; 363 (1289), Butleigh Moor; G.C. III, No. 1327 (1303), Meare Moor; G.C. II, Nos. 598 (1310), Northload Moor; 587 (1330), Pastreleshamme; and 613 (1340), Godney Moor and Hartlake Moor.

[3] G.C. II, Nos. 478 and 479; 765; 815; and G.C. III, No. 1324 respectively. See also G.C. II, No. 613 where large parcels of Godney and adjacent moors, e.g. 42 acres, 80 acres, and 12 acres, were being let out for the term of lives, which suggests the break-up of commoning arrangements and their replacement by specific occupation.

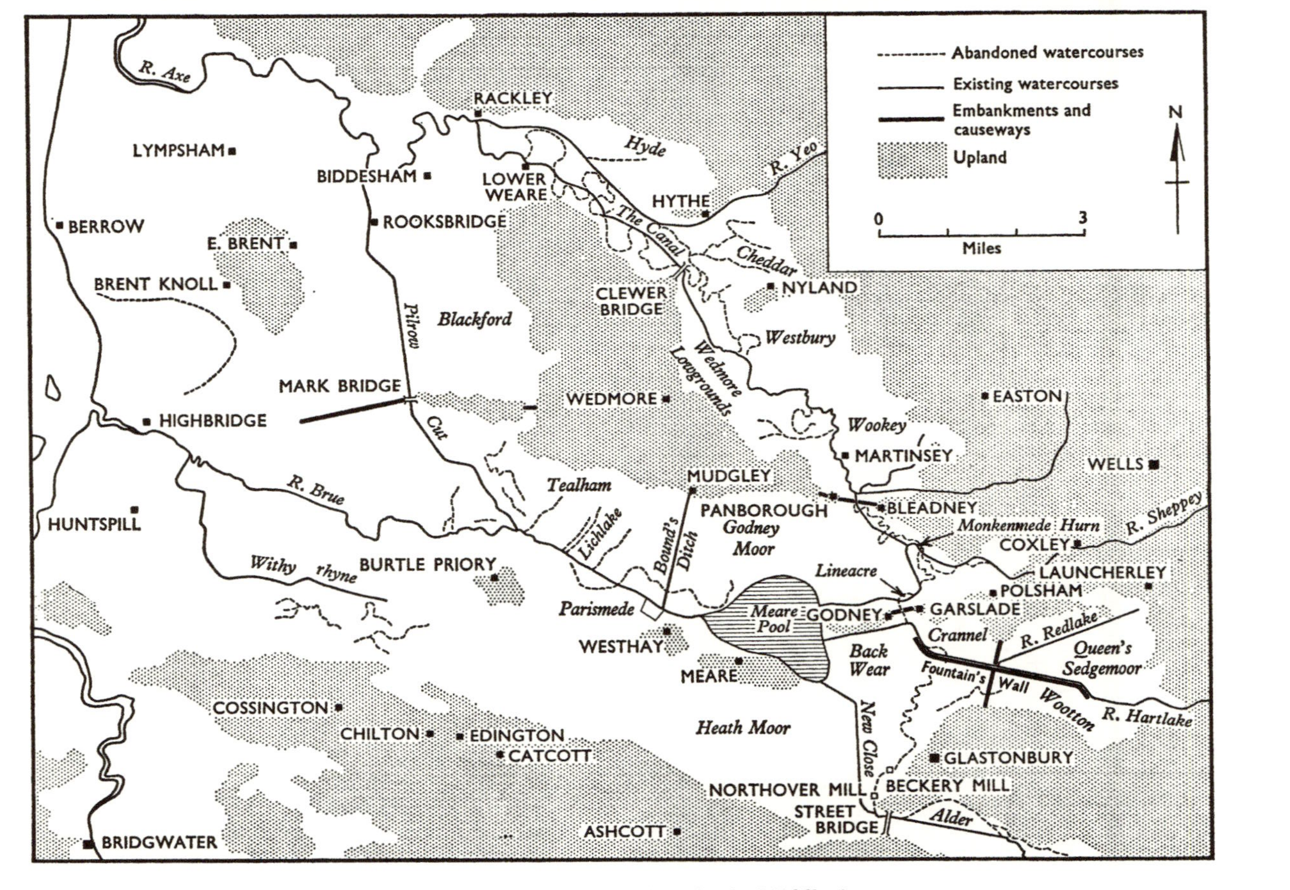

Figure 8. The Brue and Axe valleys in the Middle Ages.

1260 there were 164 acres of meadow around Glastonbury, together with 135½ acres of enclosed pasture. The morgabulum payment for the settlement was £1. 3*s*. 9*d*., which at the usual average rate of 4*d*. per acre added a little over 100 acres of meadow to the total for the settlement. There were also 103 acres of meadow around Meare 'island'. All this meadow and pasture in Glastonbury must have been reclaimed from the surrounding Levels.[1]

These piecemeal upgradings and enclosures, while being indicative of the general reclamation of the moors near Glastonbury, are of less significance than the alterations that were made to the main drainage channels of the area. It is in the tracing and elucidation of the early drainage channels, from aerial photographs and in the field, that there lies the solution to many of the questions concerning medieval drainage activity in the Brue valley. The topic bristles with problems and some of the answers are seemingly contradictory, but it would be wrong not to attempt a description and analysis of the distribution of the main watercourses, their possible origin and the known alterations to them. The low, level land in the broad expanse of the eastern end of the Brue valley allowed a complicated system of interconnected watercourses to exist; but because of possible land subsidence, peat digging, natural siltation, and drainage activity during the last 800 years, it is impossible to state categorically which way the water flowed in these watercourses. Matters are further complicated by the deliberate attempt to maintain even water-levels, for water transport was widespread in this area in medieval times in order to overcome the great obstacle of the marsh.

Although the general trend of the drainage today is in a westerly direction, it is possible that in the early medieval period there was also a northerly flow; for example, the River Brue may have once flowed north through the Panborough-Bleadney gap and united with the River Axe. Present-day conditions of slope and drainage are completely unfavourable to such a course, yet an abandoned channel can be traced from air photographs and in the field, where it is marked by an unmistakable ridge of alluvial matter which meanders through the moors,[2] via the gap between Godney 'island' and

[1] 'Rentalia et Custumaria', *S.R.S.* v, 181–7 and 206–7.

[2] The ridge-like character of this channel is very marked and is in contrast with the depressions with low alluvial flood banks which show the course of other defunct streams. Sections made by newly scoured traversing rhynes show the distinctive profile and composition of the ridge. It suggests an origin similar to that of the 'roddons' in the Fens.

Garslade 'peninsula' (Fig. 8). The existence of this abandoned channel is confirmed by the general distribution of river-derived clays, and also by the extension, from the Axe valley and along the abandoned course, of Wentloogg clays of the coastal belt variety, which suggests a continuous channel from the Axe end.

Documentary evidence is forthcoming which indicates the presence of connecting waterways between the Axe valley and Glastonbury. A thirteenth-century document states that the watercourse between Nyland, in the Axe valley, and Bleadney 'was adequate for the Abbot to take stone and lime and corn from his manor and from other places in those parts to his Abbey of Glastonbury and [they] were used to go from their Abbey to the manor of Andredesye [Nyland] in their boats'.[1] Thirty-three years earlier, the Abbot of Glastonbury had been accused of breaking three fisheries with his boats in the Axe river between Rackley and Glastonbury.[2] In the circumstances, it is not surprising that the abbey claimed that its jurisdiction extended over all the watercourses from Clewer bridge, in the Axe valley in the north, to Street bridge in the south (Fig. 8).[3] Into this north-south watercourse flowed the waters of the Hartlake and Sheppey rivers, although during periods of high floods the water from these rivers probably had a westerly flow towards Meare Pool as well.[4]

But this is not the direction of the drainage today. The rivers Brue, Hartlake and Sheppey all meet in the vicinity of Meare Pool, and the obvious straightness of their courses suggests an artificial origin. The exact date of the implementation of some of these changes is unknown; one can only make informed guesses from the fragmentary evidence. In 1294, the Brue was said to be 'embanked and run toward the

There, the natural levees of the old streams stand above the general surface-level of the peat which has shrunk and wasted owing to improved drainage during more modern times. For more information see H. Godwin, 'The origin of Roddons', *Geog. Journ.* XCI (1938), 241.

[1] *Joannis confratis et monachi Glastoniensis*, ed. Hearne, II, 337–48; also quoted in Helm, 'Somerset Levels in the Middle Ages', *Journ. Brit. Arch. Assoc.* XII, 49. As early as the eighth century Bleadney was called a harbour. See W. de Gray Birch, *Cartularium Saxonicum* (3 vols. 1885–93), I, 189.

[2] 'Somersetshire Pleas', ed. Chadwyck-Healey, No. 237 (1242), *S.R.S.* XI. Radeclive, or Rackley as it is known today, was the port of Wells where presumably sea-going ships transferred goods to the smaller barges that navigated the Levels' waterways. It is first mentioned in 1178 (*W.M.* I, 439). The stone blocks of a wharf can still be seen. Glastonbury's port of Rooksbridge on the Mark Yeo, or Pilrow Cut, had a similar function.

[3] 'Rentalia et Custumaria', *S.R.S.* V, 178.

[4] Suggested by the evidence of abandoned watercourses in air photographs of the area.

Mere' in a new channel which started from near Street bridge. The Higher and Lower New Closes, between the new and the old channels, were turned into meadow, their drainage being so good in places that some of the ground was cultivated.[1] The two courses of the Brue functioned for at least another two centuries because Leland commented in 1507 that it broke into two streams at Street causeway 'wherof the principalle goithe to Glessenbyri. The other goith thoroug low morishe grounde, and metith again with the principal streame or ever that it goith into the mere [Meare Pool].'[2] Whether the new channel of the Brue was excavated some time between 1234 and 1252, when a large meander of the river in Alder Moor was cut off, a new channel dug, and some of the water diverted into a new mill-stream to drive Beckery and Northover Mills,[3] is not certain, but the mid-thirteenth century was a time when major innovations were made in the drainage channels of the Levels.

Further north, the Hartlake river was straightened and embanked to carry the combined flow of the Redlake and Whitelake rivers through Queen's Sedgemoor and the other peat moors north of Glastonbury. This rhyne was already in existence in 1326 and formed the boundary which delimited the interests of Glastonbury and Wells in the moors; from Launcherley the new channel ran into the moors and 'straight to Heortlake bridge, and thence the water course called Heortlake shall be the boundary from the said Bridge to the bound called Batthyngwere'.[4] Hartlake bridge is part of the causeway which crosses Queen's Sedgemoor, and it is clear from the *Glastonbury Cartulary* that Batthyngwere was also known as Bacchyngwere, which can be identified with Back Wear in the alluvial lands east of Meare Pool. Fountain's Wall ran alongside the river to protect both meadows and moors to the south. Leland spoke of it as 'a marsche waulle made by mennys policy...and this waulle continuith to Hartelak bridge, and a mile lower; and then...go soone after into the mere [Meare Pool]. If this marsche waulle were not kept...al the plaine marsche ground at sodaine raynes wold be overflowen, and the

[1] G.C. II, Nos. 478 and 479.

[2] *Leland's Itinerary in England and Wales*... ed. L. Toulmin Smith, I (1907), 148–9.

[3] J. Morland, 'The Brue at Glastonbury', *S.A.N.H.S.* LXVIII (1922), 82–4. This new cut was accompanied by the construction of a culvert underneath it to drain Actis and Reed Meads to the north.

[4] G.C. I, No. 157 (1326), and for confirmation of the work done, see *W.M.* I, 324 and *W.M.* II, 617 (1352).

profite of the meade lost.'[1] But it is not known by which route the water of the Hartlake river reached Meare Pool.

Just as the Hartlake river was embanked and straightened to form a boundary in the moors between Wells and Glastonbury, so was the Sheppey river. In 1326, the part of the watercourse between Monkenmede (Hurn) and Bleadney bridge was straightened and made 12 ft wide; the two bridges, which were to be built over the new channel, were to be high enough for boats to pass underneath, which suggests contact with the Axe valley waterways and the abandonment of the old watercourse. The southern extension of this new watercourse to Lineacre (Nineacre) in the Godney-Garslade gap is indicated by other documentary evidence.[2]

The diversion of the rivers Brue and Hartlake and perhaps even the Sheppey away from the Axe valley and towards Meare Pool in the thirteenth century must have led to a great increase in the amount of water entering the pool. The observed fluctuations in its size were probably the outcome of this more direct link between the pool and the drainage system of the uplands.[3]

It is unlikely, however, that a large channel took the excess water westwards towards the sea. The western rim of the pool is a hard band of blue Lias rock and the Lias clay lies at a very shallow depth (between 9 and 10 ft O.D.); the implication is that the present Brue does not have a submerged channel or 'valley' beneath the peat as does the Parrett and the Axe.[4] Therefore, the present Brue is either an artificial channel cut through the Lias, or it is the result of erosion from the overflow of the pool during periods of high water-levels. Whichever explanation is correct, not much excavation would have been needed in order to lower the pool surface, reclaim some ground and establish a definite westward flow of water. Once a new channel was established, it is likely that the old north-south course of the

[1] *Leland's Itinerary in England and Wales*... I, 147. The ground on the north side of the wall is some 3–4 ft higher than that to the south, and this shows its flood protection function.

[2] G.C. I, No. 161 and G.C. II. No. 648.

[3] See pp. 105–6 below.

[4] Bulleid and Gray, *Meare Lake Village*, II, 8, and H. Godwin, 'Studies in the post-glacial history of British vegetation: XIII, The Meare Pool region of the Somerset Levels', *Phil. Trans. Roy. Soc.* B, CCXXXIX (1955/6), 169. A boring made north of Westhay Bridge met Lias clay at +9·8 ft O.D. and another boring made nearer to the River Brue showed 'a basal sequence of muds and peats' reflecting 'precisely the same evolutionary story as that of the raised bogs of the Wedmore-Polden [Brue] valley, and of the whole bog system, continuing northwards from this point'.

Brue would have been abandoned permanently and that the westward flowing channel would have carried most of the Brue water. The late medieval land subsidence of the fourteenth and fifteenth centuries, which is so clearly illustrated in the flooding of the peat diggings of the Norfolk Broads and the formation of the broads as we know them today,[1] may also have caused, or at least helped in causing, the process of silting in the areas of indeterminate fall in the upper ends of the Brue valley rivers; but much of this reasoning must remain conjectural until further research on this and associated problems has been carried out. These changes certainly occurred by 1327 when there is a mention of a watercourse from Ferlingmere (Meare Pool) to Lichlake, a small stream in Tealham Moor,[2] and they were complete by 1507 when 'the fresh waters descending from La Mere to Merke' (i.e. the present Brue) were observed.[3]

The cutting of the Brue west of Meare Pool naturally poses the question of where the water flowed. Two possibilities are open; one that it went via the Pilrow Cut, which appears to be the logical continuation of the Brue channel west of Meare Pool; the other that it went via the present channel, across the coastal clay belt, to Highbridge (Fig. 8). The course of the Pilrow Cut from the present River Brue to Mark looks natural in its irregularity compared with the straight channel of about four miles in length that runs from Mark to the Axe. The cut certainly contributed little to the drainage of the moors through which it passed, being excavated on slightly higher ground than the more badly drained areas on either side; once again, like other medieval cuts in this area of the Levels, its purpose would seem to be primarily one of navigation, being a connecting link between the four coastal manors of the Brents, Lympsham and Berrow, with the Abbey of Glastonbury. However, the cut, in conjunction with the Brue, must have soon developed a drainage function by carrying much of the Brue valley water to the sea.

Evidence for the existence of the cut goes back perhaps to the early thirteenth century and certainly to the early fourteenth century. In 1235, it was said that the Glastonbury waterways extended to Mark bridge, which is nearly half-way along the cut.[4] In 1316, reference

[1] Lambert, Jennings, Smith, Green and Hutchings, *Making of the Broads*, Roy. Geog. Soc. Research Series, No. 3.

[2] *W.M.* I, 226–8; see also p. 70 below, for evidence of a possible connexion nearly a century before.

[3] *Cal. Pat. Rolls, Henry VII, 1494–1509*, 322.

[4] See p. 71 below, n. 1.

was made to dykes, sluices and walls at Mark, 'for the preservation of the course of water flowing towards the sea and the safety of the lands in the Deanery of Merk, Modesly [Mudgley], Wedmore and Bydesham',[1] which must refer to the cut and the Brue because all these places lay alongside the two watercourses. As mention is made of the watercourse reaching the sea, one can infer that the cut was completed in 1316. Confirmation of this is given some time between 1327 and 1338 when the whole cut was described as being tidal, and it was still tidal as far as Rooksbridge in 1487.[2] The existence of the cut and its connexion with the Brue was confirmed again in 1358. Walter de Monyngton, Abbot of Glastonbury, 'raised a mill called Rokesmulle [Rooksbridge Mill] which stopped up all the common watercourse from Ferlyngmere [Meare Pool], whereby the land and moors were flooded annually'.[3] By the beginning of the sixteenth century the scouring of the cut by the tenants of Lympsham, Berrow and the Brents was a time-honoured practice;[4] one is tempted to suggest that the scouring of the 'Mordich', which was mentioned as a common customary service in all these manors in the early thirteenth century, was in fact the maintenance of the Pilrow Cut.[5]

The original direction of the flow of the cut was from the Brue to the Axe but the gentle incline would have allowed a reversal of flow to take place easily. This happened by the eighteenth century, if not earlier, and was caused by excessive silt-accumulation in the Axe estuary.[6] Thus, its possible use as an overspill channel taking surplus water from the Brue valley was lost and drainage conditions must have deteriorated.[7]

The alternative and obviously nearer outlet with the greatest fall

[1] *Cal. Pat. Rolls, 1313–1317*, 412–13. See also *Cal. Pat. Rolls, 1291–1301*, for evidence of a bridge over the Brue at Meare.

[2] *Joannis confratis et monachi Glastoniensis*, ed. Hearne, I, 267. 'Item apud Rokesmille, pro marinis fluxibus excludedis, magnis sumptibus opus petrinum construxit', and pp. 345–8 for the reference to 1487.

[3] Coram Rege Rolls, Easter, 32 Ed. III, m. 31*d*. Quoted in Flower, *Public Works in Medieval Law*, XL, 131. On p. 134 we also read that the Dean of Wells had two sluices at Mark 'to keep the sea water from ebbing and flowing', but from which direction the sea influence was felt is not clear.

[4] *Joannis confratis et monachi Glastoniensis*, 321–5, for example.

[5] 'Rentalia et Custumaria', *S.R.S.* v, 38, 41–3, 45, 47 and 50.

[6] R. Locke, *An essay on the Subject of Draining the Flat Part of Somersetshire...* (1800), 6–7, S.R.O. Som. Arch. Soc. Serel MSS. 113/1.

[7] The cut could only alleviate indirectly the flood waters of Meare Pool and not drain it as A. Nesbitt suggests in 'The Manor House, Meare, Somerset', *Arch. Journ.* x (1853), 130–41.

for the Brue valley drainage was the present outfall at Highbridge. The portion of the Brue west of Burtle 'island' has a conspicuously natural look compared with the river to the east, which suggests that the Brue follows the course of an older and smaller stream across the clay belt.[1] This suggestion is supported by an early thirteenth-century list of drainage duties which describes major watercourses, one of which ran 'from Burtle Pool to the land of Chilton, and from Eddington and Chilton to the land of Cossington, and from Cossington to Huntspill and from Huntspill to the sea'.[2] If this is so, then the Brue west of Meare Pool was either engineered to connect with this stream, or a natural breakthrough of flood waters occurred, a possibility which has been examined in detail already.[3] The place-name Highbridge was in existence by 1324, which indicates the presence of a deeply incised stream or pill in the clay belt at this point,[4] and by 1485, tidal doors had been erected under the bridge and it had been converted into a clyse. The manorial lords of Mark and Moor, in the peat lands, contributed towards the upkeep of the clyse because they received benefit from its existence,[5] which argues that there must have been direct access from the sea to the inland peat moors at this time.

In summary, it is probable that the Brue, Hartlake and Sheppey rivers once flowed mainly northward through the Panborough-Bleadney gap but that their diversion westward to Meare Pool, the digging of a channel west of the pool, and the excavation of part of the Pilrow Cut can all be assigned to the latter half of the thirteenth century; the evidence of watercourses linking Mark and Glastonbury some time between 1235 and 1253 suggests a mid-thirteenth century origin for some of the works. They were certainly completed by 1507.

[1] It is interesting to note that this portion of the Brue was still known as the Fishlake river in 1638 (see P.R.O. LR2/202), which suggests the survival of an older local name. The name Brue was acquired relatively recently, not appearing on general maps of the county until Cary's map of 1787. Prior to that date the river was not named on general maps of the county, though 'Brent River' was applied to it in Moll's map of 1724. See map collection in the Brit. Mus.

[2] G.C. II, No. 1015.

[3] See p. 67 above. Something similar to the first possibility was done many centuries after in King's Sedgemoor, where the final two miles of the King's Sedgemoor Drain were cut into the bed of a smaller stream which ran across the clay belt. (Compare Fig. 21 with Fig. 22.)

[4] Ekwall, *Concise Oxford Dictionary of English Place-Names*, 227–8.

[5] Brit. Mus., Harleian MSS. 433, art. 2,110, f. 209. The clyse was said to be 'now by the grete rage and tempest of water and also for the noon repayring broken and fallen to great decay: so that grete substance of Cuntry thereabouts be surflowed and likely to be finallv distroyed'.

The present outfall of the Brue at Highbridge was in existence by 1324, but it is uncertain whether or not it was connected to the rest of the Brue valley drainage system at this time.

As the new channels connected the abbey estates in the west to Glastonbury itself, the needs of navigation were probably paramount over those of drainage. Thus the diversion of river flow was not the main aim, and the old and new channels were complementary parts of an intricate system of inland water communication.[1] Nevertheless, it cannot have been long before these new channels assumed a drainage function. These works were of considerable magnitude, but their effectiveness in draining the medieval Levels was probably limited. In later years, however, they became the main lines of a radically altered drainage system, and consequently the basis of the geographical changes which occurred in the area.

The Axe valley

The earliest reference to reclamation anywhere in the Levels is for Lympsham, alongside the Axe (Fig. 8). Some time before 1126, the land was granted 'quae...ecclesiae in nullo profuerat', but when the Abbot of Glastonbury visited it later he found the land in cultivation. On one side it 'was encircled and surrounded with a wall', and on the other 'bounded by the depth of the river. In it I saw the corn was coloured red and gold, murmuring sweetly in the gentle breezes and showing a surface unbroken by weeds.'[2] The following years throughout the twelfth and thirteenth centuries are barren of information, but in the early fourteenth century there was in the Axe valley, in common with other areas of the Levels, an increasing desire to divide off the pasturing interests in the commons, which in some cases led to reclamation.

For example, in 1308, because of the difficulties and disputes over

[1] For an indication of the complexity of water transport in the Levels it is most illuminating to look at some of the duties of Robert Malerbe, a tenant of Glastonbury in the early thirteenth century. They were: 'he ought to provide a boat that can carry eight men, and be the steersman, and carry the lord abbot where he wishes...and all his men, and the cook, the hunter with his dogs, and all those who can or ought to be carried by water ...He ought to be responsible for the Abbot's wine at Pilton, after it has been put in the boat and until it has been brought to Glastonbury...To look after all waters between Clewer and Street bridges, and between Mark bridge and Glastonbury.

'And he ought to look after all the Abbot's boats in those waters and keep the waterways in Hearty Moor...' The duties concerning fisheries in the channels were even more minute and numerous. See 'Rentalia et Custumaria', *S.R.S.* v, 176–8.

[2] *Adami de Domerham*, ed. Hearne, 307–8.

common pasturing, Dean Godelee of Wells and the Bishop of Wells agreed to enclose and hold in severalty their respective moors of Wedmore Lowgrounds, which consisted of 600 acres of alluvial ground on the east side of Wedmore 'island', and Blackford moor on the west side of the 'island'.[1] There is no documentary evidence to suggest that these moors were subdivided, but an examination of air photographs does reveal a complicated prior pattern of ditches in both areas, which is orientated differently from that of the present drainage network. Whilst the pattern of ditches is highly suggestive of subdivision in the moors, its exact interpretation must await work in the field. Further examples illustrate this process of primary division and subsequent reclamation; a large area of Westbury Moor was granted to Sir Richard de Rodeneye in exchange for 9 acres of meadow in Cheddar. The lengthy agreement between Wells and Glastonbury in 1327, whereby the abbot and his tenants at Meare, Bleadney, Panborough and Martinsey were to have rights of pasture in Wookey Moor in exchange for granting similar rights to the bishop and his tenants in Godney Moor, has been explored already. Some moves were also afoot to divide Cheddar Moor because the dean was delaying a decision over the apportionment of the moor as 'it seemed to some of them to tend towards permanent alienation'.[2]

Of the very many channel diversions in the valley, little can be said. It is known that in 1316 a dispute arose between the Bishop of Wells and the Master of the House of St. Mark's, Bristol, over the latter's mills and sluices at Nether Were (Lower Weare) which were said to be causing severe flooding. The Master of St. Mark's gave the mills to the bishop in exchange for Hydemoor, which lay between Rackley and Cheddar, but throughout these negotiations mention is also made of a 'new watercourse' 15 ft wide and extending from the mills to Blakelake, by which the water 'issuing from Cheddar' could be diverted away from the mills. This suggests that much of the present course of the Axe between Lower Weare and Clewer, which cuts through many meanders and is, significantly, known as the Canal, corresponds to the new course mentioned, and also that the present

[1] *W.M.* I, 219. In *W.M.* I, 226 (1310), the other interested party in the division, Glastonbury, agreed to the apportionment, provided its tenants of Panborough, Bleadney and Clewer had common at all times of the year in the remainder of the Axe moors. Copy and confirmation in G.C. II, No. 597, and subsidiary agreement between lords of manors over tithes in *W.M.* I, 226.

[2] *W.M.* I, 198 (1328); G.C. I, No. 161, and p. 37 above; and *W.M.* I, 180 (1327).

very straight and obviously artificial portion of the Yeo between Hythe and Rackley had not been excavated in 1316, as at that time the Yeo did join the Axe.[1]

The basin of the upper River Parrett

Consideration of draining activity in the upper Parrett district has been left intentionally until last. Figure 6 shows that both the Sowy-Tone-Aller and the Glastonbury groups have three features in common; they are situated on alluvial lands, they are grouped around ecclesiastical centres which are located either on 'islands' or on areas of restricted ground, and they lie astride two of the most complex river junctions in the Levels, where the abundance of watercourses similarly provides new deposits of fertile alluvium in flood time, and facilities for transport, power[2] and fishing, during periods of more normal river flow. Each of these three features also occurs in the upper Parrett valley where the rivers Parrett, Yeo and Isle converge in the alluvial lands around the 'island' site of Muchelney Abbey. Yet, curiously enough, what experience elsewhere in the Levels has led one to expect is not paralleled here, and references to reclamation are surprisingly few. Here is a case where favourable physical and historical circumstances do not seem to have fostered much activity. Field work reveals nothing of river diversions or embankments, and it is unfortunate that in both available sets of aerial photographs of the area, extensive flooding obscures all details of surface features.

Of the few references to reclamation that exist there are some which deserve attention on account of their importance and on account of the size of the area involved. Among these are the suggestion of the enclosure of 30 acres of moorland west of Kingsbury Episcopi in 1248–9, and two enclosures in Huish and Langport North moors by 1317, one of 20 acres and one of 18 acres, and a new meadow called Lytilnye. In 1277 the Abbot of Muchelney was granted the right to appropriate and upgrade as much of 'Purimore', or Perry Moor, as he wished. Earlier, the abbey had agreed to the enclosure

[1] *W.M.* I, 188–9 and 'Bishop Drokensford's Register, 1309–1327', *S.R.S.* I, 7 and 104. For confirmatory details see 'Cartulary of St. Mark's, Bristol', ed. Ross, *Bristol Record Society*, XXI, Nos. 209 and 243.

[2] For an indication of the importance of transport and water power for the location of the woollen industry alongside the Parrett and Tone rivers in the late fourteenth century, see R. A. Pelham 'Fourteenth Century England', in *An Historical Geography of England before A.D. 1800*, ed. H. C. Darby (1936), 230–65.

and ditching of 'Prestmore' alongside the River Yeo on the condition that 'the master and brethren and their men of La Lade [Long Load] shall have no damage in Wattemore [Wet Moor] or in their moor of La Lade through inundation of water by reason of the said ditch'.[1] In the early fourteenth century the tenants of the Wells manor of Huish had succeeded in reclaiming some of the moors alongside the Yeo by an embankment.[2]

Despite these reclamations, the main rivers were neglected and the whole of the Yeo valley must have been subject to regular flooding, because in 1287, Peter de Faucunberge and the parishioners of Middleton were given leave to erect their own chapel as they could not get to Martock on account 'of the floods between', and this was still true of Kingsmoor in the middle of the fourteenth century, for it still flooded regularly and was 'under water in winter'.[3]

Although the majority of references to reclamation occur in distinct areas there are a scatter of references that do not conform to the general pattern. Most of these reclamations were around the edges of King's Sedgemoor where the downwash and alluvial deposits ameliorated the inhibiting qualities of the peat soil and where flooding was not so regular. By the fourteenth century various parcels of land had been reclaimed; there were 20 acres at Henley, 3 acres at Ham, 30 acres at Sutton Mallet, 12 acres at Moorlinch, 45 acres at an unspecified location which were 'to be sown with wheat and oats' by the Abbot of Glastonbury, as well as other areas already mentioned, to the north of Sowy.[4] A little later, a large enclosure of 200 acres was undertaken at the eastern end of the moor, but, as in so many other cases, it is not known whether this resulted in improvement sufficient to create meadowland, or if it merely remained as an enclosed moor.[5]

[1] 'Pedes Finium', No. 23 (1248–9), *S.R.S.* VI; *W.M.* I, 176–7 and 484; 'The Muchelney Cartulary', Nos. 65, 25 and 26, *S.R.S.* XIV; see also, 'The Muchelney Cartulary', No. 40, and 'The Athelney Cartulary', No. 236, for other references to drainage works in this district.

[2] 'Bishop Drokensford's Register, 1304–1327', *S.R.S.* I, 117.

[3] *W.M.* I, 450; and *Inquisitions Post Mortem, xi, Edward III*, 44 (1353–5).

[4] G.C. II, No. 384 (1302). See also p. 52 above.

[5] G.C. II, Nos. 378–81. See also *Cal. Pat. Rolls, 1317–21*, 41, and *Cal. Pat. Rolls, 1327–30*, 332.

A COMPARISON OF ELEVENTH- AND FOURTEENTH-CENTURY STATISTICS

A comparison of statistics in Somerset in 1086 with those of 1327 reveals that there had been a substantial rise in the prosperity of the Levels relative to other parts of the county between these two dates; this can only be explained by the extensive reclamation and upgrading of the moorland wastes that has already been examined. It amounted to a massive change in the economic geography of the region.

The methods and techniques of analysing Domesday statistics are well-tried and well-known to geographers.[1] Nevertheless, it is worth pointing out that the degree of reliance that can be placed on the Domesday material depends largely upon the identification of the settlements to which it refers. The early work of identification was begun by Collinson in 1791, was continued by Eyton and Whale, and largely completed by Bates.[2] He found that 600 settlements could be identified with certainty, that a dozen could no longer be traced, and that half a dozen were conjectural; there were also a number of subsidiary vills around Taunton that could not be identified. Recent work has carried even further our knowledge of the Domesday geography of the county.[3] Upon this foundation of topographical identification, the plough-teams of the Domesday vills can be plotted and expressed as a density per square mile within districts that correspond to broad geological and topographical divisions of the county.

The distribution of plough-teams is now generally accepted as a good indication of the relative prosperity and agricultural activity of regions. From the point of view of this investigation the dominant feature of Figure 9 is the overall poverty of the Levels in 1086

[1] See, for example, the works of H. C. Darby on the Domesday geography of England which include *The Domesday Geography of Eastern England* (1952); *The Domesday Geography of Midland England* (1954) (with I. B. Terrett); and *The Domesday Geography of South-East England* (1962) (with E. M. J. Campbell). See also, F. W. Morgan's valuable pioneer attempt to analyse the Somerset material geographically in 'The Domesday geography of Somerset', *S.A.N.H.S.* LXXXIV (1938), 139.

[2] J. Collinson, *The History and Antiquities of the County of Somerset*, I (1791); R. W. Eyton, *Analysis and Digest of the Somerset Survey* (1880); T. W. Whale, 'Principles of the Somerset Domesday', *Trans. Bath Field Club* (1902); and Bates, 'Text of the Somerset Domesday' in *V.C.H.* (1906) I. This transcript accompanies J. H. Round's 'An Introduction to the Somerset Domesday', *V.C.H.* (1906) I, 383–432.

[3] *The Domesday Geography of South-West England*, ed. Darby and Finn, from which Figure 9 is reproduced.

compared with many of the upland areas. The Brue valley and the lower Parrett valley shared, with the western hills, the lowest densities for the county, ranging from 0·7 to 1·0 plough-teams per square mile. The prosperity of the flood-free coastal clay belt and the Polden Hills was greater with densities of 2·4 and 2·3 plough-teams per square mile respectively; this was better than the Quantock Hills (1·7), the Mendip Hills (1·5), and the Chalk (1·5) and the Oxford Clay (1·3)

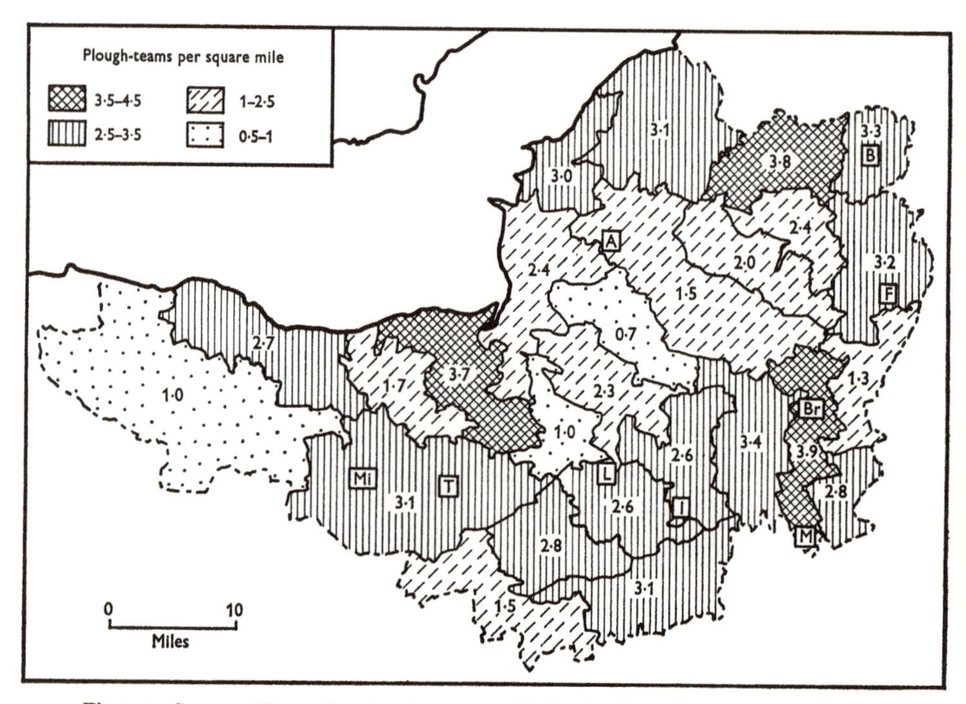

Figure 9. Somerset Domesday plough-teams, 1086 (by densities). Domesday boroughs are indicated by initials: A, Axbridge; B, Bath; Br, Bruton; F, Frome; I, Ilchester; L, Langport; M, Milborne Port; Mi, Milverton; T, Taunton. (Source: after H. C. Darby and R. Welldon Finn, *D.G. S-W.E.* (1967).)

regions in the southern and eastern margins of the county. The Northern Levels (3·0) were even more prosperous than the coastal clay belt. Although the coastal claylands and the Polden Hills were not among the poorest parts of the county, generally speaking, they fell well behind the Liassic, Triassic and Oolitic regions which stretched in a crescent-like fashion around the Levels and the Mendip Hills. On the whole, the high ground that surrounded the Levels was relatively much more prosperous than the Levels themselves.

The situation in 1086 can be compared with that in 1327 when an exchequer lay subsidy was levied (Fig. 10).[1] It was a levy of 1/20th on all movables which each lay person owned, including crops and livestock. Persons owning goods of less than 10*s.* were exempt, as, of course, were clerics, which exemptions in a county like Somerset with a great deal of clerical property, would tend to make it appear less wealthy in comparison with others than would have been the case if a complete assessment had been undertaken. This effect was offset to some extent, however, by the inclusion of the goods of the villeins of the clergy. The possibility of evasion in this, as in all taxation returns, cannot be overlooked, but it is noticeable that the amounts for many a locality concluded with a 'subtaxati', the contribution of those who had evaded the tax or who were forgotten, which suggests some effort to stop malpractices. A few totals are missing because of mutilated rolls, but it is probable that their absence would not have greatly affected the resultant distributions.[2] Thus, whilst noting the drawback of the exemption of a certain section of the population, the absence of some totals, and the ever-present likelihood of some malpractice, we can treat the return as a probably useful guide, not to total wealth, but to relative wealth in the county, and as a fairly reliable basis for comparison with the Domesday material.

A broad zone of high values extends from the coastal clay belt through the lower Parrett valley to the southern portions of the county. The tax of 16*s.* per square mile on the coastal clay belt was surpassed only by the 19*s.* per square mile of the valleys of the upper Yeo and Parrett in the Liassic regions to the south, and was closely followed by the Bath Oolitic Limestone region. The striking rise in the prosperity of the lower Parrett basin can be explained only by the extensive reclamation of moorland which we know occurred around Sowy 'island' and in Aller Moor. The two regions immediately to the south, which similarly showed some marked rises in wealth, also contained areas of lowland which had been reclaimed. There was a general levelling-off of prosperity over much of the county, in which the Levels had taken a more than average share.

A last glimpse of the agriculture of the Levels can be gleaned from a register of extents, which allows one to make a very minute

[1] The 1327 exchequer lay subsidy is printed in 'Kirby's Quest for Somerset', ed. F. A. Dickinson, *S.R.S.* III (1889).

[2] See J. F. Willard, *Parliamentary Taxes on Personal Property, 1290 to 1334* (1934).

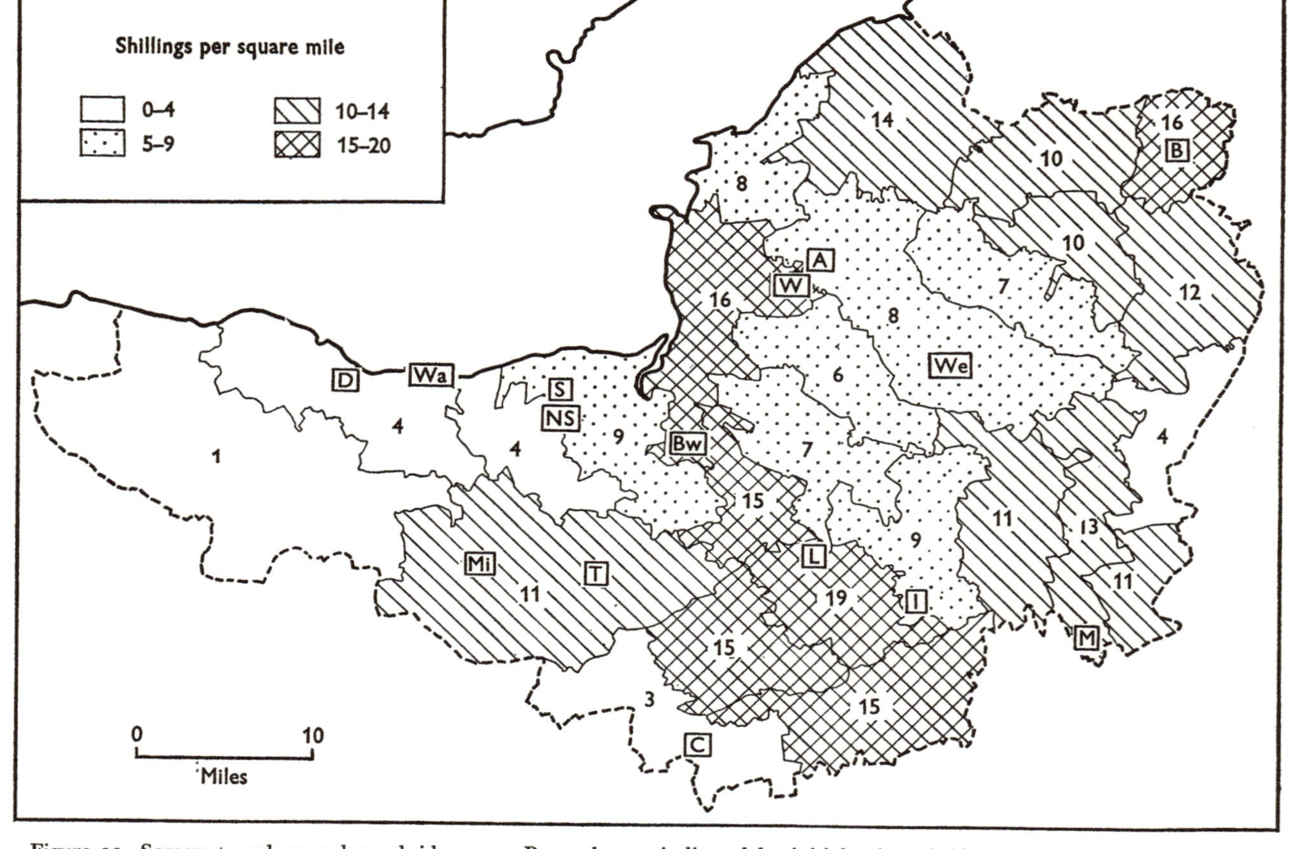

Figure 10. Somerset exchequer lay subsidy, 1327. Boroughs are indicated by initials: A, Axbridge; B, Bath; Bw, Bridgwater; C, Chard; D, Dunster; I, Ilchester; L, Langport; M, Milborne Port; Mi, Milverton; NS, Nether Stowey; S, Stogursey; T, Taunton; W, Weare; Wa, Watchet; We, Wells.

compilation of land use and its value in the Glastonbury demesne manors for dates varying between 1309 and 1324 (Fig. 11). In most manors in the uplands, along the Polden Hills, and on the coastal clay belt, arable accounted for between 70 % and 80 % of the acreage of all the land, and the only places near the Levels in which the acreage fell below one half were Meare (47·5 %), Glastonbury (41·2 %) and Withy (21·5 %). Away from the Levels similarly low proportions were recorded for Baltonsborough (33·4 %), Mells (43·9 %) and Doulting (46·1 %). Meadowland commonly accounted for only about 10 % to 20 % of all land used, but the proportion was particularly high in the restricted 'island' manors of Sowy (25·5 %), Glastonbury (34·2 %), Meare (41·0 %) and Withy (78·5 %). Away from the Levels, Baltonsborough had 49 % of all land in meadow, and Doulting 38·7 %. The other types of land recorded—woodland and pasture—generally did not enter into the total for any Levels' manor. The only sizeable proportions of pasture recorded anywhere were at Street (17·6 %), Baltonsborough (17·6 %), Batcombe (17·1 %) and Mells (30·1 %). Woodland was also absent in the Levels' manors except for Glastonbury (22·6 %) and Ashcott (13·1 %), but in the uplands to the east, Marksbury, Houndstreet and Mells had between 20 % and 25 % of their land classed as woodland. Perhaps more important to this study of the Levels was the value to the manorial economy of the various types of land use. Meadowland, in particular, was still the most valuable of all types of land, and its superior value pointed the way to the desirability of expanding pastoral activity in the Levels by the reclamation and upgrading of the moors. For example, in Glastonbury, 34·2 % of the land was in meadow but contributed 64 % of the income, in Shapwick the 16·8 % of meadowland produced 55 % of the income, and in Greinton the 20 % of meadowland produced 64 % of the income; such proportions were common.[1]

But it was the end of an era. The general downswing in economic conditions in the fourteenth century was soon to be accelerated by the Black Death. There was a quickening of the existing processes of commutation and leasehold and of the change from a demesne

[1] These proportions are based on Brit. Mus. Egerton MSS. 3321/F: a register of extents of the Manors of the Abbey of Glastonbury, compiled between 1289/90 and 1355/6. For Glastonbury, see ff. 1–41; Meare, ff. 142–64; Sowy, ff. 411–62; Shapwick, ff. 45–56, and Greinton, ff. 65–73. For other manors in the Polden Hills see Moorlinch. ff. 57–8; Hamme, ff. 76–102; Ashcott, ff. 103–18; Street, ff. 119–39; and Walton, ff. 184–94. For manors in the coastal clay belt, see Withys, ff. 61–3; S. Brent, ff. 351–64; Berrow, ff. 365–78; Lympsham, ff. 378–91, and E. Brent, ff. 392–410.

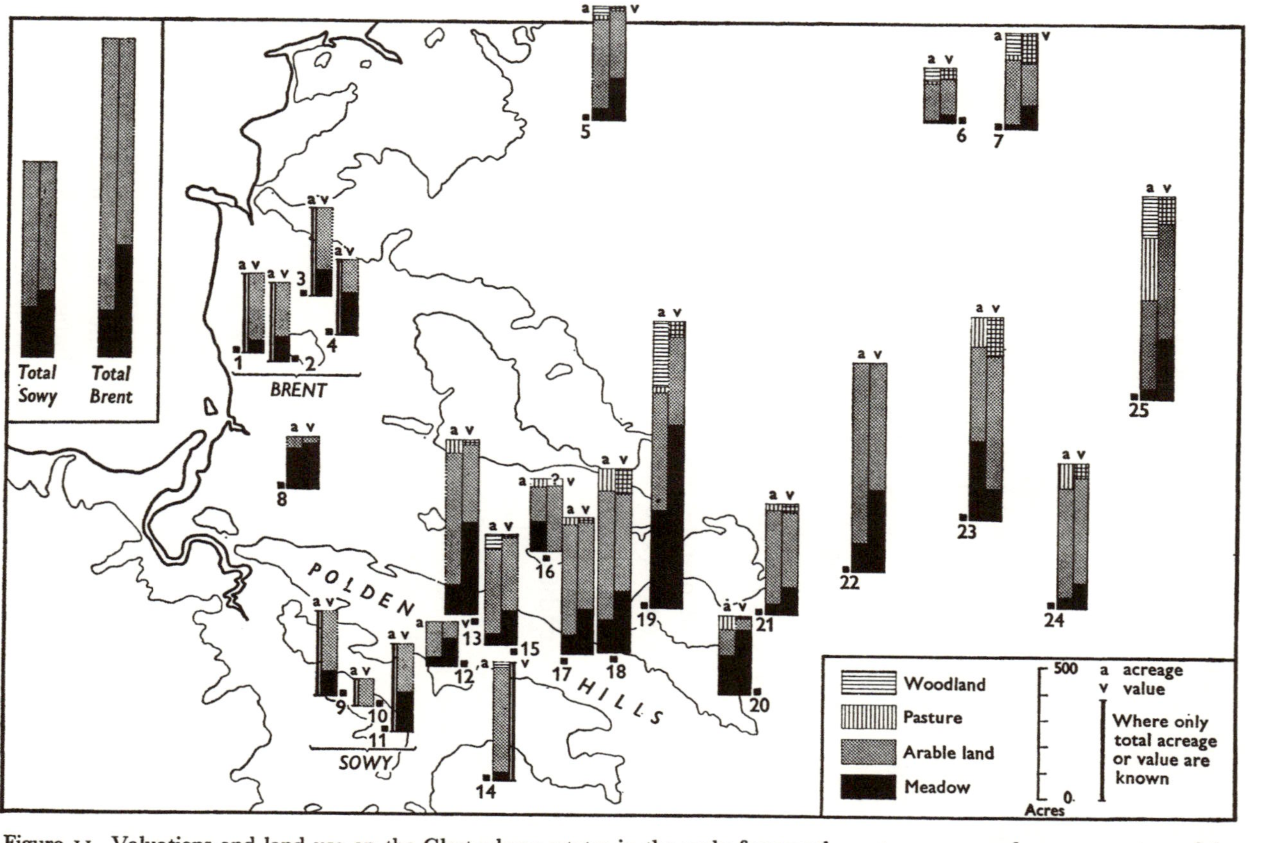

Figure 11. Valuations and land use on the Glastonbury estates in the early fourteenth century, expressed as a percentage of the total for each settlement: 1, Berrow; 2, Brent; 3, Lympsham; 4, E. Brent; 5, Wrington; 6, Houndstreet; 7, Marksbury; 8, Withys; 9, Westonzoyland; 10, Parsonage; 11, Middlezoy; 12, Greinton; 13, Shapwick; 14, High Ham; 15, Ashcott; 16, Meare; 17, Walton; 18, Street; 19, Glastonbury; 20, Baltonsborough; 21, Pennard; 22, Pilton; 23, Doulting; 24, Batcombe; 25, Mells. (Source: Brit. Mus., Egerton MSS. 3321/F.)

exploitation to a *rentier* economy, which was to continue throughout the fourteenth century and even late into the next century. The pace of reclamation seemed to pause and diminish as ownership changed and economic returns became less. As far as the Glastonbury estates were concerned, the completion of the Great Cartulary by Abbot Breynton some time between 1338 and 1340 was the last survey and act of stewardship over the great estates on the Levels and the bordering uplands.[1]

Such, then, is the evidence that indicates the progress and achievements of reclamation in the Levels during the Middle Ages. The evidence probably raises as many questions as it helps to solve; nevertheless, it does rectify some of the ignorance and vague notions about this period in the region. The drainage activity represented an impressive beginning to what was to be a lengthy task of redressing the adverse features of the physical environment of the Levels. But the contribution of the age was more far-reaching than that, for some of the drainage works themselves were to last through the ensuing centuries and became the basis of subsequent activity, for example the system of embankments on the moors around Sowy. Similarly, the river diversions in the Brue valley established a new drainage pattern based on the separate outfalls and river systems of the Axe and Brue, and this was to play a considerable part in determining the details of future drainage planning in these two areas. Above all, the new works had laid the basis for a new prosperity which was to point the way to future reclamation and improvement.

[1] See p. 40 above.

4

DRAINING ACTIVITY, *c.* 1400–1770

No common theme links together the long stretch of nearly four centuries from about 1400 to the beginning of a comprehensive approach to the problem of draining which may conveniently, and perhaps quite accurately, be said to have begun in 1770. Nevertheless, marked variations in the pace of draining activity do divide these centuries into three periods and so provide a ready basis for discussion. These periods are from 1400 to 1600, from 1600 to 1640, and from 1640 to 1770.

DRAINING DEVELOPMENTS, *c.* 1400–1600

Unfortunately there is a relative lack of information about draining activity in the Levels from about 1400 to 1600. This is partly due to the general difficulties of documentary evidence which appear to characterize these centuries of British history; sources are numerous and varied, but they are also fragmentary and hard to handle. Chief amongst the omissions, as far as Somerset is concerned, is the absence of the detailed records of the Courts of Sewers from the sixteenth to the end of the eighteenth century, which deprives us of the most important sources of information on draining. Moreover, ecclesiastical estate records, which had been such a fruitful source of information in the past, are less numerous and of less value for this period. The only documentary evidence that does hint at the flood problem, and at the steps taken to alleviate it during the fifteenth and sixteenth centuries, is concerned with what seems to be the increasing need to maintain sea defences and with problems associated with navigation and mills on the main watercourses, particularly on the River Tone.[1]

The late medieval land subsidence must have caused a considerable

[1] *W.M.* II, 184–90. In 1490, the Chapter of Wells erected a mill at Ham on the Tone, which was said to cause severe flooding upstream and to be a hindrance to navigation. In answer to the latter point the chapter gave some interesting details about the flow of the Tone: 'and all the somer season the water is so lowe and so meny shelpes and bayes in the ryver betwene our myll and Taunton, that it is not possible to convey eny bote that way; and in the wynter season the medewes be so filled and replenysshed with water, that the bootes may go over at every place, so that they shall not be lett by the myll'.

strain on the sea defences of the coastal clay belt; the fairly regular re-issuing of the Commissions of Sewers, particularly during the first decade of the sixteenth century, was, perhaps, some indication of these worsening conditions. In 1425 Cannington Priory lands were severely flooded (Fig. 14),[1] and in 1484–5 the brethren of St. Swithin's Priory, Winchester, gave up their knife-money in order to repair Bleadon sea wall which had been broken up and destroyed by the unusual 'violence and ferocity of the wind and sea'.[2] In the Northern Levels the sluice gate at Wemberham, at the mouth of the Yeo, was repaired in 1528, and twenty years later was so 'rynus and yn dekay' that the Court of Sewers directed the landowners to do something about it; the parishioners sold the silver cross of the church to raise funds for 'makyng of a sirten skluse or yere agenste the rage of ye salte water'.[3] But these are merely isolated examples of local crises and initiative in the battle against flooding, and really tell us very little. There is no substantial body of evidence from which to build a picture of conditions during these centuries, although it is certain that there must be many more examples in parochial and estate records, which have not yet been brought to light, of attempts to combat flooding.

In view of the paucity of documentary evidence, the location and extent of fifteenth- and sixteenth-century reclamation are perhaps best indicated by Figure 12. This depicts the area of lowland in the Levels, east of the coastal clay belt, that remains after the probable medieval and known late eighteenth- and early nineteenth-century reclamations are deleted. Within this area, the post-1600 reclamations (all of which are substantiated by independent evidence) are distinguished by dots and cross-shading. The area which remains is probably that of fifteenth- and sixteenth-century reclamations. The map presents difficulties, for it is easy to assume by implication the very point one is attempting to prove; if the map of reclamation during the medieval period (Fig. 6) could be claimed to be accurate and complete, then all surrounding areas would represent subsequent

[1] 'The Register of Bishop Bubwith, 1407–1424', ed. T. Scott-Holmes, *S.R.S.* XXIX (1913), 273. See also, 'Cartulary of St. Mark's Bristol', ed. Ross, *Bristol Rec. Soc.* XXI, No. 195 (1404–5) where it is stated that the causeway and ford across the Parrett at Combwich Passage had fallen into decay. There is a suggestion that it was becoming more difficult to maintain.

[2] 'The Compotus Rolls of the Obedientaries of St. Swithin's Priory, Winchester', ed. G. W. Kitchen, *Hampshire Rec. Soc.* (1892), 296.

[3] 'Pre-Reformation church-wardens' accounts', ed. Bishop Hobhouse, *S.R.S.* IV (1890), 144 and 160.

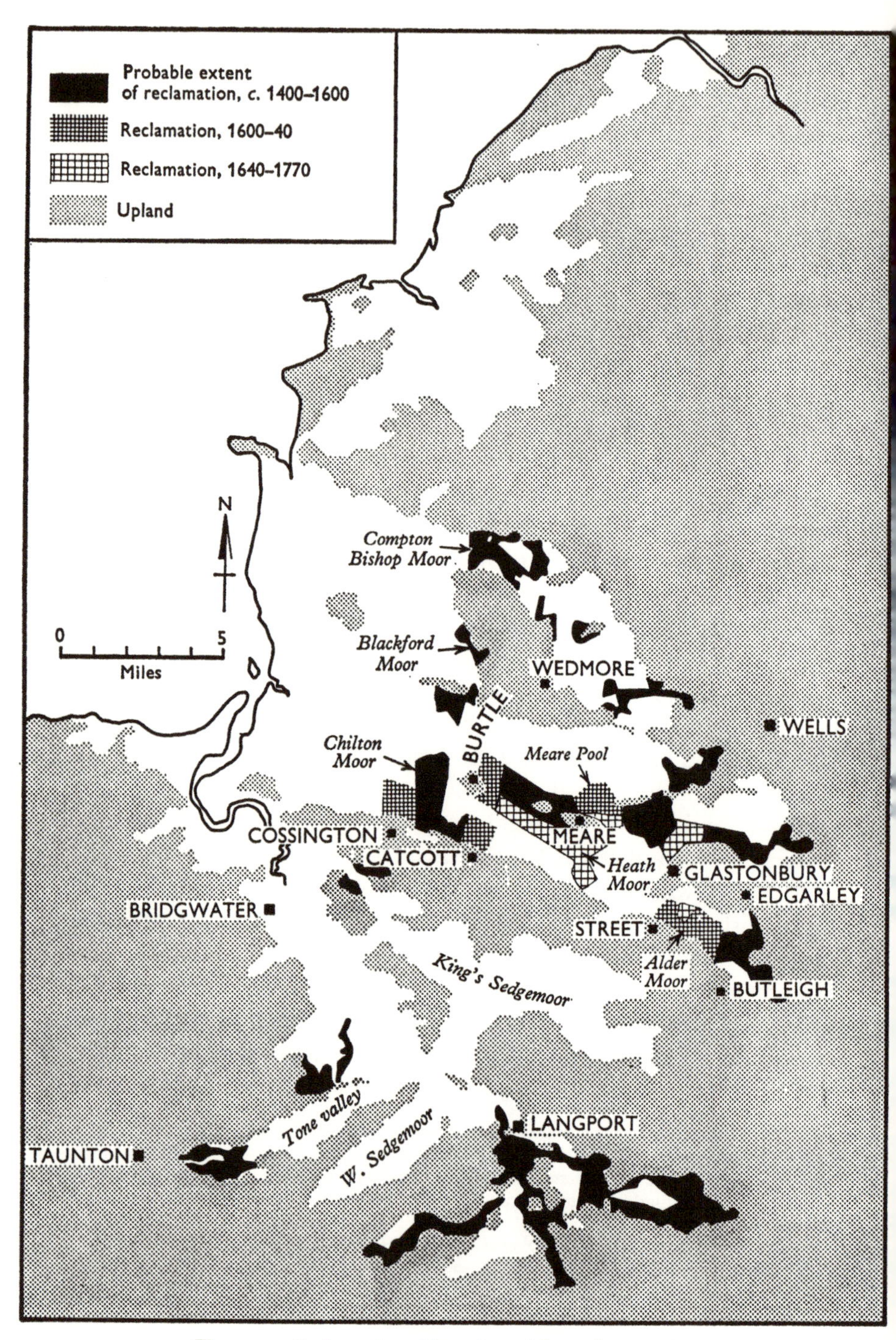

Figure 12. Reclamation: fifteenth to eighteenth centuries.

expansion of reclamation. But it is possible that the map of the medieval distribution does not accurately locate, and that it even omits, some areas. Therefore, the result in Figure 12 is a map which has a limited foundation, but which, nevertheless, offers a basis from which tentative conclusions can be drawn.[1]

The amount of reclaimed land was fairly extensive and the distribution largely what one would expect after looking at the medieval evidence (Fig. 6). Reclamation tended to be scattered and marginal, and correspond closely with the more favourable clay and alluvial soils. The concentration in the upper Parrett valley above Langport was marked; it showed the probable subsequent expansion and consolidation of the position that was hinted at by the medieval distribution in the area. Other centres of activity in the upper Brue, Hartlake and Sheppey valleys, and encircling Wedmore 'island', were consonant with previous experience and are geographically credible. More outstanding than these, however, was the tongue of reclaimed land which stretched from near Glastonbury into the middle of the Brue valley. Not only did it embrace the alluvial soils of the eastern portion of Meare Pool, and so indicate the reclamation of a part of the largest stretch of water in the Levels,[2] but it also extended over the peat soils to the immediate south and west of Meare and Westhay 'islands'. This was a significant departure from the previous pattern of reclamation, and it was repeated in the large block of land reclaimed in Chilton Moor. Altogether it seems probable that an area of approximately 22,000 acres, or about one-third of the land liable to flooding in the Levels, was reclaimed by 1600;[3] this, of course, includes the medieval reclamations. Because of the absence of documentary evidence it is difficult to say how well the land was protected from flooding, but it seems most likely that the improvement was in limiting the period of flooding in these lands rather than in totally excluding flood water all the year round.

One cannot leave this period without considering the probable effect of the dissolution of the monasteries on the draining of the Levels. The dissolution of the smaller monasteries occurred in 1536,

[1] It is gratifying to note that the one rare instance of documentary evidence indicating reclamation during this period does correspond with an area that appears in Figure 12. It is for Compton Bishop Moor, which was reclaimed and enclosed some time between 1520 and 1530. P.R.O. E.178/1927.

[2] This area had long been known as the 'new grounds' by 1638. P.R.O. LR2/202.

[3] For the evidence used in arriving at this figure of approximately 22,000 acres, see p. 109 below.

and of the larger ones three years later. Glastonbury, Wells, Athelney and Muchelney owned over two-thirds of the area of the Levels, and, as past experience had shown, they were always in the forefront of drainage matters. To deprive these lands of their traditional owners must have resulted in a slowing-down or hiatus in the progress of reclamation. At the very least, it is impossible not to feel that the dissolution led to a weakening of the maintenance of the existing drainage channels through neglect and abuse, and to a division of responsibilities which would hamper future activity. The only consoling event, from the point of view of drainage maintenance, was the increasing importance of the Commissions of Sewers.

DRAINING DEVELOPMENTS, 1600–40

The years from 1600 to 1640 were characterized by a quickened interest throughout Britain in the possibility of reclaiming and draining the lowlands around the coast. This was but one aspect of a general trend towards agricultural improvement which showed itself in the contemporary agriculturural writings of Hartlib, Blythe and Platt,[1] in the repeated petitions for the granting of patents for 'hydraulic machines for raising water and drayning marshes',[2] and in the practical experiments in draining undertaken in the Fens and elsewhere in eastern England.[3]

More locally, in the Levels, an important new stimulus to draining activity was provided by the participation of the Crown in land improvement. After the dissolution of the religious houses and the distribution of their estates, the Crown had become the largest landowner in the region. Its motive in draining was largely monetary, and arose from the financial difficulties of the Stuarts. The obvious profit accruing to the clay belt landowners from draining the coastal marshes or 'warths', as they were known locally, encouraged the Crown to satisfy its financial ambitions by attempts to drain and reclaim King's Sedgemoor and Alder Moor, and there were plans

[1] For instance, S. Hartlib in his *Legacy of Husbandry* (1651) noted that 'in Queen *Elizabeth's* dayes, *Ingenuities, Curiosities* and *Good Husbandry* began to take place and then *Salt-Marshes* began to be fenced from the Seas'.

[2] *C.S.P.D. 1547–80*, 598 and 611 (1575 and 1578 respectively).

[3] For example, Darby, *Draining of the Fens*, 23–64; Sheppard, *Draining of the Hull Valley*, East Yorkshire Local History Society, Publication No. 8; and Thirsk 'The Isle of Axholme before Vermuyden', *Agric. Hist. Rev.* I, 16; and *idem*, *English Peasant Farming*.

afoot for the reclamation of thirty other 'waste grounds' in the Levels.[1] What Michael Drayton in the rambling verse of his *Polyolbion* had called the 'plump-thigh'd moor and full-flanck't marsh' was indeed a temptation to the speculatively-minded.[2] In the search for new sources of wealth, proceedings for the dis-afforestation and enclosure of Frome-Selwood and Neroche Forests, to the immediate east and west of the Levels respectively, had already been undertaken by the Crown.[3]

An event of more local and immediate implication, which must have pointed to the need for more adequate coastal defences and a better drainage system, was the disastrous flood of 1607. On 20 January 'the sea at a flowing water, meeting with land flouds, strove so violently together that heaving down all thinges yt were builded to withstand and hinder the force of them, the banks were eaten through and a rupture made into Somersetshire'. The whole of the Brue Level, at least, was inundated to a depth of 11 or 12 ft; the flood waters reached the foot of Glastonbury Tor, 14 miles inland, and parts of the Northern Levels were similarly inundated. Huntspill, Brean, Berrow, Lympsham, South and East Brent, and many other villages were flooded.

In a short tyme did whole villages stand like Islands (compassed round with Waters) and in a short tyme were those Islands undiscoverable, and no where to be found. The tops of trees and houses onely appeared (especially there where the Countrey lay lowe) as if at the beginning of the world townes had been builte in the bottome of the Sea, and that people had plaide the husbandmen under the Waters.

Scenes that were to be described again in much later accounts of flooding were also recorded. While ricks floated away on the flood water and

a co'pany of Hogs and Pigs, being feeding upon one of the ricks... perciving it to go away more and more fro' the', they got up to ye top, and there maintained the' in eating. Nay which is more strange, conies in great numbers being driven out of their burroughes by the tyde, were seen to sit for safety on the backs of sheepe, as they swom up and down and at last were drowned with them.[4]

[1] *C.S.P.D. 1629–31*, 152.

[2] M. Drayton, *Polyolbion* (1613), 46.

[3] *C.S.P.D. 1637*, 111.

[4] Anon. tract, *A true report of certaine wonderfull over-flowings of Waters, now lately in Summersetshire, Norfolke, and other places of England* (1607). See also an article in the *Gentleman's Magazine*, XXXII (1762), 306, entitled 'The 1607 Flood'.

Eventually, when the water subsided, the Commissioners of Sewers directed 500 men to fill the breach in the sea wall at Burnham, the Commissioners and justices 'helping not with their eyes alone, but also with their hands'.[1] Granting the possible exaggeration of the narrative the detail remains, and it is an indication of the severity of the flood.[2]

While noting the circumstances which contributed towards a favourable attitude to draining during this period, we must also recognize that there was a growing body of opinion prejudicial to these trends. Events and circumstances which frustrated development in a particular area were to be of as much significance in the problem of draining the Levels as those which encouraged activity elsewhere. Opposition largely arose because of the problem of regulating the overstocking of inland peat moors, which led to a shortage of common land. A close examination of contemporary depositions leaves one in no doubt that this problem was uppermost in the minds of the agricultural population of the Levels.[3] Previous reclamation, illegal encroachments, and the growing practice of importing 'outside' cattle for fattening by the better-off farmers with unlimited rights of stockage,[4] all aggravated these difficulties. By the beginning of the seventeenth century it was said of King's Sedgemoor that

> suche a maney of hostes do inhabit neere the moore to enterteyne the beastes of forreiners...that (besides most remote parts of the same county) their common is so abundantly surcharged with greater numbers of cattel (sent to seeke fortune from the counties of Gloster, Devon, Dorset, yea, and from Ireland also), that the poore borders cattel finde their diet exceedinge shorte and often but a bare bargine to hold life and limbes togeither.[5]

[1] E. Green, 'On Somerset Chap Books', *S.A.N.H.S.* xxiv (1872), 50–66.

[2] The maximum height of the flood waters was chiselled on local churches. At Kingston Seymour church there is a mark at 25·4 ft O.D. Newlyn, and, on the other side of the Bristol Channel in Monmouthshire, a mark at 23·4 ft O.D. Newlyn can be found on Peterstone church, Wentloogg, and on Godliffe church, Caldicott. The water was said to have been 5 ft high in Kingston Seymour church and to have remained for ten days.

[3] Some selected documents which reveal the situation are: P.R.O. E.134/38 Eliz. I/Hil. 12, (West Moor); E.134/20 Jas. I/Hil. 8, (Theale Moor); E.134/10 Jas. I/Hil. 18, (Drayton Moor); E.134/2 Chas. I/Mich. 14, (Langport Moor); and E.134/34 Eliz. I/Easter 21 (Mark Moor).

[4] The scale of importation of cattle from South Wales to Somerset during this period can be gauged from a perusal of E. A. Lewis's 'Welsh Port Books, 1550–1603', *Cymmrodorion Record Series*, No. 12 (1927).

[5] Brit. Mus., Royal and Kings MSS. 17*a*, xxxvii, ff. 11–11*d*, 'An apology of the King's agents for the enclosure of Kinges Sedgemoore in the County of Somerset' (*temp.* James I), a lengthy and, probably, exaggerated account of the moor and its proposed drainage and enclosure. For confirmation of the Irish trade, see *The Agrarian History of England and Wales, 1500–1640*, ed. J. Thirsk (1967), iv, 73 and 78.

In the circumstances it was natural enough that the majority of commoners were opposed to draining, or to any alteration in the state of the commons that did not benefit them; and, what is more, because of their large numbers they could make their wishes felt. In this connexion it is well to remember that the commoners were not only the landless squatters, who had 'their greatest part of their mayntenance for the keeping of themselves, their wyfes, children and families from the commons',[1] but many were small freeholders, and many were prosperous farmers from the surrounding uplands. These were like the 'wealthy and substantiall men, though none the best bred...[who] have money in their purses to make them gentlemen when they are fitt for the degree', whom Gerard had noted in the upper Parrett valley,[2] and who could wield considerable influence when they acted together. On the whole, the evidence suggests that most commoners were content with the state of the moors, and that they were willing to tolerate the possibility of their being inundated for five or six months, and of the soil remaining waterlogged for many additional months, so long as they had some chance of de-pasturing their stock. This attitude was an important geographical factor in the subsequent draining of the Levels.

Reclamation between 1600 and 1640 occurred in two distinct areas of the Levels. First there was the recovery of the coastal lands or 'warths',[3] and secondly, there was the reclamation or attempted reclamation of portions of the inland peat moors.

Coastal reclamation and defence

The warths were reclaimed in order to overcome the shortage of pasture in the grazing lands of the coastal clay belt.[4] This shortage is reflected in the complex intercommoning arrangements that the coastal villages had with the inland peat moors. These arrangements are depicted in Figure 13, which is constructed from the evidence of two sources: from the detached portions of parishes, which preserved

[1] P.R.O., E.134/10 Jas. 1/Hil. 18.

[2] 'A Particular Description of the County of Somerset, Drawn up by Thomas Gerard of Trent, 1633', ed. E. H. Bates, *S.R.S.* xv (1900), 127.

[3] The origin of this word is obscure, but the first known reference to it is possibly in the 'Cartulary of St. Mark's, Bristol', ed. Ross, *Bristol Rec. Soc.* xxi, No. 195 (1404–5), where reclaimed land in Pawlett Hams is called 'le Wrath'.

[4] P.R.O. SP.16, 533, 3. The three clay belt hundreds of Bempstone, Brent and Winterstoke were said to have a husbandry 'consisting much in grazing and most proper for it'.

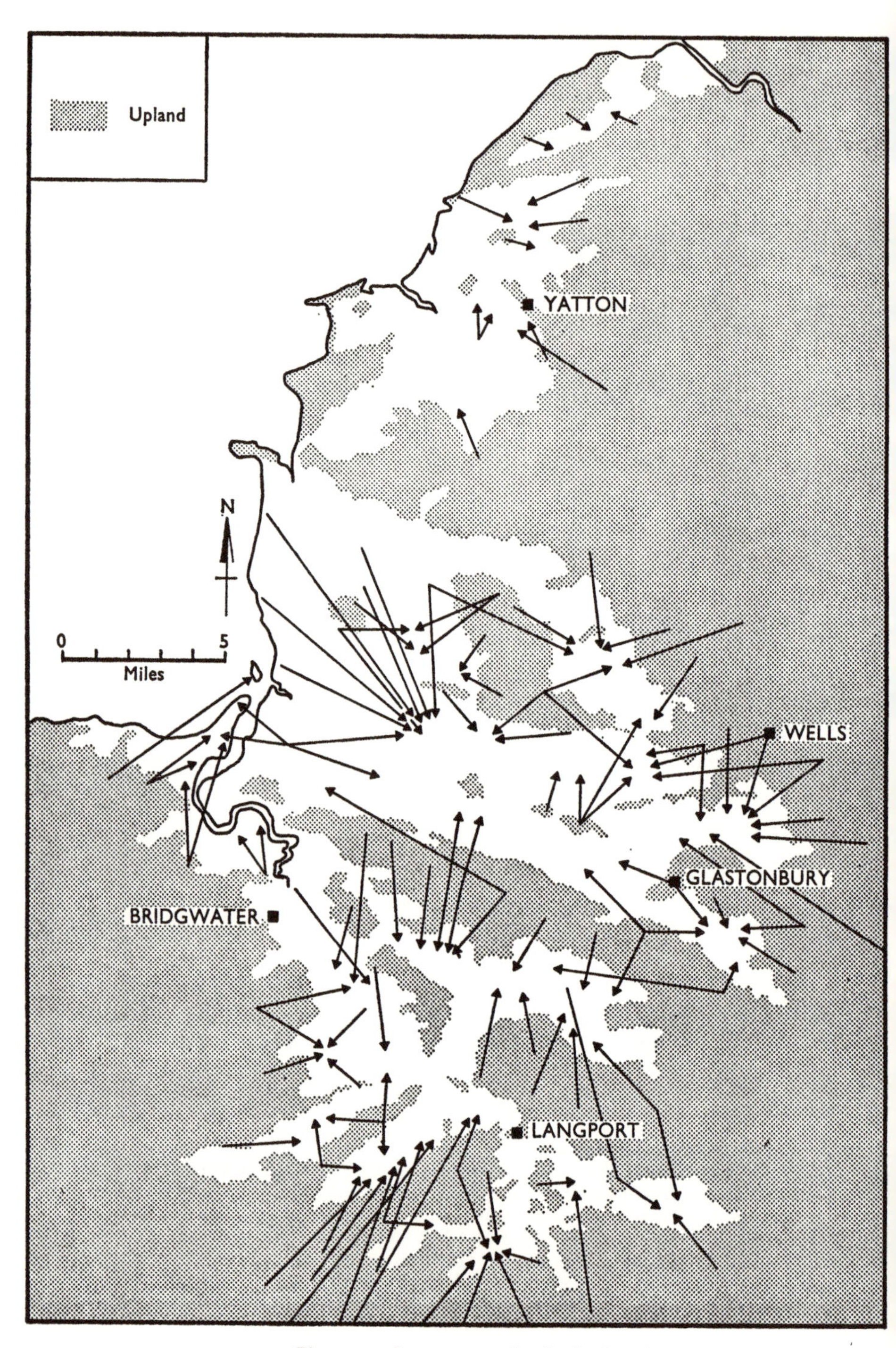

Figure 13. Intercommoning in the Levels.

in a fossilized form the complexity of these arrangements, and from an early seventeenth-century survey of commons in the Levels.[1] Some settlements such as Burnham, East and South Brent, Brean, Huntspill, Lympsham, Berrow, Badgeworth and Tarnock, drove their cattle for long distances to Mark, Tealham, and Cote Moors, in the Brue valley.[2] Some, such as Brean, even went to great lengths to increase illicitly the number of cattle which they could pasture; this is a measure of the value of the inland grazing areas to the coastal villages.[3]

An added incentive to reclamation was the value of the warths, which often equalled, or surpassed, in value some reclaimed lands in the peat moors. Very good warths at Highbridge and Bleadon commanded 3*s*. 4*d*., 3*s*. 8*d*., and even 6*s*. 8*d*. an acre. More commonly warths averaged 2*s*., though sometimes they fell to as low as 6*d*. and 4*d*. an acre. These variations were a reflexion of the composition, salinity, risk of erosion, and particularly of the liability to flooding of the individual warth when 'heavy winds and high spring tides were upon them'. In contrast, ground in the peat moors was not so valuable and new grounds in the Meare Pool area were worth only 6*d*. an acre after approximately fifty years' reclamation. The worth of virgin ground in King's Sedgemoor was reckoned to be 2*d*. per acre, with the possibility of its rising to only about 1*s*. after draining. Some better meadows at Pouland (?), near Meare, were valued at 1*s*. 3*d*. an acre.[4]

The reclaiming of the coastal warths was further encouraged by the very ease of the task. Not only was there little danger of depriving any one of his ground, and so arousing opposition as in the peat moors, but there were fewer technical or physical difficulties to overcome. All that was needed was to place hedges and brushwood on the mudbanks to accelerate deposition from the heavily silt-laden tidal and river waters. A more rapid and stable accumulation of deposits was achieved by making a low wall of stones, or by driving wooden

[1] The detached portions of a parish indicate the interest of that parish in the area concerned. The parish boundaries are those in the 1st ed. of the O.S. 1-inch Map of Somerset (1809). Documentary evidence of common rights is from P.R.O. LR2/202, but the evidence for the Northern Levels is less certain than that for elsewhere in the Levels.

[2] P.R.O. E.134/34 Eliz. I/Easter 21. For other evidence of these arrangements see *Cal. Pat. Rolls, 1547–1553*, 20.

[3] P.R.O. E.134/6 Jas. I/Mich. 12.

[4] P.R.O. E.134/6 Jas. I/Mich. 12; E.134/4 Jas. II/Mich. 16; Bodleian Library, North MSS. b. 26; and P.R.O. E. 134/26 Chas. II/Easter 23, respectively.

stakes down into the mud, weaving brushwood between them and covering them with earth turfs.[1] There can, moreover, be little doubt that the extremely wide range of the tide in the Bristol Channel greatly facilitated the growth of the warths; the spring and other flooding tides, which were heavily charged with suspended matter, were followed by long periods of low tides during which time the sea never attained the level of the warths; this enabled a closely-knit vegetation cover to gain a good foothold on the mud before the next tide reflooded it, and consequently the vegetation could hold together the existing mud and assist in the deposition of new mud.[2]

The rapidity of accretion is well attested by examples; for instance, at Lympsham, a muddy bank, visible only at low tide, rose in about ten years to form ground fit for grazing, and material which accumulated near Bleadon projected over 100 yards into the Axe estuary in only two years. This incipient warth was said to be the cause of considerable erosion elsewhere in the estuary because it deflected the river channel, and finally it was cut away. But, as local farmers pointed out, this was only a temporary solution since, because of the rapidity of deposition, 'the same will fill and grow again...within one week'.[3] These examples of the rapid rate of deposition are well supported by evidence from later centuries.

In time, a fresh warth was added to the village lands and a new wall built further out on the mudbank, or 'over' as it was known, to protect the new warth from flooding. As the new warths often grew two or three feet higher than the 'old' lands behind the sea walls, the danger of inundation was not so great as formerly, and the old wall was then soon forgotten and allowed to fall into decay.[4] Figure 14 shows the probable location and extent of the reclaimed warths; the areas depicted are those which lie beyond the succession of old sea walls which can be traced along the coast, both in the field and from air photographs. In the absence of good evidence to the contrary, these areas are assumed to represent seventeenth-century coastal

[1] P.R.O. E.134/26 Chas. II/Easter 23. The same method was used in the construction of the sea walls that were eventually built to surround the warth. Large stakes were driven into the ground and bound by thorns, and on the top was placed 'a considerable quantity of earth and turfes'.

[2] For a discussion on vegetation sequences and salt marsh development on the Somerset coast, see C. E. Moss, *The Geographical distribution of vegetation in Somerset: Bath and Bridgwater District* (1907), 7–20.

[3] P.R.O. E.134/5 Chas. I/Easter 8.

[4] P.R.O. E.134/26 Chas. II/Easter 23.

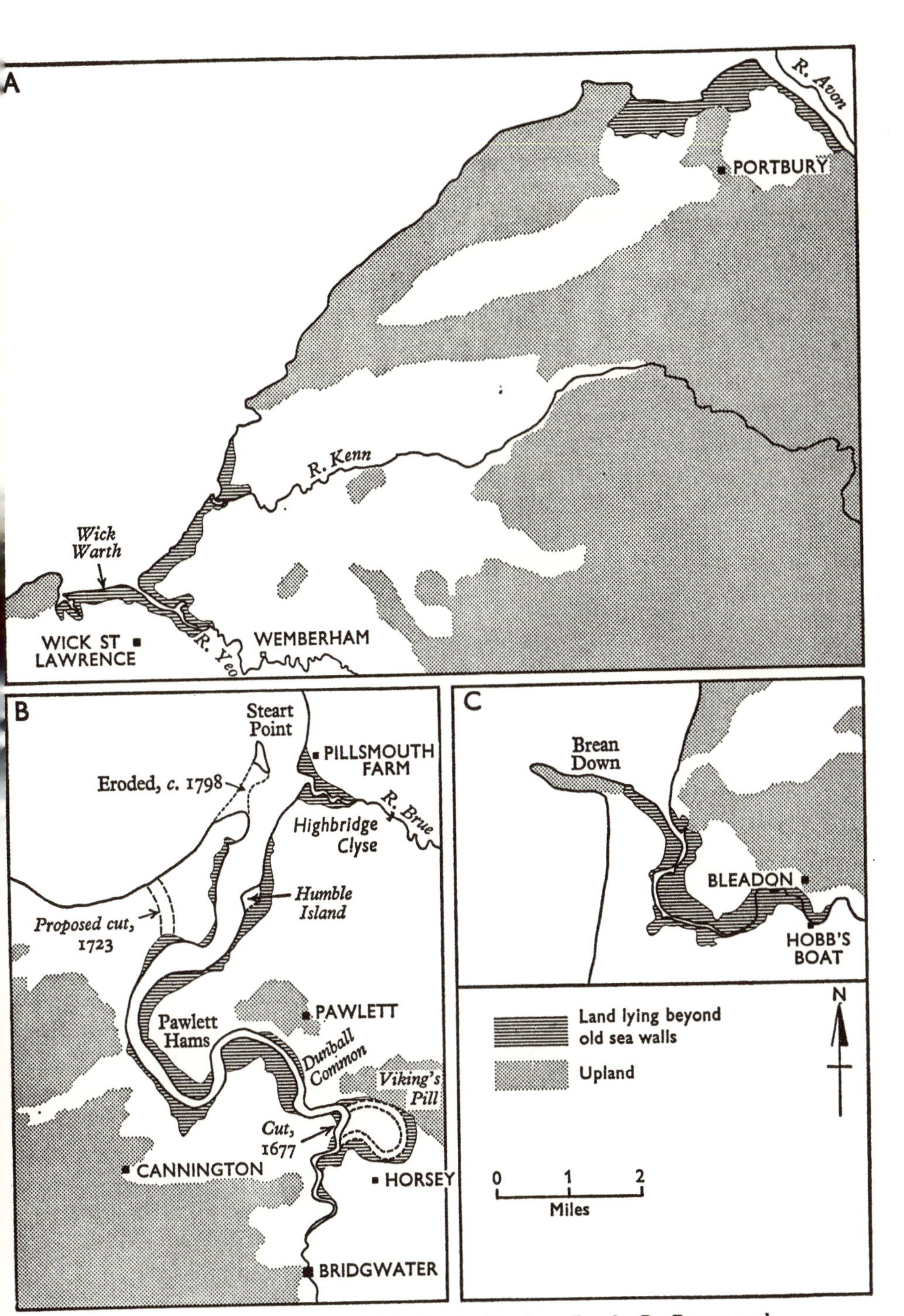

Figure 14. Reclamation of tidal lands: A—Northern Levels; B—Parrett and Brue estuaries; C—Axe estuary.

reclamation, and in many cases documentary evidence supports this view. It can be seen that the warths lay mainly in the sheltered estuaries of the main rivers where deposition was at its greatest.

The earliest recorded reclamation of a warth is in 1597, in the Parrett estuary, where a mudbank of 50 acres in extent, called 'Burlands Oads' or Ooze, lying between two channels of the river, was embanked and recovered. Thirty-six years later, 10 acres of 'wet and oozy' grounds were won from 'the old navigation' (i.e. an old channel of the Parrett) near Horsey Pignes, and a further 30 acres of unidentified warths were 'walled and taken in' from the mouth of the River Brue to the Parrett, together with 70 acres in 'the Humble' island which lay off the Huntspill coast.[1] A further 140 acres were recovered along the coast from Pawlett to Bridgwater, and there were said to be 277 acres newly embanked in Pawlett Hams alone. Part of Dunball common, amounting to 100 acres, was secured from inundation by a new wall, and another 30 acres of land reclaimed and added to it in about 1607. To the west of the River Parrett, Sharpham and Stolford commons were probably embanked and made safe from regular inundation. 'Many warths' were said to have been reclaimed in the Brue outfall below Highbridge clyse, including 15 acres near Pillsmouth Farm, and nearly 100 acres were recovered from the sea near Wick St. Lawrence in the Northern Levels.[2]

The greatest activity, however, was in the estuary of the River Axe. The documentary evidence concerning the exact location of these new warths, and of the areas involved, is a little confused and possibly repetitious, yet it seems that below Hobb's Boat separate parcels of 80, 80, 40, 20, 20, 20, 10, 10, 10, 8 and 7 acres were reclaimed. Of this possible total of 305 acres, 140 had been recently 'fenced and kept from overflowing' in the Bleadon Level. One of the warths reclaimed in 1613 was particularly noteworthy because the sea was 'kept out from overflowing the same by a windmill built at the time of the

[1] P.R.O. E.178/1983; E.178/5619. In 1739, the 'Humble' was joined to the mainland when ice blocked the small channel that separated the island from the shore, and so diverted the Parrett into its present channel. The abandoned channel quickly filled with slime and mud and became level with the surrounding land. See R. Locke, 'On the improvement of meadow land...', *Letters and Papers, Bath and West Eng. Soc.* v (1793), footnote on 205.

[2] See P.R.O. E.178/6028 and E.134/5 Chas. I/Easter 8.

said enclosure for that purpose...'. This is one of only two known instances of the use of a windmill for pumping purposes in the Levels.[1]

By their very nature, the warths were never entirely free from the danger of inundation; some in Bleadon were regularly flooded six or seven times a year; near Crottsham Pill in the Northern Levels, 200 acres of newly reclaimed land, worth 8*s*. per acre, were disastrously flooded and many cattle and sheep that were pasturing there were drowned. New houses built on the warth were in a vulnerable position; 'often times at high and spring tides, the sea water both surrounded and encompassed the houses, and both floors, about a foot above every of the same houses, and would many times flow higher were not the tide refrained by a big bank that is made above the same houses'. The same story of inundation was repeated for Steart, Otterhampton, Cannington, Dunball and Downend Commons, as well as for other lands alongside the rivers Parrett and Axe.[2] There was no such thing as success in recovering the warths, only degrees of unsuccess.

The draining of the peat moors

The area of the greatest draining activity during the years from 1600 to 1640 was in the peat moors. For the greater part of this period strenuous efforts were made by the Crown to drain the peat lands of the King's Sedgemoor. The activity associated with this project, together with the draining of Alder Moor, south-east of Glastonbury, which was closely linked with it, overshadowed the more local undertakings elsewhere in the Levels.

King's Sedgemoor. The choice of King's Sedgemoor as a centre of activity was reasonable, for in comparison with other peat areas in the Levels it possessed certain advantages which commended themselves to the would-be drainer. Its boundaries were well defined and did not merge imperceptibly into adjacent low grounds, and there were only the River Cary and some minor streams to control (Fig. 21). More important than all these natural advantages was the fact that King's

[1] Details of these reclamations are to be found in P.R.O. E.178/6028 and also in E.134/5 Chas. I/Easter 8. It is difficult to understand why a windmill was necessary for the draining of the Bleadon Level, because the fall from the Level to the low-tide line is so great that a simple sluice is all that is required to drain the area.

[2] P.R.O. E.178/6028.

Sedgemoor had been owned entirely by Glastonbury Abbey and at the dissolution of the monasteries it had passed intact to the Crown, although the twenty-five surrounding manors had been granted to various people. Fortunately, no part of the moor had been disposed of by the Crown, which meant that, in the early seventeenth century, it was the only natural drainage area in the Levels in which a co-ordinated scheme of drainage could be undertaken without the consent and co-operation of numerous landowners.

Yet, although the Crown possessed the land and the necessary legal authority to undertake new drainage works,[1] it lacked the money to construct them. Therefore, responsibility was handed out to agents who were to invest money and engineering skill in the project, and who were then recompensed by a grant of the land when drained. This was the only way to reward them, for a payment out of the rent of the newly drained land was too slow a process to commend itself to prospective investors.

Although some preliminary moves had been made as far back as 1600 to prove the Royal title to the moor, it was not until 1618 that James I decided to drain King's Sedgemoor.[2] It was described then as being 'so corse, level and subjecte to waters by raine and surroundinge, and is thereby so cold, heartless and barren, that it affordeth not in muche of it, the twentieth parte of the fruite it would thankfully yield...'. The flood waters did not only 'poison or kill the ground within, but restinge upon the very face of it the winter...and makinge a longe delaie of departure towards Sommer, doe sometimes cause it to yield little good to the commoners at all'.[3]

After preliminary negotiation between the twenty-five manorial lords of the surrounding uplands, agreement was reached whereby James retained 4,000 acres for himself, and they were allotted the residue. The manorial lords welcomed the move and petitioned James, stating, 'we believe an allotment will be more profitable than our common was [or] the whole moor has been'. The commoners took a different view, and rumour and discontent were rife;[4] consequently

[1] *C.S.P.D. 1598–1601*, 115. A General Drainage Act (43 Eliz.cll) was passed in 1601, which confirmed and expanded the articles of a proposal of 1585. It was 'an Act for the recovery and inning of drowned and surrounded grounds and the draining dry of water marshes, fens, bogs, moors, and other grounds of a like nature...' See *C.S.P.D. 1581–90*, 229.

[2] *C.S.P.D. 1598–1601*, 520, and *C.S.P.D. 1619–23*, 88.

[3] Brit. Mus. Royal and Kings MSS. 17*a* XXXVII, f. 8.

[4] *C.S.P.D. 1652*, 302. The manorial lords had been sent letters requesting their assistance, see *C.S.P.D. 1623–25*.

a commission was sent to the Levels to ascertain their attitude.[1] It is certain that the commoners objected not only to the extinction of their common rights of pasture but also to the reduction in the size of the common; the Crown had already claimed over a third of the moor and the agents might possibly do the same, leaving the commoners with a paltry section of the moor.

Nevertheless, some progress was made; during 1623 it was suggested by James's advisers that 300 acres should be divided amongst the agents, 'the inferior actors in the Sedgemoor business', to spur them on to greater effort,[2] and in July of the same year the plan for draining reached the point where one hundred pounds was allocated to commence the work. Two years later, three local men were directed to conduct 'a winter survey of King's Sedgemoor, that they may ascertain what channels are required to draw off the water without injury to other lands, and what quantities and descriptions of wood and stone are requisite for the same...'.[3]

One is not sure what it was that ultimately interrupted and stopped the drainage scheme. The opposition of the commoners may have been too strong,[4] and it is likely that James was distracted from his purpose by the pressure of more urgent outside events. In any event, James died in 1625 and nothing had been accomplished, except the creation of much bad feeling and ill-will which were to affect future drainage attempts.

On the death of James, Charles I took up the scheme and pursued it with great vigour, encouraged by suggestions that by selling and letting the moor he would gain £20,000. He informed the manorial lords that any profit he made would be used solely to support his finances, and it seems certain that a desire for the improvement of the moor was only a secondary consideration.[5] An infusion of capital into the scheme was necessary for its success. Charles found ready backers in Adam Moore and John Battalion, alias Shotbolt, who had already been the agents for James, and who were also, incidentally, carrying on an intensive lobbying campaign for exclusive rights to transport

[1] *C.S.P.D. 1619–23*, 88 and 209.

[2] *C.S.P.D. 1619–23*, 506. This proposal was never carried out and the agents then asked for 500 acres, but again without any result. See *C.S.P.D. 1619–23*, 535.

[3] *C.S.P.D. 1623–25*, 10 and 445.

[4] *C.S.P.D. 1627–28*, 256.

[5] *C.S.P.D. 1625–26*, 525. For a detailed calculation of the expected profits from draining, see Bodleian Library, North MSS. b. 26, f. 50*d*.

sea coal from the north-east coast of England to London. They were ready to deal with the commoners and all other dissenters and to arrange financial and legal matters. They also had plans for draining and improving thirty other 'waste lands' in the county, all of which would help to raise 'a powerful treasury' of nearly £200,000.[1] King's Sedgemoor was obviously only part of a much larger scheme to drain the Levels. In return for their immediate labours the agents wanted a free farm of 1,000 acres, which was to be carved out of the drained moor.[2]

The agents proposed to enclose and drain Sedgemoor; 'with convenient passages for the waters to have issue to the rivers, the very ditches of the enclosures will so drayne it, sucke out the water [and] the land will sone become dry, warm, solide and full of fruite'. They were sanguine of success and expected 'many hundreds of well fed oxen and thousands of fat sheep' to take the place of the jaded and diseased cattle when the moor was drained.[3]

The draining of the moor was to be the responsibility of a special commission working in conjunction with the agents, and between 1625 and 1639 Charles issued no fewer than nine commissions.[4] Each was couched in roughly the same terms for the draining and division of the moor, but from the great number of commissions issued it is obvious that little progress was made. Each commission was directed to a group of local men with whom rested the task of settling the claims for common, of establishing the title of the moor, and of allotting appropriate proportions. After the commission was completed, the supervision and maintenance of the drains was to be placed with the Court of Sewers, but not one commission made a final decree. Unfortunately the proceedings of the commissioners are not extant, but probably little was accomplished other than to untangle the complicated knot of claims to the common and to establish their validity.

The failure of the scheme did not arise from any technical difficulties in coping with the excess water of the moor; in fact it did not

[1] Brit. Mus. Stowe MSS. 326, f. 20, and *C.S.P.D. 1625–49*, 152.

[2] Bodleian Library, North MSS. b. 26, f. 50. Obviously there were others interested in handling the Sedgemoor business and a Thomas Jarvis petitioned the King for a licence 'to treat with the county': *C.S.P.D. 1625–26*, 525.

[3] Brit. Mus. Royal and Kings MSS. 17*a*, XXXVII, ff. 9*d*–10 and 12.

[4] 1625 (2), 1628, 1630, 1632/3, 1633, 1637/8, 1639 (2). See also P.R.O. E.178/5616 for the 'Articles of instruction for the Survey of the Moor', 8 Chas. I. The local manorial lords were again subject to general pressure from the King, as were the commissioners. See *C.S.P.D. 1627–28*, 256 and *C.S.P.D. 1628–29*, 327.

get to the stage where a rhyne was dug. Nor did its failure arise from a lack of money, for the agents were willing enough to spend time and energy on the speculation. One reason which may have hampered the progress of the draining was the strained relations between the Crown, the commissioners, the agents, the manorial lords and the commoners; they were all working for their own interests and against each other. In addition, the prospect of a farm of 1,000 acres and royal favour caused bitter jealousy amongst the agents themselves. Thus, a Mr Waldron, 'a man very criminall and defamed, for his own private ends questioned the manner of proceedinge about this work alleadginge that the Court of Lawe had exsasperated the Cuntry and that Mr Shotbolt had neither power nor friends...to sway the worke; layinge many false ymputations and aspersions on him and his adherents'. Waldron, backed by Sir Giles Mompersons, an agent of some notoriety, made representations to Charles for the right to handle the draining. In the meantime yet another group of agents attempted to do the same and to benefit by the labour of those 'who spente much tyme and expence therin'.[1] Disputes and upsets reached such a point that it was said that the draining of the moor was 'in more danger of fayling than yt was att the first undertaking [i.e. James's], if not thereby utterly foyled'.[2]

The main reason for the failure of the scheme lay with the commoners, whose rumours and opposition drained the strength and vitality of all concerned. In a detailed study of the personalities and politics of Somerset from 1625 to 1640, Professor T. G. Barnes has put forward the suggestion that 'the commissioners were in the invidious position of deciding between the claims of their powerful peers, the lords of the manors, and the claims of a multitude of small free-holders'. On the one hand, the commissioners had to serve the interests of the King and the manor lords; on the other, they had to avoid offending the mass of smaller men, who in the final analysis formed the 'good opinion of the country' and on whom those commissioners relied for their power in the county. 'It is too easy', continues Barnes, 'to underestimate the pressure that these free-holders might bring to bear upon their masters of the rural bench', the

[1] Brit. Mus. Add. MSS. 34712. Sir Giles Mompersons was satirized as Sir Giles Overeach in Massinger's play, *A New Way to Pay Old Debts* (first produced at the Phoenix, Drury Lane, in 1633). Overeach is called a 'grand incloser of what was common'.

[2] Brit. Mus. Add. MSS. 34712. See also Brit. Mus. Stowe MSS. 326, f. 20, for a petition by Battalion to Charles with a plea that he will expedite the work.

commissioners, of course, invariably being Justices of the Peace.[1] Consequently the commissioners attempted to steer a middle course in order to offend no one. This explanation is credible, for each commission had to be returned into the exchequer two terms after its issue and thus the period of delaying a decision was limited and the commissioners would be absolved of their responsibility to do anything.

These attempts to drain King's Sedgemoor can be traced a little further beyond the immediate period under consideration. As there was no immediate prospect of draining the moor, Charles sold his share of 4,000 acres in 1632. The king's agent, John Battalion, and his associates, would have welcomed the opportunity of purchasing it, for once they were the owners of over one-third of the moor their chance of success in draining the whole of it would have been enhanced. But it is probable that the sale was negotiated without their knowledge. The land went to Jeffrey Kirby, who paid £12,000 and an annual rent of £100 for it. He was a London merchant and landowner in Sutton Marsh, Lincolnshire, who had an interest in land drainage.[2] The famous Fenland drainer, Sir Cornelius Vermuyden, was also a participant in the transaction;[3] in all probability his interest was deliberately concealed, for the purchase of the waterlogged waste by a drainer of his repute would have implied his intention to drain it, and would have inevitably aroused the opposition of Battalion and the other agents.

Battalion, who still held office as the agent for draining the King's share of the moor, found the tide of fortune and favour turning against him; he was summoned before the Attorney-General in 1635 and examined on his part in the delay in the draining. In his defence he said that the whole business had become 'fylled and cloyed with many more perplexityes, questions and queres then I ever dreamed', and that it was the opposition of the commoners, not a lack of energy on his part that held up draining.[4] With conditions as they were, Battalion accused Vermuyden of fraudulently misappropriating £18,000, of being in league with the Attorney-General, and

[1] T. G. Barnes, *Somerset, 1625–1640: A County's Government during the "Personal Rule"* (1961), 154.

[2] *C.S.P.D. 1637*, 278.

[3] *C.S.P.D. 1655*, 301 for instance. Vermuyden and others always gave the date of his purchase of King's Sedgemoor, and the sums of money involved, as being identical with those of Kirby. L. E. Harris in his *Vermuyden and the Fens*, (1953) does the same.

[4] Brit. Mus. Stowe MSS. 326, ff. 20 and 21, and f. 24, Battalion to Attorney-General, Sir John Bankes, 18 June 1635 (copy).

of duplicity in dealings with himself and the other agents;[1] but the charges misfired and Vermuyden retained possession of the moor. With an unsuccessful petition by Kirby's widow and Vermuyden in 1637 for yet another commission to drain the moor and one more unsuccessful petition from the agent, Thomas Jarvis, to 'go down amongst the inhabitants' and deal with them, the Civil War finally brought all such matters to a halt.[2]

The condition of the moor remained unchanged during the next few years and a description of flooding in it during 1653 differed little from that of about thirty years earlier.[3] In 1655, Vermuyden petitioned Cromwell for a new commission 'to set out indifferently for petitioner the 4,000 acres that he may go on with so good a work ...The waste is boggy and unwholesome, but could be improved by draining.' The time was propitious for the draining of the moor; the granting of the necessary legal authority was imminent, the enthusiastic backing of many of the manorial lords was assured, and, with the agents now out of the way, Vermuyden could look forward to draining his portion of 4,000 acres and to the possibility of being asked to drain the remainder of the moor.[4] The prospect of an early start to draining diminished, however, when the old accusations of fraudulent dealings over the purchase of the moor were revived, but Vermuyden's petition was proceeded with.[5] It was finally brought before Parliament as a Bill, but rejected because 'the tenants and freeholders did not consent'; Vermuyden thereupon sold his share of King's Sedgemoor.[6] Sir Cornelius Vermuyden came nearer to draining King's Sedgemoor than anyone else had, yet the seemingly favourable combination of his skill as a drainage engineer, his ownership of one-third of the moor, and his privileged position in influential circles, could not prevail against the opposition of the commoners.[7]

1 *C.S.P.D. 1636–37*, 257; and Brit. Mus. Stowe MSS. 326, f. 25.

2 *C.S.P.D. 1637*, 275, and *C.S.P.D. 1637–38*, 55.

3 Adam Moore, *Bread for the Poor...* (1653), 19 and 26.

4 *C.S.P.D. 1655*, 301–2; and *C.S.P.D. 1655–56*, 337–8. See also Brit. Mus. Add. MSS. 6670, f. 609.

5 *C.S.P.D. 1655–56*, 132 and 337.

6 *The Diary of Thomas Burton*, ed. T. Birch (1828), I, 259; and *Seventh Report of Historical Manuscripts Commission* (1879), 76.

7 Vermuyden did not therefore reclaim King's Sedgemoor as some have said. For example, E. Cressy, *Outline of Industrial History* (1915), 48, says that Vermuyden 'having carried his project (i.e. the draining of the Fens) to a successful issue, he turned his attention to the West Country, where he reclaimed small areas of Malvern Chase and Sedgemoor'. J. Van Veen, in *Dredge, Drain, Reclaim; The Art of a Nation* (1948), 49, makes a similar claim for Vermuyden.

The great promise of improvement which the age continually held forth was never fulfilled.

Alder Moor. The successful draining of Alder Moor or Aller Moor, on the alluvium of the upper Brue valley, was in direct contrast to the failure of the King's Sedgemoor scheme (Fig. 12). It was the fervent wish of the King's Sedgemoor agents during the 1630s that 'as Sedgemore is now as Allermore once was...it may in time be as Aller Moor now is...'[1] Proceedings were begun in 1630 to drain the moor, but what was done is never mentioned.[2] Undoubtedly, the parochial and other allotments must have been divided by rhynes and the bank of the River Brue strengthened. The individual allotments were divided by the commoners as they thought fit, and it seems certain that some remained in an open, undrained condition, and were pastured in common by the inhabitants of the parishes in an attempt to retain some part of their previous economy.[3]

As in Sedgemoor, the Crown and its commission and agents were proceeding with a scheme against the will of a very large majority of the local inhabitants. When the commissioners came to Glastonbury to order the digging of ditches the commoners besieged the house where they met and stated that they 'should be undone if the said moor was enclosed'. One refractory commoner was charged with sedition and it was urged that proceedings be taken against him 'especially at the time when other designs of this nature...are generally besett with bold oppositions',[4] obviously a reference to King's Sedgemoor.

Unlike that in King's Sedgemoor, however, the initial opposition in Alder Moor was overcome, largely because of the smaller area and the smaller number of commoners involved. The commissioners did not risk offending too great a number of small tenants, thus 'forfeiting the good opinion of any appreciable proportion of their countrymen'.[5] In addition, it is probable that the procrastinations over the

[1] Brit. Mus. Royal and Kings MSS. 17*a*, XXXVII, ff. 9v–10.

[2] P.R.O. E.178/5613. Brit. Mus. Royal and Kings MSS. 17*a*, XXXVII, f. 9v. does say, however: 'in former times lying in common it was a most fenne, stagne or marishe ground, worth to the owners as much as nothinge, but it being enclosed, drayned and husbanded, it is now...and was in short space after the first enclosure, even the richest, worthiest and most notable feedinge in all these parts.'

[3] P.R.O. E.178/5613.

[4] P.R.O. E.134/12 Chas. I/Mich. 35; and *C.S.P.D. 1625–49*, 415. See also *C.S.P.D. 1631–33*, 132.

[5] Barnes, *Somerset, 1625–1640*, 155.

Sedgemoor drainage had only hardened the Crown's attitude, so that, but for the impending failure of Sedgemoor, Alder Moor might never have been drained.

There was one other reason which vitiated the effects of the opposition of the commoners to the commission. The chief of the commissioners, Sir Robert Phelips, had long been out of royal favour. Prior to the issuing of the commission, Charles wrote to him and said 'our trust and confidence in you is such that you will rather use your best endeavours and care for the preservation and increase of our ancient rent, and inheritance, than for the favour of the multitude, and doubt not but what benefit wee shall receive by your service and thankfully accept'.[1] Royal patronage was not to be neglected; Phelips did use his 'best endeavours' and expedited the enclosure and division of the moor.

Once more, the greatest grievance of the commoners was the reduction in the size of the pasture land as a result of the mode of division employed. Out of approximately 1,000 acres of the moor, Glastonbury tenants received 250 acres; Street tenants, 160 acres; Butleigh tenants, 230 acres; and 42 acres were distributed amongst sundry persons having a claim to the moor; Edgarley tenants were entirely excluded. The remainder of the moor, probably just over 300 acres, was claimed by the Crown.[2] Later, 140 acres were taken for the agents out of the total acreage allotted to the commoners.[3] Thus, the commoners were left with barely one-half of the moor. Any hopes they had of being able to pasture in the Crown share of the moor, which seems to have been left unreclaimed, were dashed when Charles eventually sold his share for £1,000 to a James Lovington, who naturally debarred the commoners.[4]

The shares which the commoners received were not only small but they were in the most frequently flooded portions of the moor; these were said to be worth only 12*d.* an acre; the Crown portion, on the other hand, was in 'the best and most commodious part...and more free from drowning than that part of the moor left out for [the] said commoners'.[5] The agents, too, had drier lands, which were said to be

[1] *C.S.P.D. 1625–26*, 495; and *C.S.P.D. 1639*, 447. See also *C.S.P.D. 1639*, 448, for a similar letter from Secretary Coke to Phelips.

[2] P.R.O. E.178/5613.

[3] P.R.O. E.134/12 Chas. I/Mich. 35. For a particular instance see P.R.O. E.134/1654/Trinity 2, where one agent, a Mr Louthwaite, was given 60 acres 'for his paines in solisittinge the business'.

[4] *C.S.P.D. 1631–33*, 458.

[5] P.R.O. E.134/12 Chas. I/Mich. 35; and E.134/1654/Easter 15.

worth 20*s*. per acre. Another complaint was that access to the moors was difficult because the old droveways were severed by the new rhynes, and the new droveways, which passed over the waterlogged allotments of the commoners, were useless in wet weather. One grievance of the commoners, which was peculiar to that area, was the loss of the alder groves which grew on the moor and provided fuel, as well as feeding, for cattle and swine. Indiscriminate felling under the direction of the agents had ruined the groves.[1] Taken together, it was not surprising that 'it was the general voice of the greatest part of the tenants of the said several parishes there, that they did disagree and dislike the said inclosures and allotments'.[2]

As a consequence of draining and enclosure the condition of the commoners deteriorated; the number of beasts that could be kept on the moor declined, it being said that some commoners kept only a quarter of the number previously grazed. Arable land in the uplands was turned over to pasture to accommodate the displaced cattle, and 'some that kept whole ploughs cannot now keep about halfe a plough'. In time, many houses were in decay in Glastonbury, and the poor of the four parishes had 'much increased...for the said moor was formerly a great reliefe to the poor'.[3]

Anger and resentment at these events found their ultimate expression in acts of violence and destruction. These occurred after Lovington had undertaken to recompense Edgarley, for its loss of common, with approximately 100 acres of the allotment he had purchased from the Crown, but the agents who handled the transfer then appear to have expropriated the ground.[4] The commoners broke down the walls and fences, filled in the rhynes and stopped the flow of water, so that most of the moor, which had been the 'most notable feeding in all those partes', reverted to its original state. For the next nine years at least, the commoners of the four parishes pastured their cattle as of old, without hindrance.[5]

[1] P.R.O. E.134/12 Chas. I/Mich. 35.
[2] P.R.O. E.134/9 Chas. I/Mich. 8.
[3] P.R.O. E.134/1654/Easter 8; E.134/1654/Easter 15; and E.134/1654/Trinity 8.
[4] See P.R.O. E.134/1654/Trinity 8, and E.134/12 Chas. I/Mich. 35.
[5] P.R.O. E.134/1654/Easter 8 and E.134/16 Chas. I/Mich. 30. D. G. C. Allen, 'The Rising in the West 1628–31', *Econ. Hist. Rev.* v (1952), 76–83, states that there was no violence connected with enclosures in Somerset like the disturbances at the reclamation by Lord Berkeley of the 'New Gained Grounds' at Frampton and Slimbridge on the River Severn in 1631, and the Skimmington riots over the enclosure of the Forests of Gillingham in Dorset, Braydon in Wiltshire, and Dean in Gloucestershire, but there is evidence to the contrary here.

Meare Pool and other parts of the moors. In contrast to the protracted negotiations over King's Sedgemoor and Alder Moor, the difficulties in dealing with the other areas in the Levels between 1600 and 1640 seem insignificant. These areas were all in the Brue valley and they are depicted in Figure 12. In 1606, the manorial lord and tenants in Cossington co-operated to reclaim part of the moor. Every commoner was given an acre in recompense for each beast-lease which he held. The 'New Grounds', as they were called, were then divided by ditches which facilitated the drainage of the moor.[1] The Polden edge of Catcott Moor, to the east of Burtle 'island', was said to be 'now all severall' in 1638, which strongly suggests that it too had received some measure of reclamation.[2]

Evidence suggests that reclamation also took place in the Parrett valley during the period 1600 to 1640, although the reclaimed areas cannot be located precisely. Thus, in 1633, Gerard noted that if the peat moors in the Tone valley and West Sedgemoor were reclaimed, they would 'yield a vast revenue, which is ye cause the takeing of them in hath bin aymed at but with little successe...', which clearly indicates reclamation activity. But greater success attended the efforts of the local inhabitants near Langport, who were embanking the river and straightening bends in its course 'whereof they recovered a great quantity of very riche land'.[3] Other unspecified improvements were carried out in West Sedgemoor in 1637, which appear to have been so successful as to induce some commoners to attempt to expropriate portions of the moor.[4]

More outstanding than these, however, was the draining of Meare Pool. It had been described in 1535 as 'a fysshyng...which is in circuite fyve myles and one myle and a half brode, wherein are great abundance of pykes, tenchards, roches and jeles and of divers other-kind of fysshes...'[5] Leland's description of 1537 was similar; the pool was 'at high waters in winter a 4 miles in cumpace, and when it is lest a 2 miles and an half and most communely 3 miles'. These fluctuations

[1] S.R.O. Glebe Terriers, 199 (1606).

[2] P.R.O. LR2/202. This seems more than substantiated by a document in S.R.O. Som. Arch. Soc., Parochial MSS. 147 entitled 'A Presentment...of all the Lands, Rights and Duties which belong to Mr Abell Lovering' (1634). This gives details of new enclosures from the moor.

[3] 'A Particular Description of the County of Somerset' ed. Bates, *S.R.S.* xv, 220 and 131.

[4] *C.S.P.D. 1637*, 215.

[5] Report to the King on the 'Late Attainted Monastry' (1539) by Pollard and Moyle. Printed in *Peter Langtoft's Chronicle*, ed. T. Hearne, (1725), II, 343–88.

in size were due to variations in the rainfall which caused the pool to expand and contract, mainly in the area known as the West Waste.[1]

Because of the paucity of documentary evidence, one can do little more than attempt a tentative reconstruction of events from the few scant pieces of information that are available. The date of draining is uncertain, but by 1630 a Mr William Freake 'had drayned manie hundred acres of ground there'. Eight years later these 480 acres of new ground (still owned by William Freake) were described as being 'lately a fish pool'.[2] Further investigation reveals that, during an enquiry in 1684 concerning the tithes due from the new land, three deponents stated that the draining had taken place before their recollection; two were 60 years old, and one about 81 years old.[3] Thus, although one can say little that is definite, the implication of the evidence is that the pool was reclaimed not very long before 1630, a time when so much attention was being paid to the problem of draining elsewhere in the Levels.

The new grounds on the floor of the former pool were largely devoted to meadow. Yet, in 1620, a Christopher Cockerell, who was well known 'for the skill hee had attayned in sowinge, dressinge, and orderinge of flax and flax-seed was sent for by divers gentlemen into theis westerne-parts'. By 1632, Cockerell had rented parts of the former pool and employed people from Glastonbury for weeding, preparing and planting the ground with flax. Unfortunately, the experiment was short-lived, for during 17 August 1637 or 1638, 'a fierce and longe raine' caused severe flooding and 'the flax and seed were so spoiled and carried away that very little was saved and that little nothing worthe'.[4] Even as late as 1685 the value of the land still depended upon the state of the drainage; in some years it was tolerably dry and its value was believed to be '6*d*. an acre, one acre with

[1] *Leland's Itinerary in England*, ed. Smith, I, 149. For another description of the pool see P.R.O. 315/420. The expanding pool frequently inundated 75 acres of land to the west; see *Joannisconfratis et monachi Glastoniensis*..., ed. Hearne, II, 318.

[2] *First Appendix to Seventh Report of the Historical Manuscripts Comm.* (1879), 694, and P.R.O. LR2/202. Some confirmation of this general date for reclamation is given in a document in Brit. Mus. Add. MSS. 24821, f. 202 which is a copy of a fee farm roll of 1651. It says that the rent for the pool had been 'constantly answered from time to time until Michas. 1641, but by what right or title I cannot certifie since which time it hath not been paid by reason, as I am informed, that the said Meere is drayned soe that there is neither Fish nor Swannes in the same'.

[3] P.R.O. E.134/4 Jas. II/Mich. 16. For confirmation of these estimates, see Brit. Mus. Add. MSS. 24821, ff. 206–7.

[4] *First Appendix to Seventh Report of the Historical Manuscripts Comm.* 694.

another', while in other years it was 'very subject to be drowned and flooded in great rains' and liable to revert to its virgin state.[1]

Again, what was done to drain the pool in about 1630 is not clear, but if 480 acres of land in the northern half of the pool were recovered, then the rivers Sheppey and Hartlake must have been embanked, straightened, and connected with the River Brue (Fig. 15). From the deposition of 1685 we know that the 'New Cutts', or Decoy rhyne, had been dug about twenty-five years earlier and its channel must

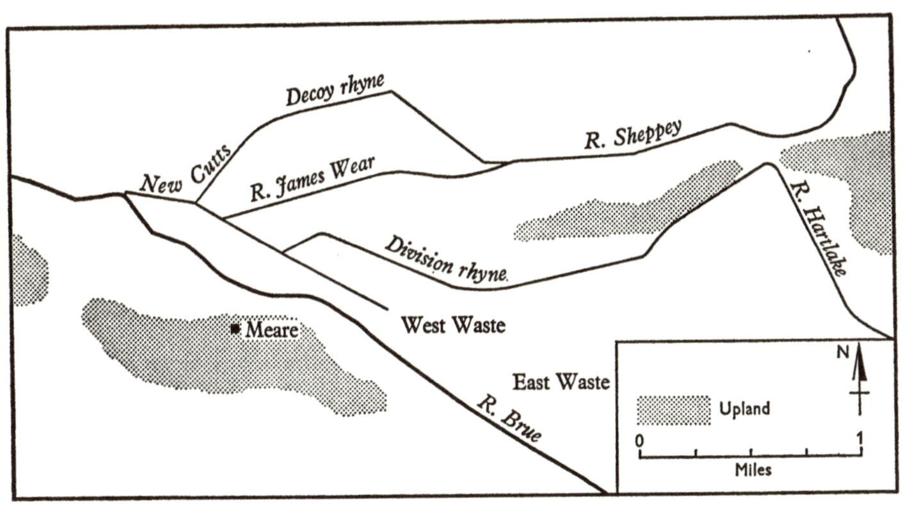

Figure 15. Meare Pool, 1630.

have provided a new and improved outlet for the river running through and near the pool. It is more than probable that the southern half of Meare Pool remained unreclaimed throughout this period for it was not until 1712 that the whole of the pool was said to have been covered with a network of rhynes, as it is today.[2] This conclusion can be verified, to a certain extent, by looking at general maps of the county of Somerset published during the seventeenth and early eighteenth centuries. Meare Pool is depicted as a stretch of open water of varying shape on the maps of Saxton (1575), Speed (1610), Bloome (1671), Sellers (1695), Moll (1723) and Strachey (1736); but it is not shown on Bladeslade's map of 1741, or on later maps.[3]

[1] P.R.O. E.134/4 Jas. II/Mich. 16.

[2] W. Phelps, *Observations on the Great Marshes and Turbaries of the County of Somerset, with Suggestions for their Improvement* (1836), 12.

[3] J. Cowley's map printed in 1744 also shows Meare Pool, but it is likely that this map is an inaccurate copy of an earlier edition; see T. Chubb, *A Descriptive List of the Printed*

But these works did nothing to alleviate the basic difficulties of the unrestricted entry of the tide up the rivers and of the insufficient capacity of the river channels. Flooding was regular and widespread. A glimpse of conditions in one district is provided once more by Gerard who travelled through the Parrett valley in 1633. The moors at Kingsbury Episcopi, Muchelney, Aller and around Burrow Bridge, were 'soe covered with water you would rather deeme them Sea than lande', and the inhabitants of the uplands surrounding Aller Moor were forced to come to church in boats 'and in them also carry their dead corpses to buriall'. Not until the 'watery spectacle', which yielded 'a great commoditie to the fowlers', had subsided with the onset of summer, were the moors of use.[1] Then the unsuspecting traveller looking out from the eminence of Glastonbury Tor could only see 'those fertile and pleasant Moores and Meades' and other 'large and fertile grounds' that were said to produce the largest and best cattle in England.[2] There were two vastly different faces to the Levels; and which one the traveller saw depended upon the time of year in which he passed through the region. Camden's observation, nearly half a century before, that 'as in summer time it may really be termed a summer-country so no less may it in the winter season be called a winter-country; so wet, moist and marshy it is for the most part; which makes it very troublesome to travellers',[3] still held true. The seasonal variations in the aspect of the Levels and their economy were well summed up by Bloome in 1673 when he said that 'What is worst for the rider is best for the abider'.[4]

The progress of reclamation c. 1640

A summary of the progress of reclamation after the first few decades of the seventeenth century can be obtained from a survey made in

Maps of Somersetshire, 1575–1914 (1914). These changes in Meare Pool are discussed in M. Williams, 'The draining and reclamation of Meare Pool, Somerset', *Thirteenth Annual Report of the S.R.B.* (1963), 52–3.

[1] 'A Particular Description of Somerset, Drawn up by Thomas Gerard of Trent, 1633', ed. Bates, *S.R.S.* xv, 220, 215, and 63. For example, Gerard commented on the River Tone (p. 231), which passed 'through deep and dirty moores in the winter time most covered over with water, but at Summer affording greate plenty of grass'.

[2] 'A relation of a short survey of the Westerne Counties...August, 1635', ed. L. G. W. Legg, *Camden Miscellany*, xvi, 3rd ser. vol. LII (1936), 80 and 78; and G. E. Fussell, 'Cornish Farming, A.D. 1500–1910', *Amateur Historian*, IV (1960), p. 339.

[3] W. Camden, *Britannia*, Bishop Edmund Gibson's translation (1695) of the last Latin edition of 1607, I, 177.

[4] T. Bloome, *Britannia* (1673), 195.

1638 of the 'Moores and Lowe Grounds' of the Levels.[1] This survey included all land liable to flooding that lay to the east of the coastal clay belt, together with some marine clay lands west of Wedmore 'island', which were habitually flooded. A detailed measurement was made of each unreclaimed moor and also of the meadows which were 'not very hurtfully surrounded'. These are tabulated below.

Table v. *The 1638 survey of 'Moores and Lowe Grounds'*

	Commons (acres)	Meadows (acres)
Parrett (above Langport)	2,167	3,346
Parrett (below Langport)	6,351	6,516
Brue valley	17,014	10,358
Axe valley	4,994	4,574
King's Sedgemoor	10,696 (est.)*	—
	41,222	24,794

* King's Sedgemoor was omitted from the survey, probably because the survey was Crown-sponsored and aimed at ascertaining what ground in the Levels could be reclaimed. Consequently, as the drainage of King's Sedgemoor was only then being considered, it may well have been omitted. The figure of 10,696 acres is taken from a survey of the moor made in 1791. There is no reason to suppose that it does not represent the area of the moor at an earlier date, for the boundaries of the moor are well defined by higher ground, and will not have changed. A contemporary estimate of 14,000 acres in Bodleian Library, North MSS. b. 26 is a gross overestimate.

From the above table, it can be seen that by 1638, 24,794 acres were classed as meadow and therefore regarded as reclaimed; these accounted for just over one-third of the 66,016 acres of land liable to flooding in the Levels,[2] that is assuming, of course, that an acre of the mid-seventeenth century is of the same dimension as that of today. Of this reclaimed meadow, 2,711 acres are known to have been recovered between 1600 and 1638, which leaves approximately 22,000 acres as the medieval and sixteenth-century contribution to the draining of the Levels. After 1638, on the other hand, nearly two-thirds of the Levels were still left to be reclaimed; this land was almost wholly in the peat moors, the greater part of it lying in the vast unbroken tracts of peat in King's Sedgemoor and the Brue valley (Fig. 12). Consequently, the period 1600 to 1640 is noteworthy

[1] P.R.O. LR2/202.

[2] This figure of 66,016 acres of floodable land in 1638 compares reasonably well with the 69,072 acres which were inundated during the great floods of 1872–3. See Grantham, *Report on the Floods in Somersetshire in 1872–73* (1873).

because it marks the transition between two distinct geographical phases in the reclamation of the Levels: that on the clay and alluvium before 1600, and that on the peat after about 1640.

Unfortunately, it is very difficult to check these conclusions on reclamation activity for the period from the end of the Middle Ages until 1640 by comparing the relative prosperity of the Levels with that of the uplands. The subsidies of 1636 (Ship Money), 1641 (£400,000 for Ireland), and 1649 (£90,000 per month for six months) were not genuine assessments at a local taxation level; rather they were arrived at by the division amongst the counties of a certain total sum to be levied on the whole country. The county totals are therefore allocations rather than assessments, and probably no more than a rough estimate based on some older local assessment, so that the subsidies are 'probably more indicative of the distribution of wealth at the beginning of Elizabeth's reign than in the seventeenth century'.[1] The Hearth Tax of 1644–5 is also of little use, for, in addition to the loss of eight out of the twenty-seven books for the county,[2] any conclusions based on the density of hearths (dwellings) would give an erroneous impression of the relative importance of the Levels, where, despite some draining activity which had led to an extended grazing season, there were few habitations until the end of the eighteenth century because of the recurrent autumn and winter floods. Until more evidence is forthcoming, one cannot be too dogmatic about this period.

DRAINING DEVELOPMENTS, 1640–1770

In marked contrast to the activity from 1600 to 1640, the remaining years of the seventeenth century, together with the greater part of the eighteenth century, saw a near stagnation of endeavour to improve the Levels. There was little difference between the area of the unreclaimed moors of the 1638 survey of 'Moores and Lowe Grounds' and the same moors during the years from 1770 to 1830 when they were finally drained. The areas of these moors at both periods can be compared, and the results are set out below in Table VI. Again, assuming the dimensions of acres have remained constant between the two dates, the only really significant differences in area occurred in

[1] R. S. Schofield, 'The geographical distribution of wealth in England, 1334–1649', *Econ. Hist. Rev.* XVIII (1965), 483–510.

[2] R. Holsworthy, *The Hearth Tax for Somerset*, E. Dwelly's National Records, I (1916), where some books are printed, and P.R.O. E.179/354 for the rest.

Table VI. *A comparison of the area of unreclaimed ground in the Levels,* 1638 *and c.* 1800

	Acres		Differences	
	1638	*c.* 1800	1638 more than *c.* 1800	1638 less than *c.* 1800
The upper Parrett				
Kingsmoor	610	610	—	—
Wetmoor	503	530	—	27
West Moor	1,054	1,100	—	46
The lower Parrett				
Langport Common	65	50	15	—
Week Moor	160	160	—	—
Tone valley	1,794	1,793	1	—
West Sedgemoor	2,875	2,900	—	25
North Moor	1,415	1,466	—	51
Place Drove	42	—	42	—
Brue valley				
Common Moor	543	—	543	—
Crannel Moor	407	369	38	—
Queen's Moor	1,618	1,630	—	12
Tealham Moor	798	800	—	2
Tadham Moor	1,802	1,830	—	28
Godney Moor	1,543	1,520	23	—
Westhay Moor	1,616	1,660	—	44
Ashmoor Moor	217	200	17	—
Mark Moor	1,946	1,967	—	21
Huntspill Moor	642	620	22	—
Little Cote Moor	88	75	13	—
Edington Moor	700	740	—	40
Heath Moor	4,218	2,838	1,380	—
Upper Brue Moors	965*	—	165	—
Axe valley				
Knowle Moor	648	645	3	—
Westbury Moor	416	420	—	4
Panboro' Moor	119	130	—	11
Stoke and Drayton Moor	750	761	—	11
Wedmore Moor	601	595	6	—
Ewe Moor	114	115	—	1
Blackford Moor	514	357	157	—
Allerton Moor	106	102	4	—
Cheddar Moor	1,727	1,763	—	36

* At the time of the 1638 survey, 800 acres of Alder Moor were being reclaimed.

Blackford Moor (157 acres), Alder Moor (165 acres), Common Moor (543 acres) and Heath Moor (1,380 acres); these are depicted in Figure 12. Thus, the total of 2,245 acres represents the possible contribution of 130 years' effort in the draining of the Levels between 1640 and 1770.

Of the reclamations themselves little can be said. Details of the Heath Moor and Blackford Moor reclamations are unknown. The draining of Common Moor was completed by 1722[1] and was particularly noteworthy because a pump or 'engine', as it was termed (which was, in all probability, a windmill), was used to lift the water.[2] The 165 acres remaining in Alder Moor were treated under the same Act as Common Moor.

Throughout this period repeated attempts were made to drain King's Sedgemoor. Some time between 1660 and 1685 over one hundred freeholders petitioned Charles II to divide and drain the moor according to the agreement made nearly fifty years before,[3] and in 1669 an Act was obtained for 'opening the ancient and making new roynes and water courses in and near Sedgemore'.[4] The works envisaged were not fully specified, but they appeared to deal mainly with the western end of the moor near the Rowing Lake, a stream which ran from the north of Sowy 'island' and alongside Lake Wall to the Parrett (Fig. 21). Although the Act was passed, nothing came of it. Past experience suggests that the commission that went to execute the works once more ran up against the opposition of the majority of the commoners. Eighteen years later, yet another scheme was proposed by a Philip Bertie, but, again, nothing further is heard of it.[5]

During the early eighteenth century, Celia Fiennes[6] and Daniel Defoe,[7] both of whom had accurately observed new drainage works in the Fens, saw nothing comparable in the Levels. Indeed, Defoe's main observations on the Levels were to note the susceptibility of the coastal lands to inundation, and the effects of the great storm of November 1703, which had caused widespread flooding.[8] Dr Claver

[1] This was the first Parliamentary Enclosure Act in Somerset. The Act is No. 79 in the S.R.O. numeration. The date of the Act is 1721, and the date of the Award is 1722. In future, only the number of the Act, together with the two dates, will be given in references.

[2] *The Diary of a West Country Physician, A.D. 1684–1726,* ed. by E. Hobhouse (1934). Entry for 5 November 1723. See also, for the enclosure of the moor, entries for 12 November, and 19 and 24 December 1724.

[3] Brit. Mus. Add. MSS. 35251, f. 21 (*temp.* Charles II).

[4] 10–11 William III C.26 (1699). See N. Luttrell, *A Brief Historical Relation of State Affairs from September 1678 to April 1714* (1852), IV, 495, for the introduction of the bill in Parliament.

[5] 'Propossals afforded for the dividing of Kingsedgemore in the County of Somerset', *S.R.O. Kemeys-Tynte MSS.* (1716), 49.

[6] C. Fiennes, *The Journeys of Celia Fiennes* (*c.* 1702), ed. C. Morris (1949).

[7] D. Defoe, *A tour thro' the whole island of Great Britain, 1724, 1725 and 1727,* ed. G. D. H. Cole (1927).

[8] *Ibid.* 270, and *idem, The Storm* (1704).

1 Glastonbury Tor

2 The terror of a flood: a breach in the River Tone, 1950

3*a* Sowy 'island'. A view from the moors along the road to Othery, showing the small but marked break of slope which raises the 'island' above the level of flooding

3*b* A typical moorland scene in the Southern Levels: a droveway, a rhyne, pollarded willows and cows

4*a* Lake Wall, which runs from Westonzoyland to the River Parrett. The road runs along the top of this massive medieval embankment and the houses on the left of the photograph occupy the raised northern edge of the wall

4*b* Peat digging in the raised bog of Ashcott Heath, in the southern portion of the Brue valley. The cut turfs are stacked in beehive-shaped mounds to dry. Coarse grasses and birch trees are the characteristic vegetation of these acid peat soils. Glastonbury Tor can just be seen in the far distance

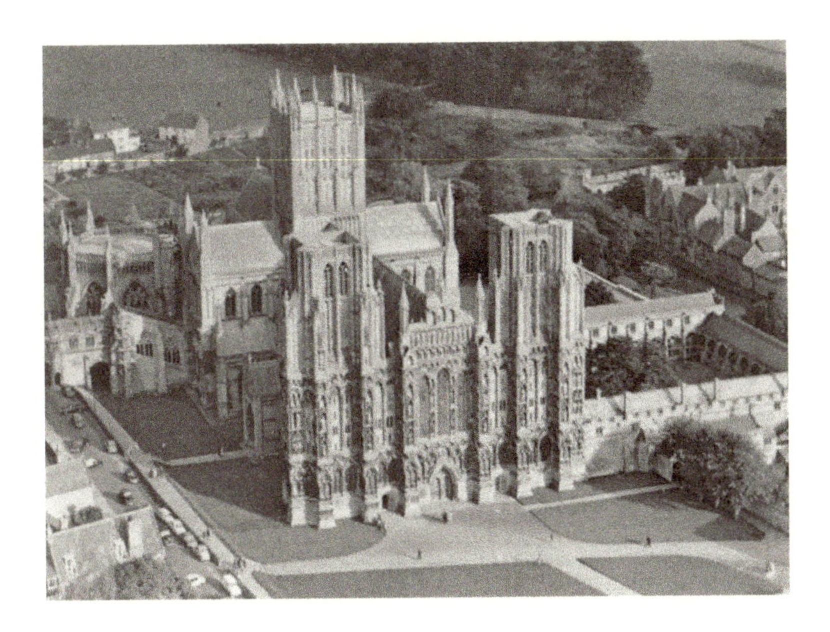

5*a* All the glory and wealth of one of the greatest ecclesiastical establishments in medieval Britain, and one of the greatest landowners and reclaimers in the Levels, is summed up in the richness of the architecture of the west front of Wells Cathedral

5*b* The large parish church of Wedmore, with its characteristic tower, is typical of many of the churches of the settlements on the 'islands' and fringes of the Levels

6*a* Highbridge Clyse at high tide

6*b* Highbridge Clyse at low tide

7*a* Athelney 'island' during a flood, 1960. The main Taunton-to-Glastonbury road runs north of the 'island' and is covered with water. The monument on the 'island' (arrowed) commemorates King Alfred's exploits in the ninth century

7*b* Flooding in Southlake Moor, 1951, looking north-west along the River Parrett from Stathe towards Burrow Mump and Burrow Bridge (arrowed) alongside. The damming effect on the floods of Burrow Wall, which connects the Mump to Sowy 'island', situated to the right of the picture, is evident

8*a* The King's Sedgemoor Drain, looking east from Greylake Bridge during a flood

8*b* The lower end of the King's Sedgemoor Drain, where it cuts deeply through the coastal clay belt, showing the timbering on the drain bottom. This photograph was taken in 1938, before the drain was widened

9 Langport at flood time, looking north-west towards Aller, *c.* 1935. This early photograph shows the River Parrett in the foreground with only the tops of the embankments showing above the flood waters. The river curves around the high ground on which Langport stands and passes through a constricted channel on the left of the photograph, which is crossed by Langport Bridge. The Portlake rhyne, which was excavated to relieve the Parrett of some of its water before passing under Langport Bridge, cuts through the centre of the 'peninsula' and the town

10*a* Cutting the Huntspill river near Puriton, 1940

10*b* Widening the King's Sedgemoor Drain, 1939

11*a* The Huntspill river and clyse after heavy rain. The main Taunton-to-Bristol road crosses the river by the second bridge up the river. A portion of the reclaimed land or 'warth' of Humble Island lies on the coastal side of the sea walls which run either side of the clyse. Those on the left are clear, but those on the right run inland and are very degraded through neglect

11*b* The Edithmead meander, looking west towards Burnham and the Bristol Channel. The heavy rain of October 1960 for the first time delineated the outline of this ancient river course. The pattern of underdrains and gutters in the clay fields can also be seen

12*a* Westonzoyland main rhyne and pumping station. The steam pump house was erected in 1830, and is on the right. The new diesel-driven pumps are in the building located in the rhyne

12*b* The Huish Episcopi pumping station (1964). One of the most modern of the new electrically-driven pumping stations in the upper Parrett valley. This illustration must stand as an example and symbol of the extensive and numerous engineering works undertaken in the Levels during the last twenty years

Morris, who was a commissioner of the Court of Sewers from 1707 to 1726, described nothing of very great importance done by the commissioners during that time. Only the Brue was widened and Bason bridge rebuilt in 1725, the Brue bank at Alder Moor strengthened, and a clyse provided for a rhyne near Kennard Moor,[1] but these were minor alterations.

The piecemeal reclamation of the moorland blocks and minor drainage works were only nibbles at the edge of the problem of flooding in the Levels. The central core of the problem of unsatisfactory main river channels and outfalls, complicated by tidal entry and siltation, remained untouched. These difficulties had long been known to be the major reason for flooding and they were stressed more fully by Dugdale in 1662. He observed that the problem arose partly from 'the flatness of those parts at least 12 miles eastward from the sea, which gave way for the tides to flow up very high; as also that the filth and sand thereby continually brought up did not a little obstruct the outfalls of those fresh waters which descend from Bruton, Shepton Mallet and several other places of this shire'. The seriousness of the situation in the Brue valley was partly averted because Highbridge Clyse enabled the fresh water to be penned up during flood periods, and 'evacuated with such a force as grinding out the silt which would otherwise choak up that ostiarie'.[2]

Conditions in the outfalls of the other rivers of the Levels were not so favourable as those at Highbridge; the sinuous course of the River Parrett and its estuary below Bridgwater, in particular, had long proved a problem, both to the draining of the Levels and to river navigation. As a consequence of these difficulties, steps were taken in 1677–8, by Sir John Moulton, to cut a new channel through the neck of a large meander, known as Viking's Pill, in the lower course of the Parrett (Fig. 14).[3] Although the cut was designed primarily to facilitate navigation, it also aided drainage by ensuring a more rapid

[1] *The Diary of a West Country Physician. A.D. 1684–1726*, ed. E. Hobhouse. Entries for 1 September and 7 October 1724; 12 May and 9 November 1725; and 3 August 1726.

[2] Dugdale, *History of Imbanking and Drayning*, 104.

[3] Bodleian Library, Gough MSS. Somerset, 7, f. 28. That Horsey Level is the site of this cut-off meander there is no doubt, and 'The Old River' is shown on a map illustrating the 1723 scheme of river cuts that are considered below. The map hangs in the committee room of the Somerset River Board. These two schemes are dealt with in M. Williams, 'River diversions on the Parrett in the 17th and 18th Centuries', *Thirteenth Annual Report of the Somerset River Board* (1963), 53–5. At an inquisition taken at Bridgwater in 1687/8 on 'Concealed lands' in Somerset, a deponent testified that the meander neck was breached by natural forces, which, of course, is quite possible. P.R.O. E.178/6028, f. 33.

flow in the river, thus ridding the hinterland of floods more quickly, and helping to scour out the tidal deposits in the main channel. In the process of cutting the new channel, about 150 acres of new land were gained through sedimentation, and the abandoned bed of the river filled so rapidly that

> in about 10 or 12 years [it] was never level wth. the other lands and was so hard...that men could walk on it in Dry Weather and neap Tides in ye Summer Seasons and in 20 years was Levell with the other lands and worth about 20*s*. per acre and has been esteemed for many years past some of the best land in the contry.

A proposal to implement a similar scheme lower down the Parrett estuary was put forward in 1723. This proposal is documented in detail and it provides an insight into contemporary opinion, not only on the draining of the Levels, but also on engineering practice in the early eighteenth century. The advantages and disadvantages of the proposed scheme also foreshadowed and summarized many of the arguments for and against the implementation of similar schemes for cuts across meanders in the lower Parrett in the early nineteenth century.

Councillor Harrison of Bridgwater, together with Mr Payne, 'A Quaker Merchant', were two of the promoters of the scheme and they proposed that a new cut be made through the Steart Peninsula. The advantages that would accrue to the public from the scheme were felt to be as varied as they were numerous. One benefit was that the journey of ships to Bridgwater would be shortened by six miles, and the difficulties of tacking up the channel, south of the Gore Sand and around Steart Point, obviated. This would mean that ships could be at Combwich before they could be at Steart Point, and arrive at 'Bridgwater key at one Tide Whereas now ships after they are come into the River by the help of one Tide are forced oftentimes to wait another Tide and sometimes 2 or 3 tides before they get up to the said Key'.[1] This, concluded Councillor Harrison, must inevitably

[1] Bodleian Library, Gough MSS. Somerset 7, f. 12. These documents comprise the following: (i) Letter from Councillor Harrison, Bridgwater, to the Duke of Chandos, 1 January 1723 (ff. 11–20); (ii) Objections to the scheme by the Duke of Chandos, 13 January 1723 (ff. 21–5); (iii) 'Answers by Mr Payne...to the forgoing Objections made by H.G.', Bridgwater, 16 January 1723 (ff. 26–9); (iv) 'A More particular Calculation of the Expence of this Cut', Councillor Harrison, Bridgwater, 17 February 1723 (ff. 30–51); (v) Letter from the Rev. Mr Lowthorp to the Duke of Chandos, 17 February 1723 (ff. 52–5); (vi) A copy of a letter from Mr Payne, Bridgwater, to the Duke of Chandos, 10 February 1723 (ff. 56–8).

stimulate and increase the trade of the port of Bridgwater and of the surrounding country.

From the point of view of the draining of the Levels, the benefits of the scheme were calculated to be equally good. Harrison expounded at length:

By thus Shortning the Course of the River the Fresh water that lyes upon the Vast flats of Levels on each side of the River (which may be Observ'd, by the by, to be all very good Pasture grounds) to the Depth of three foot or more in many places during the Winter Seasons and sometimes to the middle of April and is Dra'ned off from thence by severall large Ditches, Rhoines or Draines which Empty themselves here and there into the said River and now fetch along [the] Course as far as Start before they are discharged into ye Sea, will now when this new Cut is made be Drained off (as it is supposed) from the said fflats or lowlands and they will be eased of that great Bulk of Waters that do now lye on the same so long and do so greatly damage the Herbage thereof much sooner than they are at present whilst the River continues to take so long a Course and it is so long a time before the said Waters can now be brought into and clearly discharged into ye Sea.

A third advantage of diverting the Parrett lay in the possibility that a large amount of land could be reclaimed by damming the ends of the old course of the Parrett with hulks filled with stones and allowing the natural silting process of the river to fill in the intervening depression. Thus 'about 1,400 acres will...be filled up wth. sediment of the Tide and become good land wch. may be adjudged to be worth 20*s*. per Acre per Annum'. Further, the oft-recurring proposal for the reclamation of the tidal flats between the Parrett estuary and Brean Down was put forward as another probable benefit that would ensue from the scheme. Not only would new land be won from the sea, but the closing up of the old river course would reduce the amount of land which had to be protected from high tides by embankments, the maintenance of which, Harrison estimated, was costing about £350 per annum.[1]

Councilor Harrison felt that the only strong and valid objection that could be made against the scheme was that the new cut would enable the tides to flow

stronger and longer and consequently the waters rise higher at Bridgwr. and everywhere else all up along the sd. River and so overflow the low lands on each side thereof, unless the Banks be raised some feet higher to prevent

[1] Bodleian Library, Gough MSS. Somerset 7, ff. 13–14.

the same, wch. will Cost so much money as to render this project altogether impracticable.[1]

Yet, he consoled himself on this point by the thought that if the tides came in sooner, they would ebb sooner and the danger would then be averted.

The total cost of the new works was hopefully estimated by Harrison to come to about £6,400, of which only £1,000 was for the excavation of the cut, the remainder being spent on an Act of Parliament, obtaining land, surveying, and providing a ferry to Steart. The very small proportion of the total amount to be spent on the new channel through Steart Point arose from the ingenious, but unrealistic method of engineering proposed.

Two cuts, in the form of two frustums of triangular pyramids of unequal length with their lesser ends meeting, were to be excavated. The cut leading from the River Parrett was to have an opening 150 ft wide and 20 ft deep, diminishing to 20 ft wide and 6 ft deep at its lesser end. The total length of this cut was to be 2,500 ft. The other cut, which opened into the Bristol Channel, was to be of the same dimensions but only 1,500 ft long. Half of the earth excavated from the two cuts was to be thrown up on the sides to make protective embankments. Then the Parrett was to be diverted into the two cuts by damming the old channel with a hulk filled with stones. By its own erosive power the river would scour away the remaining earth, which would be loosened and 'at Times cast and thrown down towards the Bottom of the said Cut to be wash'd away by the Tides and Currents'.[2] Of a possible total of 444,440 cubic yards to be removed from the two cuts to make them a serviceable navigation channel of equal width and depth (150 × 20 ft) along their whole length, only about 93,100 cubic yards, or under one-quarter of the total, was to be taken out of the channel. The removal of the remainder depended entirely upon the erosive power of the river and the tide.

In the face of such hopeful calculations it was little wonder that other people were ready to point out some of the probable difficulties that might arise. The Rev. Mr Lowthorp realized that the success of the scheme depended upon the rapidity of the current through the

[1] *Ibid.* f. 16. These fears have been vindicated by modern experiments. See Du Cane, *River Parrett Estuary Scheme.*

[2] Bodleian Library, Gough MSS. Somerset 7, ff. 30–2.

intended cut, which would depend in turn upon the fall between the existing Parrett river and sea level, three-quarters of a mile away; such calculations had not been made. Another potential difficulty was the possibility of the growth of a bar at the sea end of the cut, formed from the soil scoured out of the cut and from suspended marine silt. If formed, it would be a serious obstruction to shipping. Perhaps more serious than this was the possibility that the daily movement of the tide might bring more of the suspended material back into the Parrett estuary and towards Bridgwater than the river could take out to sea, thereby worsening rather than improving conditions at the port. Finally, Lowthorp pointed out an important fact overlooked by Harrison. He said, 'If I remember right there are considerable sluices at High Bridge for the draining the marshes there: it ought therefore to be examined how these Drains will be affected by shutting up the Mouth of the Old River'.[1]

The Duke of Chandos was equally outspoken in his criticism, saying that 'all the Estimates I have ever yet known for undertakings of this Nature have prov'd abundantly short of what the real Expense has amounted to and indeed there are so many unforseen accidents which unavoidably happen in the Execution of these sort of Designes that most exact computors have always to their own cost found themselves abundantly mistaken'. In his opinion, the scheme would cost nearer £30,000 than £6,400. As to the quality of the land reclaimed from the old river bed, a point on which the promoters of the scheme had placed so much emphasis, the duke thought that it 'must be a perfect Hole and very difficult to come at and manure and cultivate'.[2]

There was little hope that the scheme would commend itself to local landowners and merchants while its benefits were so uncertain and the probable cost so great, and so it went no further. Perhaps the last words lie with the Rev. Mr Lowthorp: 'The Gentleman...that drew up that Scheme, I am persuaded, is much better acquainted with his Pen than a Spade or he would have found a great many other expensive Articles to Account for.'[3]

During these years an effort was also made to reduce tidal influence near King's Sedgemoor by building a clyse at Dunball. At this time a small stream ran off the clay belt to Dunball and it was not connected to the main drainage of King's Sedgemoor (Fig. 21).

[1] Bodleian Library, Gough MSS. Somerset 7, ff. 53–5.
[2] *Ibid.* ff. 21, 23.
[3] *Ibid.* f. 55.

In 1699 this stream was said to be open to the tide, but by 1764 a clyse had been built across its outfall.[1]

Taken as a whole, the years from 1640 to 1770 were barren years for the draining of the Levels, and the solid advances made before were wasted in over a century of inactivity. By 1770, nearly two-thirds of the area of floodable land in the Levels was still unreclaimed; lowland Somerset permanently fell behind other comparable areas in England in the stage of development of its drainage system and consequently in its geographical evolution. The inactivity was characterized as much by the lack of new rhynes and especially of windmills[2] (a common feature of the landscape of other marsh areas) as by the absence of any proposals, schemes or plans. The piecemeal approach to draining prevailed throughout the period. The obvious neglect to improve the main river channels and outfalls was ultimately going to vitiate the success of the piecemeal reclamations. As a consequence, they were not only limited in extent but also temporary in the benefit they conferred.

The removal of Crown sponsorship of draining, due to the opposition of the commoners and ultimately to the disturbances of the Civil War, was a fatal blow from which the region did not recover for over 130 years. An indication of the opposition and difficulties faced by the Crown was the failure to follow up the 1638 survey of 'Moores and Lowe Grounds'. The end of the Civil War did not bring peace to the Levels. Unsettled conditions still prevailed and draining was in abeyance. In 1685 the abortive rebellion of the Duke of Monmouth, which culminated in the Battle of Sedgemoor, had received great support from the inhabitants of the Levels. But after the defeat the most cruel repressive measures were taken against the people of the region under the direction of Judge Jeffreys. It was little wonder that men's minds did not turn to draining or other peaceful pursuits for many years.[3] Despite these very real interruptions, however, one is constantly reminded as much of what might have been done as of what was done. It was an age of lost opportunities.

[1] In 1764 the Bawdrip Jury of Sewers were asked to view the channel from Crandon bridge to Dunball Clyse, 'it being necessary to take it into view because it bays back the water'. S.R.B. Comm. of Sewers, Western Div. 1 Oct. 1764.

[2] With the exceptions in the Bleadon Level (p. 95 above) and in Common Moor (p. 112 above).

[3] See Bodleian Library, Ballard MSS. 48, f. 74 for a manuscript plan of the battlefield at Westonzoyland. For correspondence concerning the battlefield with topographical detail of Sedgemoor and its watercourses, see *Ninth Report of the Historical Manuscripts Commission* (1884), pt. III, 5.

THE MAINTENANCE OF THE DRAINAGE SYSTEM

The new drainage works that came into being from the fifteenth to the eighteenth century were of little use without some system of maintenance. The medieval system in which ditching, wallwork etc., were customary villein services[1] was no longer applicable because of the disintegration of the manorial system during the fourteenth and fifteenth centuries.

In its place, the system of maintenance *ratione tenurae* became common and was used widely throughout the Levels from the fifteenth century onwards.[2] Briefly, the system of maintenance 'by reason of tenure' meant that the repair of drainage works was to be carried out by those who owned or occupied the land adjacent to such works. The acceptance of this responsibility was often the result of oral agreement between owner and occupier and was enforced by the Court of Sewers, but in order to avoid any misunderstanding this obligation was sometimes written into new leases; thus tenure and responsibility were complementary.[3]

The great defect of the system arose from the continual subdivision of estates, which resulted in the subdivision of responsibility and in the lack of comprehensive maintenance. The position was aggravated considerably by the fact that those who were responsible were no longer 'frontagers' and lived many miles away. They were unlikely to do many repairs, let alone be on the spot if an emergency occurred. For instance, by about 1640, two miles of the Huntspill sea wall were divided into nearly 200 portions varying from between 900 ft to one foot in length, and wall work in the Northern Levels was divided into lugs (15 ft) and spade lengths (4 ft).[4] With so many subdivisions there was a constant possibility that non-performance of duty by one person would jeopardize the work of all.

[1] For examples of medieval ditching duties, see *Liber Henrici de Soliaco*, ed. Jackson, 168. Also 'Rentalia et Custumaria', ed. Elton, *S.R.S.* v, 44, 47, 49, 149, 153, 154, 156, 160, 217 etc.; and pp. 43–5 above.

[2] See *Joannis confratis et monachi Glastoniensis*, ed. Hearne, I, 324–6; also *Royal Comm. on Coast Erosion and Afforestation*, Appendix IV, 306, using De Banco Rolls, 12 Hen. IV, Mich. No. 599, and 'The Cartulary of St. Mark's, Bristol', ed. Ross, *Bristol Rec. Soc.* XXI, No. 195 (1404–5) for a good example of maintenance and landownership in Pawlett Hams and Stockland Marsh. For seventeenth-century examples, see P.R.O. E.134/5 Chas. I/Easter 8; 26 Chas. II/Easter 26; and 9 Chas. I/Easter 23.

[3] For instance, P.R.O. E.134/26 Chas. II/Easter 23.

[4] S.R.B. 'Survey of the Yatton Jury of Sewers in the Parishes of Kingston Seymour, Kenn, Yatton, Brockley, Chelvey and Blackwell taken in 1848'. This contains documents of wall work responsibility for as far back as 1637.

Despite its many drawbacks and shortcomings, the system of *ratione tenurae* worked reasonably well in the coastal lands and meadows because all the land was owned or occupied by someone, but this was not the case in the commons which accounted for about two-thirds of the low-lying land east of the clay belt. Here tenements, known as austre or ancient tenements, had the sole burden of drainage duty because of their exclusive right of 'common sans nombre' in the moors. All other tenements were known as overland tenements and were new creations won out of the waste of the moors. They had neither rights of pasture (however valuable or extensive they might be) nor drainage duties.[1] It seems highly probable that the rigid delegation of duty to austres only was an attempt to avoid the subdivision of responsibility which had occurred elsewhere.

The working and origin of the 'Austre System' of drainage maintenance have hitherto been unknown, but they are amply illustrated by documents found in the Somerset Record Office.[2] These details, however, are of less concern to us here than are the effects of the system on the draining of the Levels. Generally speaking, it gave rise in time to the same abuses of subdivision and neglect as did the system of *ratione tenurae;* for example 5,436 ft. of the West Sedgemoor common wall (which protected the moor from inundation of the River Parrett) were divided into 791 work sections. Again, like holders *ratione tenurae*, austre holders lived many miles from the works they maintained; this was a result of the intricate intercommoning arrangements which then prevailed (Fig. 13). Against these disadvantages one must place the unity and identity of interest that arose from the scattered settlements intercommoning within a particular moor. By the mid-seventeenth century, if not earlier, austre holders co-operated in some areas and commuted their individual burden to a money payment; people were employed to do the duties and a uniform maintenance of the whole work was achieved. For example, in West Sedgemoor, one John Small, was appointed to collect money from the intercommoning parishes and tythings of Fivehead, Broadway, Capland, and Beercrowcombe 'and collect of every commoner a like some for his wall work for each common...and...separately of

[1] For a discussion on ancient systems of tenure in Somerset, see R. Locke, *The Customs of the Manor of Taunton and Taunton Deane* (1816). For sixteenth- and seventeenth-century confirmation of these duties, see P.R.O. E.134/41 Eliz. I/Trinity 4 and E.134/10 Jas. I/Hil. 8.

[2] S.R.O. Dodson and Pulman MSS. Boxes 1 and 2 of 94.

every person occupying ancient tenements...considered of having common over Sedgemoor...one shilling and sixpence'. The other parishes with rights of common in the north of West Sedgemoor, i.e. North Curry, West Hatch, Hatch Beauchamp, Wrantage, Lillesdon and Knapp, also employed such a person.[1] When extraordinary damage was done by autumn or winter floods double the normal sum was collected, and on such occasions the owners of austre tenements would send out their own labourers, as well as carts of material, to stem any breach.[2]

Generally speaking, both systems of maintenance, i.e. *ratione tenurae* and austre, were neither satisfactory, nor conducive to perfect drainage, and, from what can be ascertained, they barely kept pace with decay. The more detailed evidence available for the early nineteenth century makes it obvious that by that time maintenance was totally inadequate for the needs of the more perfect drainage then required.

Finally, no discussion on the maintenance of the drainage of the Levels would be complete without mention being made of the Court of Sewers. In 1532 an Act was passed to consolidate the somewhat haphazard proceedings of the *ad hoc* commissions of the Middle Ages. Commissions could now be issued when needed, courts could be held, and those charged with neglect or wilful damage could be punished, taxed or fined by verdict of a jury of local men.[3] From the evidence that does exist of the court's activities, it seems possible that the maintenance of existing works was fairly well done, given the elementary and piecemeal drainage system of the Levels before the late eighteenth century.[4] The widespread viewing of works by local juries, and the system of fines for neglect, had long settled down into a regular and hardened system by the end of the seventeenth century.[5]

[1] S.R.O. Dodson and Pulman MSS. Box 2 of 94, F.L.9.

[2] S.R.O. Dodson and Pulman MSS. Box 1 of 94, O.B.14. See also P.R.O. E.134/26 Chas. II/Easter 2. For other examples of drainage maintenance during the eighteenth and nineteenth centuries, see H. P. Olivey, *North Curry: Ancient Manor and Hundred* (1901), 129, 130, 230, 235, 238, 246, 251–3, 256, 257 and 259 (West Sedgemoor and the Tone valley); and F. A. Knight, *The Heart of Mendip* (1915), 299 (Uphill), and 351–2 (Cheddar).

[3] See S. and B. Webb, *English Local Government: Statutory Authorities for Special Purposes* (1922), IV, 13–106, for details of the legal powers of the Court of Sewers. Other Acts had preceded this for at least a century.

[4] 'Pre-Reformation church-wardens' accounts, 1349 to 1560', ed. Bishop Hobhouse, *S.R.S.* IV, 133, 139, 157, 160, 162, 165, 171 and 172 for examples of the action of the Court of Sewers in Yatton.

[5] See *Diary of a West Country Physician, A.D. 1684–1726*, ed. E. Hobhouse, 134.

The one major defect of the court was its legal inability to make new drains or embankments, and so co-ordinate and improve the many individual works that had been created in the Levels; in this respect the court was entirely handicapped and could merely maintain the *status quo*.[1]

[1] See H. W. Woolrych, *A Treatise of the Law of Sewers including the Drainage Act* (1849), 94–5. This legal restraint was sometimes questioned in other lowland areas of Britain but generally accepted in Somerset.

5

EXPECTATION AND DISAPPOINTMENT

THE BASIS OF THE NEW ACTIVITY

In 1770 there was an unprecedented outburst of draining activity in the Levels.[1] A start was made to reclaim the last remaining wastes of both peat and alluvial soils, and the co-ordination of the drainage systems of the Brue and Axe valleys, King's Sedgemoor, and part of the Northern Levels was begun. It looked as though the solution to the problem of flooding was in sight.

This activity was in line with general agricultural improvement and expansion throughout Britain.[2] From the middle of the eighteenth century agriculture slowly rose from its depressed state and responded to the new stimulus of growing population and expanding markets in the new industrial areas, this upward trend being accelerated by the scarcity of food and by the rise in the price of wheat during the Napoleonic Wars. These boom conditions did not subside until the termination of the wars in 1815.

It was not long before Somerset and the west were caught up in the general trend of improvement which was sweeping through the Midlands and the east of England. The founding of the Bath and West of England Agricultural Society in 1777 was one of the numerous local manifestations of the new interest and spirit in agriculture. From the point of view of the development of the draining of the Levels, the most important fact was that rents and returns were high because of the need to increase food production, and consequently this put a premium on marginal land. Thus economic conditions were favourable for the reclamation of the remaining moors of the Levels, as indeed they were for reclamation schemes for other low-land areas in Britain. When one recalls, however, the opposition and

[1] For an outline of the main advances and features of draining and agriculture during this period see M. Williams, 'The draining and reclamation of the Somerset Levels, 1770–1833', *Trans. Instit. British Geog.* XXXIII (1963), 163–79.

[2] General trends of depression and prosperity in British agriculture here and elsewhere in this work are based on those set out in Lord Ernle's *English Farming, Past and Present*, (1936 ed.). But see G. E. Mingay, 'The Agricultural Revolution in English History: a reconsideration', *Agric. Hist.*, XXXVII (1963), 123–33, for some cautionary views on Ernle.

hostility that baulked draining attempts in Somerset in the preceding century, it is evident that the general background of expansion and improvement in agriculture alone does not fully explain the new upsurge of activity. There were, in addition, local factors which helped to develop a more favourable attitude towards draining amongst all sections of the population of the Levels.

Foremost amongst these additional factors was the old problem of over-stocking, which had reached unprecedented proportions.[1] Amongst the worst offenders were the more prosperous farmers of the peripheral settlements who, it was said, 'not only take the advantage of keeping their stock on the driest and best parts thereof and taking them out in time of flood...but from their contiguity to the common they are enabled and do keep thereon great numbers beyond their fair proportion to the manifest injury of other commoners'.[2] These grievances were given added weight by the fact that the austre holders (i.e. the commoners) still had to perform their drainage duty irrespective of their inability to common their cattle successfully. Many commoners wanted a fairer and more uniform distribution of land, and felt that a drained allotment, albeit small, was more desirable than unlimited rights of stocking on moors, which 'for many months of the year are covered with water and very frequently lame the sheep which are turned on them'.[3] Indeed, 10,000 sheep were said to 'have been rotted in one year' on the moors of Meare parish before enclosure and draining took place.[4] Thus attitudes changed and opposition to the draining and enclosure of the moors tended to weaken.

Secondly, the efforts of certain people to point out the benefit and practicability of reclaiming the moors aroused interest amongst the larger landowners and tenants in the possibility of improving their properties. Prominent amongst these people was John Billingsley, a farmer from Sutton Mallet, who wrote *A General View of the Agriculture of Somerset* (1794). He was a founder member of the Bath and West of England Agricultural Society and he served on many enclosure commissions, including those of Crannel Moor (1793) and Queen's

[1] The number of cattle brought into King's Sedgemoor alone, for one year, had reached approximately 30,000 by 1775. See *Journals of the House of Commons*, XXXIII (1775), 650.

[2] S.R.O. Dodson and Pulman MSS. Box 1 of 94, O.B.10. See also Billingsley, 195, for similar examples.

[3] S.R.O. Dodson and Pulman MSS. Box 1 of 94, O.B.11, 'West Sedgemoor Inclosure Letter Book', 1816.

[4] Billingsley, 168.

Sedgemoor (1794). There was also William White, who surveyed many of the draining schemes; and Anstice, Jessop and the Easton brothers, all of whom were engineers, and put many of the draining schemes into operation with commendable enthusiasm and ability.

Of greater importance than these, however, was Richard Locke of Pillsmouth Farm, Burnham, who pursued with a missionary zeal his aim of introducing local landowners and gentry to the benefits of improving the Levels. Throughout the years from 1750 to 1770, Locke and his relatives had carried out many experiments in improving the meadows on the coastal clay belt, which had been impoverished by constant mowing. These experiments included manuring, levelling and spreading ditch refuse, and, above all, assiduous attention to surface draining techniques, thereby doubling and sometimes trebling the value of the land in a short time.

In 1769 Locke turned his attention to the peat moors, and, when purchasing an estate from the Rev. Sir George Stonehouse in the Churchland Moors, realized that an austre tenement valued at £20 could be made worth £100 if drained and enclosed. 'He hereupon encouraged me to promote inclosing', writes Locke, 'insomuch that I actually measured and mapped twenty thousand acres of moors and commons upon speculation; and persevered in writing pamphlets to prove the utility of inclosing, till the publick mind became convinced.'[1] This was not done without opposition from the commoners and Locke said that he was 'stoned, bruised and beat by the mob till the blood has issue from my nose, mouth and ears', and his effigy was burnt 'by the owners of Geese'.[2] Such behaviour merely stimulated interest in his projects for reclamation and 'opened the eyes of the blind—removed the prejudice of the weak and prepared the publick mind for increasing their property by inclosing and agricultural improvements'. His articles in the *Bath and West of England Society Journal* were read eagerly, and he had the satisfaction of congratulating himself later 'that the neighbourhood is two millions the richer

[1] R. Locke, 'A historical account of the marsh-lands of the County of Somerset', *Letters and Papers, Bath and West Eng. Soc.* VIII (1796), 283. See also Locke Survey, f. 210, for a similar example of improvement in Godney Moor. It is possible that Locke wrote other pamphlets on draining and enclosure which have never been found or identified because of a lack of a signature. See J. G. Locke, *Book of the Lockes* (1853), 344–5.

[2] Locke Survey, f. 475. See also f. 526 where the effects of publishing a pamphlet in 1769 on the draining and enclosure of Yarrow Moors are discussed.

for sheding...[my] blood'.[1] Modesty was not one of Locke's virtues; he was even hopeful that his success in promoting reclamation in the waste lands of the Levels might influence the agricultural policy of the country, for, he argued, 'if these forty-five thousand acres of waste land in Somersetshire only are worth when enclosed an additional forty thousand pounds per annum and no person injured, will it not have a tendency to promote a General Enclosing Bill, specially when we consider that there are in England, Wales and Scotland, twenty millions of acres of waste land'.

Indeed, Locke was typical of the second tier of the local eighteenth-century 'improvers', people like the Sykes family in the Yorkshire Wolds, and the Knights of Exmoor, and he had no less effect on his region than they had on theirs. He was, said the editor of the *Bath and West of England Society Journal*, 'well known and respected for his abilities in the improvement of grassland', and, what was more important to some, 'it was the result of practice founded on his own reflexions, unaided by the perusal of agricultural authors; and therefore...having the recommendation of the greater originality'.[2]

Thirdly, many graziers and dairy farmers in the fertile clay belt, especially around Huntspill and Burnham, had become increasingly wealthy throughout the eighteenth century and had money to spend on the reclamation of the peat moors. The breaking-up of the larger estates had allowed a class of freeholders to emerge whose 'industry nursed by frugality' had caused a marked rise in prosperity. In Burnham, in 1750, only one man was said to be worth £1,000 and there were only five jurors. But by 1796, fifty were worth £10,000 or more, and ten were worth £100,000. Thirty-five farmers were styled 'Gentlemen' and served on the Grand Inquest of the County at Quarter Sessions. Gone were the days when the population of these coastal parishes consisted of 'poor renters and cottagers who existed without hot dinners, silk clothing, carriages of pleasure, mahogany furniture, clocks, watches, and even Tea kettles'.[3] The increase in

[1] Locke Survey, f. 300, and f. 475. It is interesting to compare Locke's experiences with those of William Fairchild who attempted to survey King's Sedgemoor in 1775. He was met by commoners who told him 'he should receive damage' if he continued his levelling. He was then 'threatened with his life; and they went so far as to dig his grave'. It was said 'that there were an hundred or two met him, and a reward of a hogshead of cider offered to anyone who could catch him'. *Journals of the House of Commons*, XXXIII (1775), 649.

[2] *Letters and Papers, Bath and West Eng. Soc.* V (1793), 180.

[3] Locke Survey, ff. 96–8, and 'On the improvement of meadow land...', *Letters and Papers, Bath and West Eng. Soc.* V (1793), 208. Collinson had also noted the prosperity of

wealth could be gauged by the amenities which came to these hitherto backward and isolated marsh villages. Locke said that there was 'a shop, a butcher, a baker, a barber, a surgeon, an attorney and some giggs in almost every parish; but when I remember first, none of these existed between the two market-towns of Bridgwater and Axbridge'.[1]

Thus, with ready capital, a rising national demand for food, and the inspiration of Locke's example, the graziers enclosed and improved the last remaining wastes on the clay belt, many of these being merely roadside strips.[2] They then turned their attention to the peat moors where hundreds of acres of waterlogged peat could be rented for as little as 1*s*. per acre per annum, and drier portions from 2*s*. to 5*s*. For 20*s*. the drier moor could be purchased outright. When one grazier purchased many acres near Meare, and began to 'exert himself in making this land lye plain and dry...his neighbours followed his example and without the assistance of any ashes, dug soil or compost of any kind whatsoever, part of their lands have sold for £40 per acre and many hundreds per annum, the remaining are at present worth 40*s*. per acre per annum'.[3] There was a local demand for the improvement of the moors, and capital for this purpose was available, a situation which had not existed before.

Finally, the process of reclamation in the Levels was now easier and more satisfactory because the Parliamentary Enclosure Acts could be used to effect drainage improvements. This was not a purely local phenomenon, it is true, but it was a marked characteristic of the late eighteenth-century and early nineteenth-century activity in the Levels. The Enclosure Act provided a way around the absurd position of the Commissioners of Sewers who were unable to construct new works. Furthermore, it was no longer essential to reward the Crown or agent each with as much as one-third of the newly-reclaimed moor; portions could now be sold to anyone (including the

these coastal grazing lands, their fine herds of cattle and general well-being: Collinson, *History and Antiquities of the County of Somerset*, 179–80, for Burnham; I, 182, for Mark; II, 389, for Huntspill; and II, 100, for Pawlett.

[1] Locke, 'Historical account of the marsh-lands of the County of Somerset', *Letters and Papers, Bath and West Eng. Soc.* VIII, 267–8.

[2] S.R.O. E.A. Nos. 68, 40, 66, 126 and 30. These are not depicted in Figure 16.

[3] Locke, 'On the improvement of meadow land', *Letters and Papers, Bath and West Eng. Soc.* V, 209–10. See also W. M. Acres, *A Brief History of Wedmore* (1954), 51, where it is recorded that the value of an estate in Wedmore rose from £3,850 in 1787 to £15,000 in 1808, largely through the inclusion of the newly-drained moors within it.

commoners) to defray costs, and a direct tax could be levied on the area according to the benefit received. This mollified some of the objections to draining which the commoners had voiced in the past. In all ways, conditions were favourable for a new attack on the last remaining floodable waters of the Levels.

THE REGIONS OF DRAINING

The various moors of the Levels can be divided conveniently, and quite accurately, into five regions which correspond to major river catchment areas. They are, from north to south, the complex area of the Northern Levels consisting of the catchments of the rivers Kenn and Congresbury; the Axe valley; the Brue valley; the Cary valley and King's Sedgemoor; and the Southern Levels, which include the Parrett valley and the valleys of its tributaries the Isle, Yeo and Tone. The work of draining was divided and was very different in each of these five regions; but before proceeding with a detailed examination of new reclamation and draining activity in each, we should glance at the overall pattern of reclamation.

The areas reclaimed between 1770 and about 1830 were the last remaining vestiges of common and waste which were the 'hard core' of difficult land. These areas are depicted in Figure 16 and their boundaries are taken from the maps accompanying the Parliamentary Enclosure Awards. If the distribution of the reclaimed areas is compared with the distribution of soils shown in Figure 1 it can be seen that reclamation occurred almost wholly on the peat moors, which had been so consistently avoided during the previous centuries. Exceptionally badly-drained alluvial and clay areas, such as the Tone Moors, Mark Moor and the moors in the Parrett valley above Langport, were also reclaimed, but they amounted to barely one-sixth of the total area of land recovered between 1770 and 1830.

The pace of reclamation varied, and moved in well-defined stages from one catchment area to another. There was a nucleus of activity in the Brue valley from about 1770 to 1780, and this activity spread into the Axe valley and then into King's Sedgemoor during the next two decades, but barely affected the Northern Levels until after 1800 or the Southern Levels until after 1810. These regional variations were important, and the lateness of reclamation in the Southern Levels should be noted.

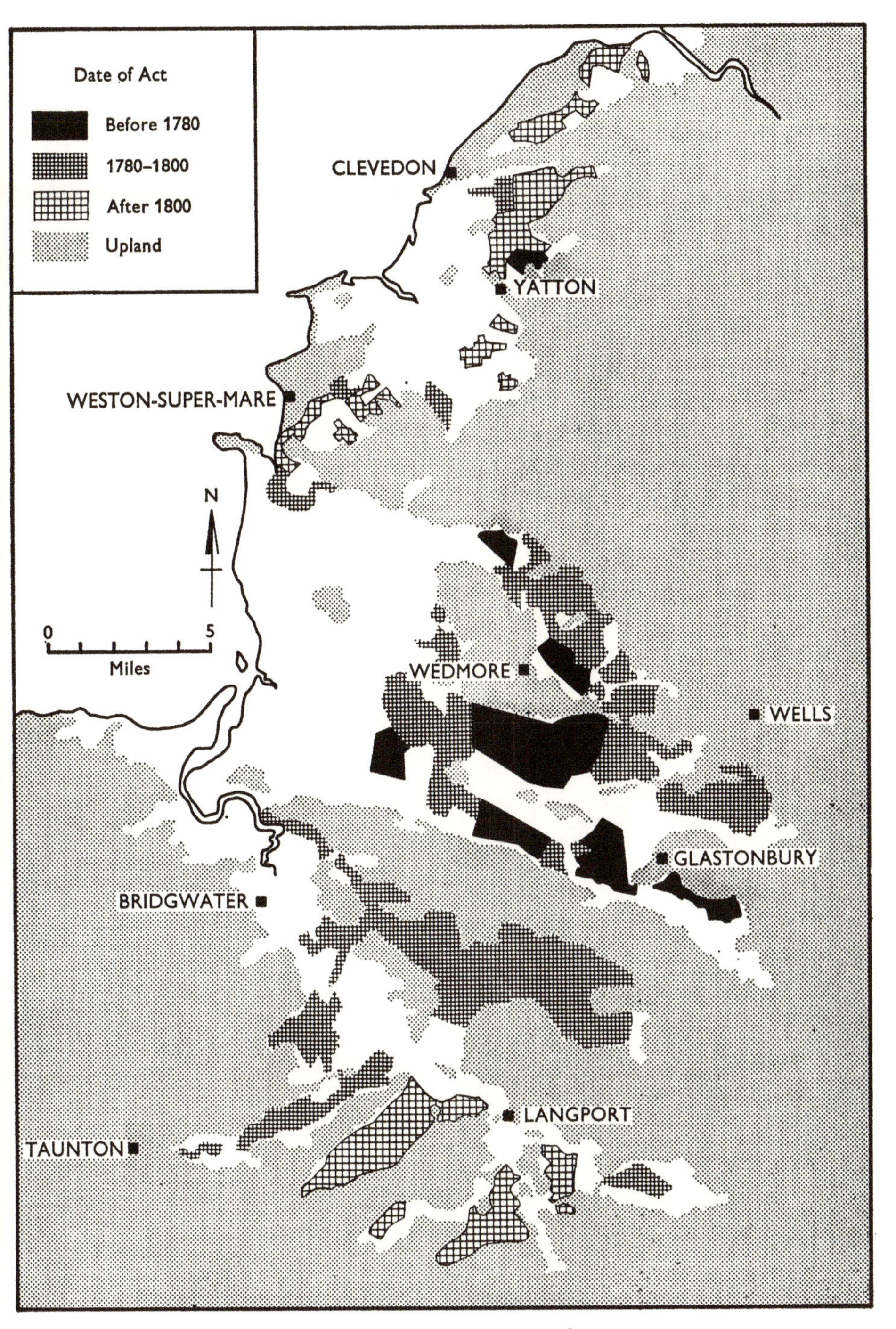

Figure 16. Reclamation: 1770–1833.

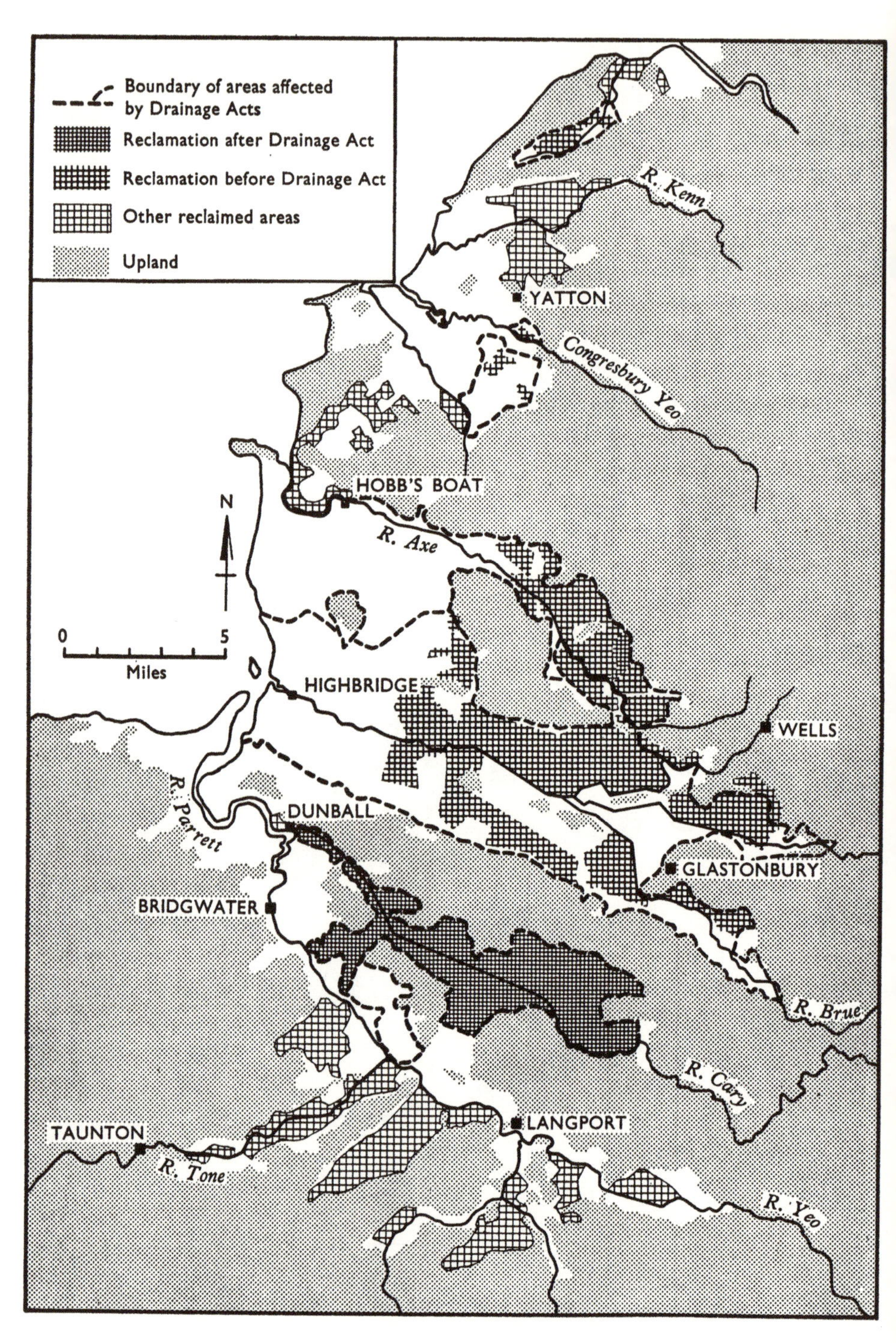

Figure 17. Reclamation and comprehensive drainage: 1770–1833. (Source: S.R.O., Enclosure Award and Drainage Act maps.)

The regions of draining

When the reclamation of most of the open commons had been completed, the whole character of drainage activity changed and comprehensive drainage schemes or Drainage Acts, as they were often called, were started. These schemes were aimed at co-ordinating the entire drainage within a catchment area and they represented the first real attempt to plan the drainage of the Levels on a regional basis. Figure 17 shows the boundaries of the areas affected by the schemes for co-ordinated drainage; these are superimposed on the distribution of the local moorland blocks reclaimed between 1770 and 1830. The schemes for the Brue valley (1801), Axe valley (1802), Weston valley (1810), the Congresbury area (1819), and the Middlezoy, Othery and Westonzoyland lowlands (1819), postdated the individual reclamations which had failed to achieve a lasting improvement in drainage and lessen floods. On the other hand, the scheme for the co-ordinated drainage of King's Sedgemoor (1791) preceded the reclamation of the individual blocks of moorland, and, as a result, it was more successful than its northern counterparts in alleviating floods. The Southern Levels remained untouched by this movement for a more integrated drainage, and the lack of a comprehensive scheme for the Parrett valley and its tributaries was a distinctive feature of the region that has lasted to the present day. In this way there were significant and sometimes subtle differences between the regions in the timing and type of reclamation, and in their agricultural emphasis once the moors were relieved of excess water. These five regions of activity will therefore be treated separately.

The Brue valley (1770–1810)

The reclamation of the remaining areas of open peat land in the Brue valley preceded that of the other regions of the Levels. The reason for this is uncertain. Perhaps it could be attributed to the presence of Highbridge Clyse which gave the valley protection from tidal inundation, an advantage which no other area of the Levels had. A more likely reason was that the valley was the nearest peat area to Huntspill and Burnham, where the wealthy graziers, and particularly Locke, provided much of the impetus in matters of draining. One thing is certain: the valley possessed no particular natural advantages which would have aided draining; on the contrary, this unbroken stretch of peat lowland presented many problems.

Notwithstanding these difficulties, an Act for the reclamation of

Tadham Moor was obtained as early as 1774.[1] (See Fig. 20 for the location of these moors.) The basis of the scheme was a main rhyne 2½ miles long, running through the centre of the moor and parallel to the Brue. With the later draining of Westhay and Godney Moors, the rhyne was continued eastwards for another four miles, the whole rhyne being the precursor of the North Drain. In 1777 there followed the reclamation of Huntspill Moor, and, a year later, of Kennard Moor and the Glastonbury portion of the southern raised bog.[2] The year 1781 witnessed the reclamation of the clay moors near Mark and also of Tealham Moor.[3] During 1793 and 1794 there followed the reclamation of Crannel Moor and Queen's Sedgemoor, and in 1797, the heaths of the raised bog in the parishes of Street, Ashcott, and Shapwick were reclaimed.[4] In addition to these schemes there were others for Kennard Moor, Chapel Allerton Moors, and Knowle Moors.[5] In the short space of twelve years there was no area of the valley that did not have some form of drainage facilities. There is an enormous mass of detail about rhynes, clyses, banks and other new works that were constructed under these Acts inspired by local district enterprise. A glance at a map of the area tells the story better than words can.

The immediate improvement in the moors was promising. Billingsley said that 'the rhynes and ditches necessarily cut to divide the property, together with the deepening of the general outlets, discharge so much of the superfluous water, that many thousand acres which heretofor were overflown for months together, and of course of little or no value, are become fine grazing and dairying lands'. He further estimated that out of 17,500 acres so improved,

[1] E.A. No. 82, 1774 and 1778. See also S.R.O. Sexey MSS. 20, A and C, and 21 F for many interesting details on the pre-enclosure moorland, the claims of commoners and the subsequent division of the moor.

[2] E.A. Nos. 125 (1778 and 1783), 83 (1788 and 1791), 126 (1777 and 1782) and 18 (1778 and 1783) respectively.

[3] E.A. No. 115, 1781 and 1784 and E.A. No. 136, 1781 and 1785. S.R.O. Wharton MSS. 49, contains a petition by sixty commoners against the enclosure of Mark Moor. See also S.R.O. Kemys-Tynte MSS. 49, Letter, Bampfylde to Sir C. K. Tynte, 3 January 1776, in which the proposed draining of Mark Moor is discussed.

[4] E.A. Nos. 81 (1793 and 1795), 8 (1797 and 1798), 5 (1797 and 1798) and 12 (1797 and 1798) respectively. Arthur Young in *A Farmer's Tour through the East of England* (1771), IV, 20, noted that there were 'very large tracts of flat marshy land, half poisoned with wet' in Queen's Sedgemoor, which needed 'nothing but draining to be rendered the best meadow in the country'.

[5] E.A. Nos. 51 (1786 and 1797), 37 (1794 and 1796) and 134 (1782 and 1786) respectively.

'six parts out of seven are cleared of stagnant water and rendered highly productive'. The improvement, however, was shortlived and limited, and, in spite of his sanguine account, Billingsley had to admit that all was not well.[1] The reclamations had never been tested under severe flood conditions, and when such conditions occurred in 1794 the whole valley was flooded (Fig. 18). Warner found the area in the same condition at the end of September some six years later and said that 'one wide sheet of water spread itself over the flats which not three weeks before I had traversed on foot'.[2]

The flooding was of a twofold origin. First, the local reclamations were not having 'the most perfect influence on the country' and merely succeeded in ridding one moorland block of water at the expense of another area. By their very nature they were unrelated and unco-ordinated; no attempt had been made to integrate their drainage with that of the major outlet of the Brue. Consequently, the waterlogging of the peat increased and its drainage condition deteriorated. Secondly, the major channel, the Brue, was in a poor condition because of its indirect course, lack of fall, and shallow and narrow sections. The most important defects were those of the outfall and of Highbridge Clyse which may be catalogued as follows:

1 The clyse was situated so far up the Brue estuary that the Parrett could not scour away the accumulated mud that hindered the working of the sea doors (Fig. 19); neither was the flow of the Brue itself always sufficient to scour it away, except during the height of a winter 'fresh';[3]

2 The three arches of the sluice were too small to allow the free evacuation of the landward flood waters. In 1801, Sutcliffe estimated that the sectional area of the combined sluice openings was only 3/8th of that of the river behind, and it was 'no matter of surprise that the drainage of the...river was so extremely bad'.[4] Locke also attributed the great damage done by the floods, that had lately swept through the valley, to the inadequacy of the openings;[5]

3 The sill of the sluice was too high. An additional fall of ten feet could have been gained by more judicious siting;

[1] Billingsley, 167 and 168.

[2] R. Warner, *A Walk through some of the Western Counties of England* (1800), 59.

[3] Billingsley, 169–70.

[4] S.R.B./A.D. 2, Sutcliffe's Second Report on the state of the Axe, 18 May 1801.

[5] Locke, *An Essay on the Subject of Draining the Flat Part of Somersetshire*, S.R.O. Som. Arch. Soc. Serel MSS. 113/1.

4 The cut through the clay belt was narrow and curving in some places. It was also obstructed by mud shoals and extruding masses of peat, locally known as 'pill-coal', which were pushed up into the river bed by the pressure of the banks;

5 Through the silting of the Axe outfall the waters of the Pilrow Cut had changed direction and now flowed into the Brue, worsening conditions there rather than alleviating them, as before.

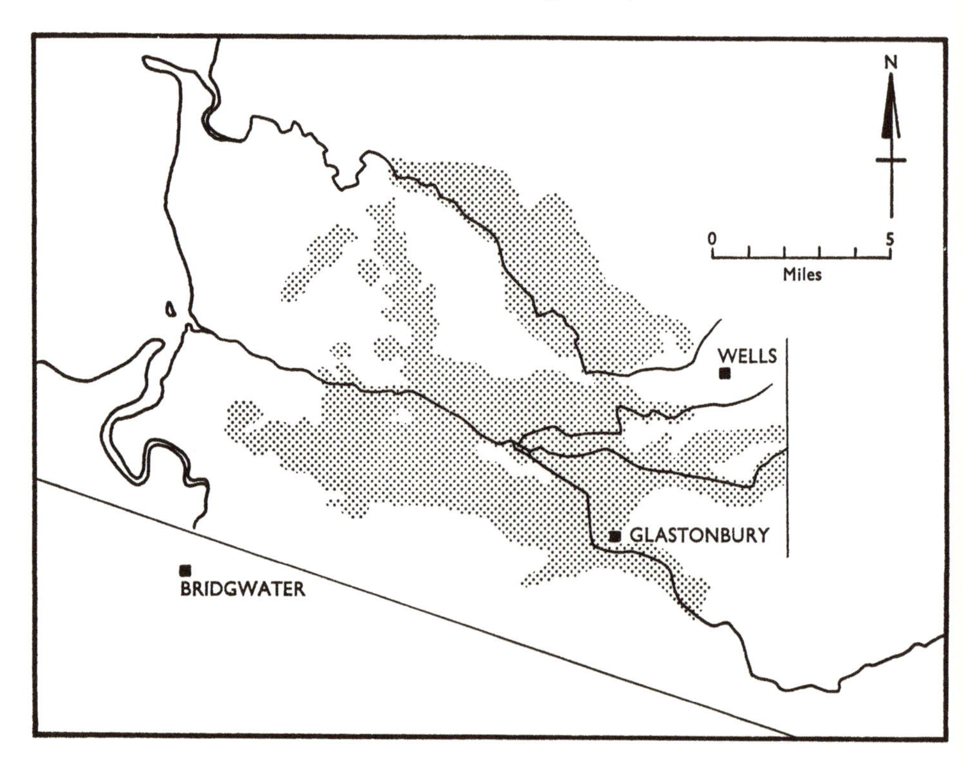

Figure 18. Flooding in the Levels, 1794. (Source: after a map showing 'flooded lands' by Wm White and entitled 'A plan for more effectually draining the turf bogs and flooded lands near the rivers Brue and Axe in the County of Somerset', in J. Billingsley, *A General View of the Agriculture of the County of Somerset* (1798), opposite 169.)

All these defects tended to produce a high water-level in the Brue during periods of heavy rainfall or tide-lock, which in turn caused overspilling in the surrounding moors. The high water-level also produced a flow-back in the majority of incoming streams;[1] consequently, water in the moors themselves could be discharged gravitationally

[1] Locke, *Essay on the Subject of of Draining the Flat Part of Somersetshire*, 6–7.

only when the level of the main river subsided. Obviously the lowering of the water-level throughout the whole course of the River Brue was the key to solving the problem of successful drainage. There was one other difficulty which was peculiar to the area, and that was the fluctuating level of the raised peat bogs due to precipitation and flooding. Until the bogs were stabilized the uninterrupted flow of water in the rhynes could not be assured.[1]

All these problems were given an added urgency by the breaching of the sea wall at Huntspill during September and November 1798. The sea flowed through the gaps across the clay belt and remained locked in the peat moors. This breach was caused by the erosion of Steart Peninsula about 1783, this peninsula having previously protected the Huntspill Wall from the full onslaught of westerly gales and high seas (Fig. 14).[2]

To alleviate the defects in the Brue, Billingsley suggested that Highbridge clyse should be completely rebuilt and enlarged; the sill should be 10 ft lower and the bed of the river reduced to an inclined plane of one foot to one mile from Highbridge to beyond Meare. This would facilitate the more rapid discharge of flood waters, and promote the sinking and consolidation of the raised bogs. He also proposed that the new sluice should be placed in a new cut, one mile long, which would by-pass the old sluice and outfall to the south. Four other new cuts were suggested by him: one near Mark Moor; one to help the North Drain to discharge its water more effectively into the Brue; one across a bend in the River Sheppey at Hurn; and a cut of 12½ miles from near Cripps's bridge, through the southern raised bog, to tap the headwaters of the Brue in Alder Moor.[3] Locke came forward with other suggestions; he advocated a new sluice at Highbridge, a catchwater drain along the foot of the Polden Hills, a new outlet across the clay belt for the Brent area, and also cuts in Blackford Moor and the River Brue.[4]

With all this discussion on proposed improvements[5] it was inevitable

[1] For information on these fluctuations, see Billingsley, 169, and Phelps, 'On the formation of peat bogs and turbaries...', *S.A.N.H.S.* IV, 101.

[2] S.R.O. Som. Arch. Soc. Serel MSS. 24/77 and 113/44 for an account of the events leading up to the breach, and copies of the subsequent legal proceedings. See also S.R.B./P.C.S. Bridgwater, 22 January and 11 July 1799. For a discussion on the changing shape of Steart Peninsula and Island, see A. P. Carr, 'Cartographic record and historical accuracy', *Geography*, XLVII (1962), 135–44.

[3] Billingsley, 166, 169–70, and map opposite 169.

[4] Locke, *Essay on the Subject of Draining the Flat Part of Somersetshire*, 1–7.

[5] For other proposals see S.R.B./B.D. 3. 'Account of the Highbridge River; what should be done'.

that there were those who objected to any change in the area. Clay belt farmers, who were vitally interested in the improvement of the peat moors, nevertheless opposed the lowering of the outfall because it would deprive them of water for summer irrigation and for fencing.[1] This was yet another example of the conflicting attitudes of clay belt and peat land proprietors. Everyone objected to being rated for new works,[2] especially at a time when an acre rate had been levied on the whole valley by the Commissioners of Sewers in order to raise money for the repair of the Huntspill sea wall. This levy was without precedent in the Levels, for formerly liability had been *ratione tenurae* with the 'frontagers', but the commissioners argued that the whole of the valley benefited from the maintenance of the wall and therefore ought to repair it. A bitter and protracted legal battle ensued from 1799 to 1807, which was finally won by the commissioners.[3]

Ultimately, worsening conditions goaded the farmers into action, and it was said that 'smarting under the affliction of those late unusual rains [they] are now resolved to drain these land upon the best principles possible'.[4] They obtained an Act, the preamble of which stated that 'there were several lands and grounds...lying on or near to and draining into the River Brue and which...from their low and flat situation are frequently overflowed and greatly injured by the waters'.[5] William White, the surveyor, was employed to survey the whole of the Brue catchment and prepare, probably for the purposes of rating, a map showing the extent of the land that was flooded

[1] Billingsley, 170. This could have been overcome by hatches.

[2] S.R.B./B.D. 3 contains the following doggerel verse entitled 'Highbridge New Needless River' which summarizes most of the the objections voiced at that time:

> Here's anger, discord, fury, fear, defate,
> and all besides that Can man's Ruin make.
> Europe's disturbances now Seems half undon
> and Some may wish the Game had neen begun.
> It will Ruin some Thousands as most people say
> Therefore you Gentlemen the Tax take away.
> The Freeholders do all Think the same
> That the land survoyers and masters will get gain.
> The expence of cuting the Needless New River
> will ruin some Gentlemen's family for ever.
> The freeholders lands they will mortgage to pay,
> the lesshold estates they must sill, and then Run away.

[3] R. Locke, *A Letter to George Templer*..., pp. 21–2, S.R.O. Som. Arch. Soc. Serel MSS. 113/1. See also S.R.B./*P.C.S.* Bridgwater, 20 September 1800; 17 and 29 September 1801; 27 June 1803; and 4 May and 6 October 1804. S. D. Cole, in his *Sea Walls of the Severn* (1912), 29, comments on the breach in the Huntspill Wall.

[4] Locke, *Essay on the Subject of Draining the Flat Part of Somersetshire*, 1.

[5] 41 Geo. III, c. 72.

during the previous ten years, and showing the depth of water on the land.[1] The aims of the works envisaged under the Act were to co-ordinate the drainage of previous individual reclamations, and to improve the conditions of the main channels throughout the catchment area of the Brue. All Billingsley's proposals, which had been widely circulated in the county, were embodied in the Act, with the exception of the one dealing with the course of the new South Drain.

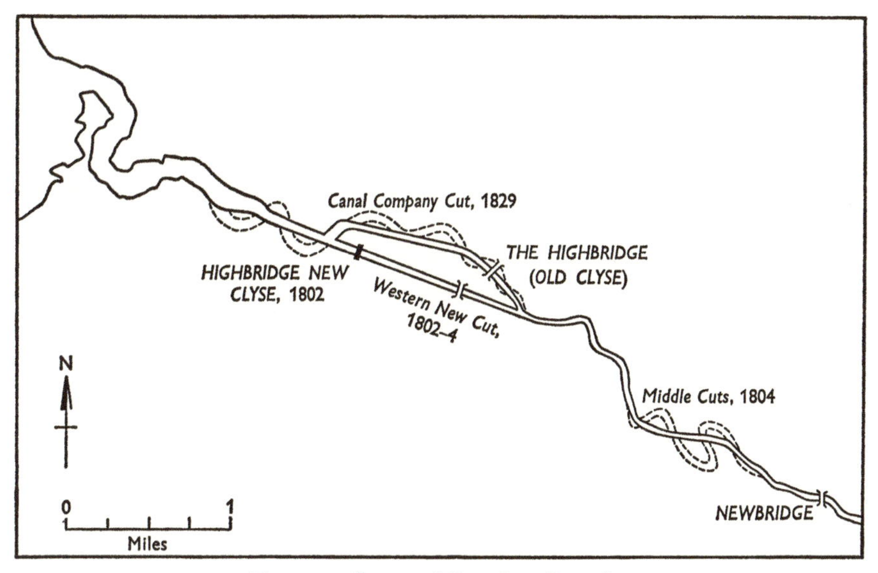

Figure 19. Brue outfall works: 1800–26.

Work started with the excavation of the new outfall cut at Highbridge and with the clyse and bridge across it (Fig. 19), but progress on the sluice-pit and cut was seriously delayed when flood water flowed into them in 1802.[2] In February of the following year, heavy seas beat down the newly-erected sluice doors, probably because they had not closed when a stone or some other object became lodged beneath them, and the sea flooded the construction works and peat moors inland. The broken doors were replaced by new ones but these were never effective; they continually leaked and caused the new cut

[1] S.R.B./*P.B.C.* 19 August 1801. It is interesting to note that White had already prepared a similar map in 1794 for Billingsley. See Figure 18.

[2] S.R.B./*P.B.C.* 18 August 1801, for survey of new outfall cut, and 21 April 1802, and 24 March 1803 for the new works. See also 22 April 1803 for a report on the damage.

to fill with mud.[1] To guard against the possibility of another flood arising from the failure of the sluice, additional doors were erected under the arches of the new Highbridge. The old outfall was abandoned and subsequently filled with river and marine alluvium, and rose 20 ft in twenty-five years; it was said to be 'producing fine crops of corn' in 1825.[2] Further cuts were made in the main channel of the Brue at Highbridge and beyond Cripps's bridge in 1803 and 1804;[3] during the latter year work started on the South Drain, on the new exits for the North Drain and on Hurn Weir (Fig. 20).[4] By the spring of 1805, the Pilrow Cut was improved and sluices placed at either end; this made it possible to divert the Brue into the Axe in order to regrade and 'bottom' the Brue channel.[5] A great variety of other cuts, sluices and improvements to existing rhynes, which can be traced from the Proceedings of the Commissioners and from the Award of the Commission, were also carried out.[6] The most important of these are depicted in Figure 20.

It seems obvious that the commissioners had learnt something from the mistakes made in the construction of the King's Sedgemoor Drain (see p. 150 below). The sluice was built with the sill as low as possible, the cut through the clay belt had gradually sloping sides that would not slip, the refuse from the new cuts was spread at least six feet away from their banks, and drains in the peat lands were 'bottomed' with timber to stop their bed welling up.[7] On the other hand, it is apparent that the commissioners had little idea of what to do after the improvements were made to the main river channel. They proceeded in a somewhat haphazard manner, advertising in the Press for committees of landowners, who knew the problems of certain areas, to suggest improvements and draw up plans.[8]

The Brue scheme was badly conceived and, in spite of the expenditure of approximately £60,000, floods still incommoded the valley. The problem lay in the re-conditioned North Drain and in the new South Drain. If these two lateral drains had worked successfully

[1] S.R.B./*P.B.C.* 16 July 1803.

[2] Phelps, 'On the formation of peat bogs and turbaries...', *S.A.N.H.S.* IV, 97.

[3] S.R.B./*P.B.C.* 13 August 1803, and 8 January 1804, for details of cuts from Highbridge to the sea, and S.R.B./*P.B.C.* 7 and 11 July 1807, for the Little Moor Cut.

[4] S.R.B./*P.B.C.* 24–29 June and 15 August 1804; 11 January 1806.

[5] S.R.B./*P.B.C.* 20 March 1805.

[6] Both in the S.R.B.

[7] S.R.B./*P.B.C.* 11 January 1806.

[8] S.R.B./*P.B.C.* 13 April 1804, and 1 April 1805, for example.

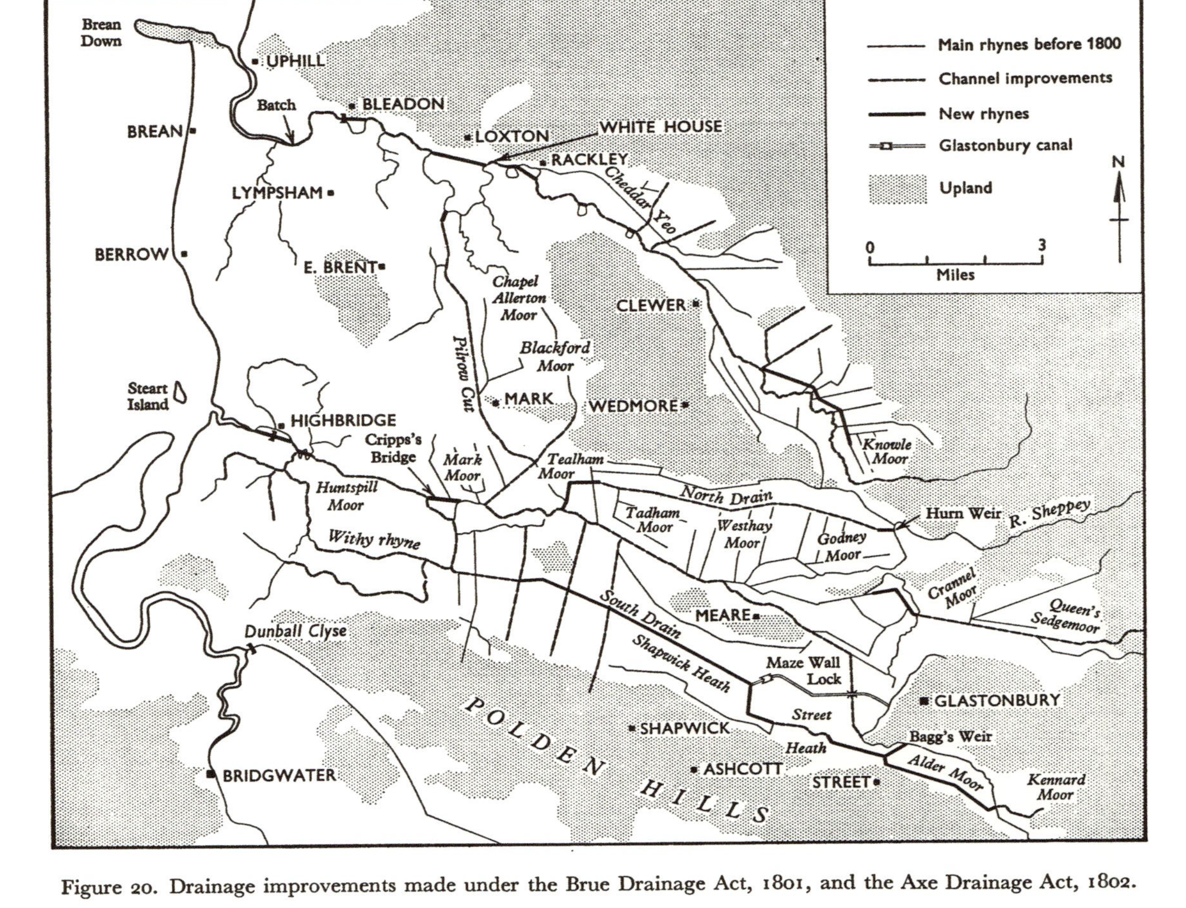

Figure 20. Drainage improvements made under the Brue Drainage Act, 1801, and the Axe Drainage Act, 1802.

then much of the surface water of the Brue valley would have been evacuated. The drains might have been successful if they had been provided with independent outfalls through the barrier of the clay belt and into the sea, or at least into the lower Brue estuary, as was the case with the King's Sedgemoor Drain; but, instead, they joined the Brue between four and six miles inland, at a point where the main river would be running at a higher level than the incoming streams during any period of heavy rainfall or tide-lock. When these conditions prevailed, a flow-back into the two drains occurred and flooded the moors. Thus, at a time when the North and South Drains were most needed they were of least use; differences of only a few inches in the water-level of the Brue rendered them ineffective.[1] Much later, self-operating doors had to be provided to prevent this.

In addition to these basic defects some of the works were imperfectly finished because construction was delayed by floods, and also because the time allotted under the Act for their completion expired in 1806. As a consequence, the contractor made many of the cuts too small in order to complete the work in time; 13,883 cubic yards were not excavated in the upper end of the Brue alone, which consequently reduced the ability of the rhynes to pass off the flood waters which they were designed to carry.[2] Provision was made to hold £700 in reserve for the completion of the work after the commissioners had handed over their powers in 1806, but the work was never done owing to further flooding.[3] Further, £1,652 of the income of the commissioners was said to have been applied 'to purposes foreign to those for which it was legally applicable'.[4] No strict account was kept of the large sums of money that passed through their hands, and consequently the drainage suffered. Many people felt that the drainage could have been much better than it was;[5] their disappointment was justified by subsequent events.

The Axe valley (1770–1810)

On the whole, flooding in the Axe valley was neither so frequent nor so damaging as that in other regions of the Levels. This was largely

[1] Differences of up to 5 ft were common.

[2] S.R.B./*P.B.C.* 11 August and 12 September 1806.

[3] Phelps, *Observations on the Great Marshes...*, 13.

[4] 'Brue Drainage Account, 1807–1834', *S.R.O. Chilton-upon-Polden Documents, Civil Records.*

[5] Phelps, *Observations on the Great Marshes...*, 14, for example.

because of the smaller area of lowland involved,[1] and, in addition, the relatively narrow valley meant a greater slope towards the river, which facilitated the gravitational drainage of the moors. Moreover, the possible early medieval diversion of the rivers Brue and Sheppey away from the Axe would have resulted in the River Axe possessing a channel larger than it normally needed, and therefore capable of carrying moderate flood waters. Perhaps more important than these reasons was the fact that both the Axe and its major tributary, the Cheddar Yeo, originated in areas of Carboniferous Limestone where steady percolation of rainfall tended to give the rivers a flow that was fairly constant and not subject to such great fluctuations as in the other rivers of the Levels.

In spite of these more favourable physical factors some 2,500 acres of low-lying peat and clay land in the upper end of the valley were regularly flooded when high tides coincided with a high water-level in the river. Petitions and agitation throughout 1800 leave one in no doubt that the extensive reclamations of the late eighteenth century had been only of temporary benefit. Similar in nature to, and contemporary with, those of the Brue valley, these reclamations were not integrated with the drainage of the main channels (Fig. 16).[2]

With characteristic acumen, Billingsley pointed out that flooding arose from the unrestricted entry of the sea in the outfall and lower reaches of the river.[3] Both the influx of sea water and the vast amounts of silt it deposited tended to raise the level of the river in the upper regions of the valley and caused flooding. The evacuation of landward floods was also impeded by many meanders and stretches.[4] Thus, after the unusually bad winter of 1799, when the valley was under water for five months, the local landowners felt that some improvements were necessary in the main channel and its outfall, and joined with the proprietors of the Brue valley in an endeavour to obtain an

[1] The catchment area of the River Axe is only 81 square miles compared with 217 square miles for the Brue, 555 square miles for the Parrett and its tributaries, and 81·8 square miles for the Cary.

[2] It is not proposed to enumerate the enormous mass of detail about sluices, cuts and rhynes, which accompany the Acts. The relevant references are: E.A. No. 82, 1774 and 1778 (Wedmore Moor); E.A. No. 83, 1778 and 1791 (Panborough Moor); E.A. No. 117, 1777 and 1779 (Compton Bishop Moor); E.A. No. 134, 1782 and 1786 (Knowle Moor); E.A. No. 22, 1788 and 1791 (Bleadon); E.A. No. 21, 1788 and 1791 (Westbury Moor); E.A. No. 74, 1791 and 1793 and E.A. No. 80, 1791 and 1793 (Stoke and Draycott Moors); E.A. No. 38, 1795 and 1801 (Cheddar Moor); E.A. No. 33, 1811 and 1821 (Stoke Moor).

[3] Billingsley, 168.

[4] S.R.B./*P.C.S.* Axbridge, 9 April, 21 October and 20 November 1800.

Act for the draining of both areas. They advocated that the Axe river should be widened and deepened and that a clyse be erected across its mouth at Uphill.[1]

These suggestions were embodied in a report by Josiah Easton, an experienced land-surveyor of Taunton, who was employed by the landowners. He found that the river had

> no barrier to the tide and from...Hobb's Boat...[to] White House...the river is not more on an average than fourteen feet wide and three feet deep, whereas the river above White House...[to] Rackly Bridge is twenty-four feet wide and six feet deep; when the tide flows into this part of the river the straight below will not admit the waters to return into the Bristol Channel. This straight is an obstruction to the water above and these are the principal causes of the drowning of this level.

Consequently, these portions of the river needed attention. In his opinion, a clyse near Hobb's Boat would not give sufficient fall, nor would the capacity of the river channel behind it be large enough to contain the flood waters during tide-lock. Thus the outer estuary near Brean Down and Uphill was the best spot for a clyse, which would then also protect the coastal parishes of Brean, Berrow, Bleadon and Lympsham.[2] In November 1800, the Axe landowners decided to pursue their planned improvement independently of the Brue valley scheme. In order to avoid the great expense of obtaining a special Act of Parliament the proposals were submitted to the Commissioners of Sewers for their approval and action.[3] The commissioners naturally delayed making any decisions until they were sure that they had the necessary legal authority to undertake the construction of new works.[4] After taking legal advice, they felt that they had this authority, and began to draw up a schedule of lands to be rated, itself a new departure from accepted practice and open to some dispute.[5] But they were still reluctant to go ahead, and asked for a second opinion from Mr Sutcliffe, of Leeds, on the works proposed

[1] 'A Case for the Owners and Occupiers of the Axe Area for better Drainage', 11 October 1800. It was stated that 'there were some thousands of acres of land frequently inundated owing to the imperfect drainage of the Axe'. (S.R.B./A.D. 4). For the position in 1794, see Figure 18.

[2] J. Easton, 'To the Proprietors of the Lands affected by the imperfect drainage of the River Axe', 1 October 1800 (S.R.B./A.D.4). He estimated that the cost of the works would be £11,703.

[3] S.R.B./*P.C.S.* Axbridge, 21 October 1800.

[4] S.R.B./*P.C.S.* Axbridge, 20 November 1800.

[5] S.R.B./*P.C.S.* Axbridge, 12 and 26 January 1801. See page 136 above for problems of general rates in the Brue valley.

by Easton, largely, one suspects, from motives of delay, as well as from genuine doubt as to their exact powers and the practicability of the engineering proposals.

Sutcliffe drew up a supplementary report.[1] He condemned the proposal to build the sluice at Uphill on the grounds that unnecessary expense would be involved and time wasted in building it in such a wide part of the river. In addition to this, he did point out three basic defects in Easton's original plan. They were:

1 To save considerable expense, Easton did not want to make the sill of the Uphill Clyse level with the river bed, but 8 ft above it. This, however, was nearly level with the sill of the proposed Hobb's Boat Clyse further upstream, and consequently no advantage of fall would be gained by placing the clyse at Uphill;

2 Sutcliffe noted the increasing use of the river for navigation, especially with the opening-up of the South Wales coalfield. He felt it inevitable that additional expense would soon be incurred by having to provide a lock;[2]

3 Easton's provision for an opening of only 100 sq. ft in the clyse was unrealistic, especially when the effects of tide-lock had to be considered. Sutcliffe felt that 480 sq. ft was nearer the mark.

Therefore, with these considerations in mind, he urged that a clyse be constructed at a point intermediate between Hobb's Boat and Uphill, and that the river should be merely widened and deepened, and no cuts made across the meanders.[3]

In all fairness, the delaying tactics of the commissioners had brought to light some serious defects in Easton's plan; yet, even with Sutcliffe's suggestions in mind, they did not act, but once more asked for legal opinion on their obligation to provide for the needs of navigation.[4] Between 18 May and 27 November, the commissioners met and delayed their final decision four times.[5] Finally the landowners were exasperated by the unwillingness and inability of the commissioners to act, and took matters into their own hands by applying to Parliament for an Act, which was obtained in 1802. Jessop was appointed engineer. He rearranged both Easton's and

[1] S.R.B./A.D. 4, Sutcliffe's Second and Third Reports, 18 May 1801 and 6 July 1801. See also S.R.B./*P.C.S.* Axbridge, 11 February and 1 May 1801. Originally, Rennie had been approached to comment on Easton's proposals, but could not undertake the work.

[2] He notes this again in his Third Report of 6 July 1801 (S.R.B./A.D. 4).

[3] S.R.B./A.D. 4, Sutcliffe's Second Report, 18 May 1801.

[4] S.R.B./*P.C.S.* Axbridge, 20 July 1801.

[5] S.R.B./*P.C.S.* Axbridge, 19 August, 20 October, 13 and 27 November 1802.

Sutcliffe's plans only to turn eventually to Billingsley's suggestions. The clyse was to be built at Hobb's Boat, and three major cuts made on the river at near Loxton, Rackley, and Bleadon (see Fig. 20 for these cuts and other works undertaken).[1]

Work started on the clyse in 1803, and, in order to excavate the sluice-pit, the river was diverted down the Pilrow Cut, which had been improved lately under the Brue Drainage Act. Five feet of riverine clay and twenty feet of fine sand were excavated before a reasonable base for the clyse was found. But even this proved too unstable for such a heavy structure. It was said that 'two methods only have occurred for rendering of the work safe—piling of the piers and abutments and planking the whole extent of the sluice—or sinking deeper for a foundation and laying a temporary adequate weight to consolidate the substratum'.[2] Construction was further hampered by springs in the base of the pit.[3]

Despite the many set-backs the sluice was finally erected in 1810.[4] The improvement was immediate and lasting, the sea no longer flowed inland choking the outfall, and landward floods were dispersed more quickly. By 1826 it could still be said that 'the flat which was formerly covered with water now affords some of the richest land in the kingdom'.[5] Taken as a whole, the expectation of successful draining was upheld through later years; consequently the Axe valley proved to be an exception to the general rule which held fast for the other areas of the Levels throughout the nineteenth century.

King's Sedgemoor (1770–1810)

In contrast to the position in the Brue and Axe valleys, a scheme for the comprehensive drainage of King's Sedgemoor preceded the reclamation of individual blocks of the moor. Because the main drainage pattern was already in existence, the individual reclamations by

[1] S.R.B./A.D. Letter: Conway to Dickinson, 24 February 1803. In the meantime the commissioners had taken yet more legal advice on their obligation to construct a lock, but found it was unnecessary. S.R.B./A.D. 2, Letters: Jessop to commissioners, 8 February 1803, and Conway to Jessop and White, 18 February 1803, and 11 May 1803 respectively.

[2] S.R.B./A.D. 2, Letter: Anstice to Conway, 18 May 1805. The first solution was tried. See Anstice to Conway, 8 February 1806.

[3] S.R.B./A.D. 2, Conway's Letter Book, Jessop to commissioners, 8 February 1803. Jessop advised the use of a small steam engine to save time in pumping the water out of the pit. This is the earliest reference concerning the use of a steam engine for pumping in the Levels.

[4] The Award to the Commissioners is deposited in the S.R.B.

[5] Brit. Mus. Add. MSS. 33691, ff. 248–9.

parliamentary Act which followed, contributed less to the draining of the moor than was the case elsewhere in the Levels. The early inception of a comprehensive scheme in King's Sedgemoor is not altogether surprising when one recalls the strenuous efforts that were made to drain the moor during the preceding century. Moreover, it was still physically the most favourable of the peat areas in the Levels for which co-ordinated drainage could be planned.

The condition of King's Sedgemoor before draining was particularly bad. In 1771 Arthur Young found the moor covered with water, which, he said, 'has no way to get off but by evaporation; in winter it is a sea and yields scarce any food except in very dry summers. What a disgrace to the whole Nation it is.'[1] The state of the moor remained unaltered, and four years later it was said to be 'a morass by reason of the waters standing on it...the whole is nearly on a level, and is low in comparison to the adjoining lands...it is seldom fit for cattle more than two or three months in a year with safety'.[2] It was estimated to be worth little more than 18*d.* per acre, but on talking to local farmers Young felt assured that it wanted 'nothing but draining to be made well worth from 20*s.* to 25*s.* per acre on an average'. The possibilities of draining also struck an observant traveller, Shaw, who thought that there was no question that 'by proper management it might be rendered infinitely more valuable'.[3] Because of the prospect of such profits and the generally more favourable attitude to agricultural improvement prevailing at this time, new attempts were made to drain the moor in 1772, 1776, and 1788, the whole movement probably being spurred on in its later stages by the publication of a map of the moor made by Richard Locke some time during the 1780s.[4] The map showed not only the existing rhynes but also the new rhynes necessary for draining the moor, the way in which the parochial allotments could be marked out, and a computation of the cost of draining and of the cost of converting the drains to navigable canals. This was another of Locke's speculative surveys and substantiates his claim that he 'measured and mapped twenty thousand acres of moors and commons upon speculation'.

[1] Young, *A Farmer's Tour through the East of England*, IV, 13.

[2] *Journals of the House of Commons*, XXXIII (1775), 648 and 649.

[3] Young, *A Farmer's Tour*, IV, 14, and S. Shaw, *A Tour of the West of England in 1788* (1789), 323. For confirmation of these figures for the expected improved value of the moor after draining, see *Journals of the House of Commons*, XXXIII, 649 and 652.

[4] S.R.O. Dickinson MSS. Box 66.

All these attempts to drain the moor were unsuccessful because of the opposition of the commoners and some of the manorial lords, but more particularly because of the fraudulent way in which the sponsors attempted to present different Bills to the local inhabitants and the House of Commons. This was especially true of the 1776 attempt which was no more than a financial speculation by Lord Bolingbroke in order to raise some ready money to settle a gambling debt. These attempts are fully documented and throw considerable light on the methods, both legal and illegal, employed in enclosing and reclaiming the waste land of Somerset; but as they did not succeed in bringing about the draining and improvement of the moor and a change in the geography of the area they are not considered in detail.[1]

Eventually, in 1791 an Act was passed for 'The Draining and Dividing of King's Sedgemoor'.[2] By this time all the manorial lords appear to have approved of the measure, although the commoners petitioned by a majority of two to one against it.[3] This did not reflect the true magnitude of the opposition, for, as a local solicitor and owner of High Ham manor explained, 'the apprehension of disobeying their landlords must have deterred very many from signing the counter petition'.[4] Nevertheless, opposition was overcome, and the bill was passed. Over 4,063 claims for the land in the moor were put forward, but when all the bogus claims of overland settlements and squatters had been disregarded only 1,796 remained and were allowed.

The local inhabitants felt that the draining of the moor could be accomplished by merely widening and opening old outlets, i.e. the River Cary and Rowing Lake; but the existing channels did not provide a satisfactory basis for improvement since they were of inadequate dimensions, susceptible to tidal influence at their outfalls, and of insufficient gradient. (See Fig. 21 for the pre-drainage pattern

[1] *1772 attempt*: S.R.O. Kemys-Tynte MSS. 49, Letters E. Burrell to Sir C. K. Tynte, 5 September 1772, and Sir C. K. Tynte to 'A member of the House of Lords', 26 October 1776.

1776 attempt: see J. L. and B. Hammond, *The Village Labourer* (1911), 58–65. Great use is made here of Carlisle MSS., ed. R. E. G. Kirk, *Historical Manuscripts Commission*, XLII, 301–3. See also S.R.O. Kemys-Tynte MSS. 49, and *Journals of the House of Commons*, XXXIII, 648–52.

1788 attempt: Billingsley, 191–2.

[2] E.A. No. 116 (1791 and 1795).

[3] *Journals of the House of Commons*, XXXXVL, 193, 207, 236, and 267–8.

[4] S.R.O. Meade King MSS. Sect. D.3., Document entitled 'Mr Mildmay's opinion concerning inclosing the moor'.

of rhynes in the moor.)[1] For example, Cowhouse Clyse, at the Rowing Lake outfall, was at a higher level than the land in the northern part of the moor which it was supposed to serve, and Bennett Clyse was not working. With the existing drainage system unsuitable for remedy or extension, the commissioners took a bold approach, and, on the advice of Jessop and William White, sought a completely new outfall at Dunball.

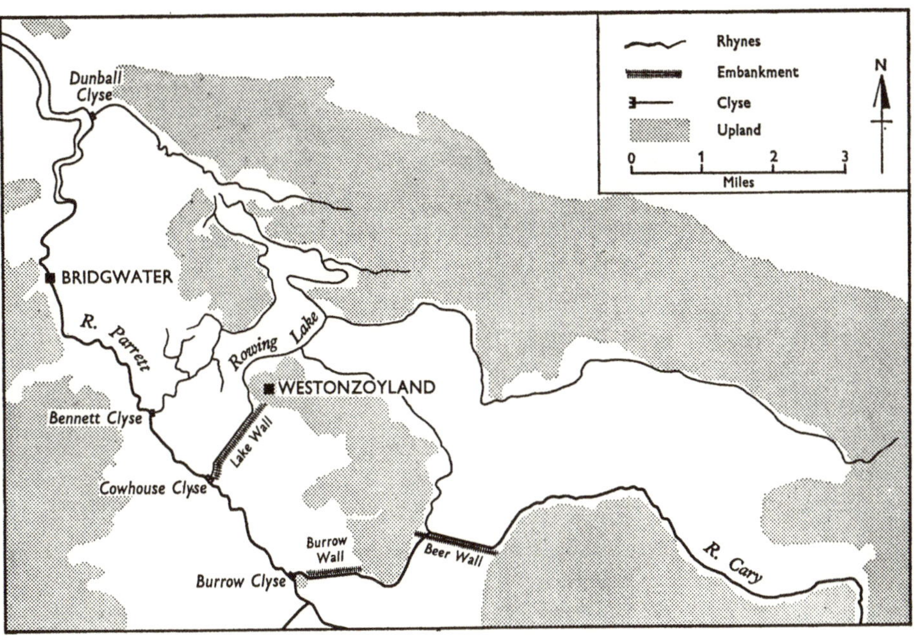

Figure 21. King's Sedgemoor, 1791. (Source: based on a map by Wm White (1791), S.R.O., Strangeways MSS., Map 54.)

The River Cary was diverted at Henley Corner and the whole pattern of drainage re-orientated towards the new Dunball outfall in a new channel of 26 ft 6 in. top width and 5 ft 6 in. depth, increasing to 55 ft wide and 14 ft deep at Dunball (Fig. 22).[2] The new channel extended for approximately 10 miles through the centre of the peat moor and for a further 2 miles through the clay belt to Dunball. The course of a stream that already ran off the clay belt and towards Dunball was utilized in the last two miles of the new cut. At the Dunball

[1] Based on S.R.O. Strangeways MSS. Map 54, a map of Sedgemoor (1791) by William White, and the evidence of air photographs.

[2] The increase in top width was merely to maintain the bed width of 10 ft where the drain passed through high ground of the clay belt.

outfall an old clyse was demolished and a new one built with a 10 ft 8 in. square opening. The construction of the clyse was greatly impeded because of unstable foundations which resulted from the fact that it was situated in the middle of what was once the bed of a meander of the Parrett that had been abandoned in 1677.[1] It seems fairly certain that the location of the clyse at Dunball was forced on the commissioners by landowners near Down-End and Walpole who

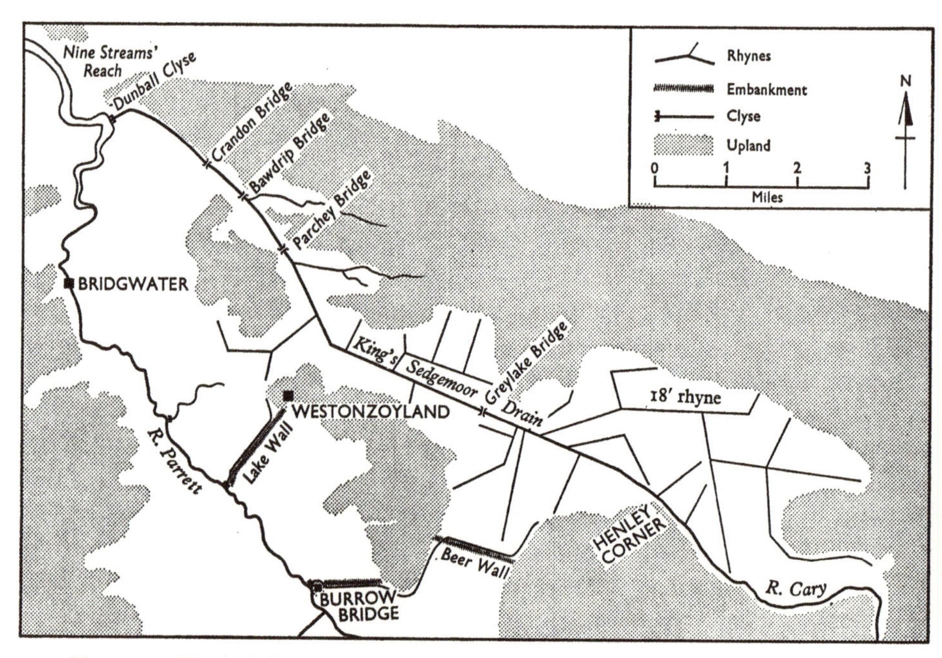

Figure 22. King's Sedgemoor, 1796. (Source: based on a map by Wm White (1796), S.R.O., Strangeways MSS., Map 55.)

opposed the course of the original cut which was to meet the Parrett at Nine Streams' Reach.[2] New bridges were built across the drain at Crandon, Bawdrip, Parchey and Greylake, as well as at many other places, the Greylake and Bawdrip bridges being fitted with penning sluices for irrigation purposes. In addition to the Main Drain, the 'Eighteen Feet Rhyne' was cut to drain the north-eastern portion of the moor; Shapwick, Street, Aller and Chedzoy rhynes were similarly 18 ft wide, and all lesser parochial rhynes were 15 ft wide.[3]

[1] See p. 113 above and Figure 14.
[2] See *Journals of the House of Commons*, XXXIII, 648.
[3] Details and information of the bridges and rhynes are taken from the Award.

With the exception of the Main Drain, the pattern of the other major rhynes in the moor was not dictated by the physical needs of the drainage; the course of the 'Eighteen Feet Rhyne' was particularly anomalous (Fig. 22). The reason for this lay in the practice of allotting reclaimed common land exclusively to those who had rights of common and performed the drainage maintenance, that is to say, the austre holders. They received a proportion of the moor approximately similar to the proportion which their own number within a parish bore to the total number in the whole moor; for example, Shapwick had 3·49 % of all austre tenements and was allotted 3·43 % of the moor. The combined parochial allotment was then divided off by one of the main rhynes. (These proportions are shown below in Table VII.) The drainage pattern of King's Sedgemoor bears little direct relationship to the physical geography of the region; it can be understood only in the light of the historical context of the time and of the circumstances when the pattern was first formulated.

Despite the many merits of the scheme, it did have faults which quickly made themselves felt.[1] First, because of the difficulty of

Table VII. *King's Sedgemoor: austre tenements and land allotted*

Parish	% of austre tenements	% of land allotted	Parish	% of austre tenements	% of land allotted
Bradney	0·50	0·36	Horsey	0·50	0·42
Bawdrip	1·16	0·87	Bower	1·00	0·89
Stawell	1·33	0·97	N. Petherton	0·61	0·48
Sutton Mallet	1·78	1·27	Middlezoy	5·50	5·69
Cossington	1·90	1·46	Othery	5·95	5·89
Chilton	2·16	2·12	Westonzoyland	7·11	6·05
Edington	1·83	1·64	Aller	4·56	5·20
Catcott	2·28	2·00	High Ham	5·17	5·39
Moorlinch	1·89	1·64	Huish Episcopi	1·39	1·84
Shapwick	3·49	3·43	Long Sutton	3·83	5·66
Greinton	1·83	1·76	Low Ham	2·33	2·42
Ashcott	3·28	3·28	Pitney	2·44	2·45
Butleigh	3·61	5·80	Compton Dundon	5·06	3·99
Walton	2·50	4·67	Somerton	12·79	11·52
Street	3·28	3·94	Droves and		
Chedzoy	4·28	3·22	rhynes		0·39
Woolavington	3·00	2·11	Total	100·00	100·00
Dunwear	1·66	1·18			

[1] There was flooding in the moor in 1791 and 1794. See W. Marshall, *The Rural Economy of the West of England* (2nd ed. 1805) II, 271.

building Dunball Clyse on such unstable foundations the sill was 4 ft higher than it should have been. Consequently, the fall in the drain was not adequate, and this encouraged excessive sedimentation in its central sections and in the side rhynes, which eventually caused flooding. Secondly, the new bridges were too narrow and bayed back the water to a height of 18 in. or more during severe flooding, when unimpeded egress was most needed. Henley Corner, Greylake's Fosse, Parchey and Crandon bridges were the worst, the damming effect being most apparent with the last two bridges, which were situated in the lower reaches of the drain where it passed through the clay belt and where the water could not overflow.[1] Thirdly, the deepness of the cut through the clay belt between Crandon bridge and Dunball (Fig. 22) meant that great masses of the excavated earth were deposited on the edge of the drain; the earth sank into the ground and the peat subsoil rotated on a flat arc, forcing up the soft peat in the drain bottom. The bottom of the drain had to be timbered at great expense, but this was no cure for the trouble which recurred incessantly. It was not finally alleviated until the spoil banks were flattened out and spread over the adjoining fields. The steepness of the sides of the cut also resulted, during rainy weather, in serious sliding and slumping of the bank, which again was repaired only at great expense.[2] Thus, despite an expenditure of £32,632 (which did not include the cost of the parochial subdivisions) and despite great optimism as to the outcome of the scheme, these defects tended to reduce the value of any improvement made.[3]

The cut through the clay belt did at first sight seem foolhardy, and contemporary comment was most disparaging. Billingsley wrote:

It is impossible to describe the ridicule which this undertaking excited. Some thought the Commissioners mad, others and by far the majority ascribed the boldness of the plan to the liberality of the proprietors in allowing the Commissioners three guineas per day for attendance and management; and drew this sage conclusion, that the work would never be finished but would be protracted till the expenses *would equal* the value of the moor.[4]

[1] J. Aubrey Clark, 'On the Bridgwater and other Levels of Somersetshire', *Journ. Bath and West Eng. Soc.* II (1854), 105–6.

[2] Billingsley, 195–6.

[3] See S.R.B./*P.C.S.* Bridgwater, 10 June and 6 October 1796.

[4] Billingsley, 94.

Others expressed the opinion that the expense of digging the Main Drain would be so great as to throw the moorland proprietors into debt for the next twenty years.[1]

The only practical standpoint from which one can judge the work of the commissioners is to ask what would be done today. It seems fairly certain that the same scheme would be proceeded with as the only logical solution to the problem of draining the moor. The principle of a new outfall passing through the barrier of the coastal clay belt and of straight cuts to give maximum fall, was surely correct.[2] The high level of Dunball Clyse was something which would not have happened if the commissioners had been able to adhere to their original plan. The steepness of the cut through the clay belt and the narrowness of the bridges were not fundamental defects in the plan, but merely points which better engineering knowledge and ability would have ultimately rectified. It is significant that the features of the scheme were exactly the same as those of Locke's plan of the 1780s, and it bore a marked likeness to his proposals in all its major aspects. Locke's vision as a drainage engineer was but another facet of the ability of this energetic and remarkable man. Unlike other regional drainage schemes implemented during this period, the 1791 King's Sedgemoor scheme was bold, imaginative, and correct in its basic conception of an independent outfall, and of a long straight cut to achieve maximum fall; but like them its success was limited.

Some mention must be made of the parochial reclamations which followed and which put the final touches to the drainage pattern of the moor. Fifteen parishes enclosed and recovered their lands between 1797 and 1798, immediately after the completion of the main drainage scheme, and three other parishes did so at a later date (Fig. 16).[3] Little is known of the remaining fourteen parishes; Acts were obtained for Othery and Woolavington, but whether an Award was signed for them is not known; the Sutton Mallet allotment was

[1] S.R.O. Meade King MSS. Sect. D.3.

[2] It is interesting to note that the same principles were adhered to in the construction of the channel later known as the Huntspill river in 1940–41. See p. 238 below.

[3] E.A. No. 49 Cossington 1807 and 1807; E.A. No. 110 Chilton 1798 and 1800; E.A. No. 98 Westonzoyland 1830 and 1834; E.A. No. 75 Middlezoy 1798 and 1800; E.A. No. 109 Edington 1814 and 1816; E.A. No. 56 Catcott 1798 and 1799; E.A. No. 106 Aller 1797 and 1798; E.A. No. 108 High Ham 1797 and 1799; E.A. No. 131 Huish Episcopi 1797 and 1799; E.A. No. 5 Ashcott 1797 and 1798; E.A. No. 43 Long Sutton 1797 and 1806; E.A. No. 34 Butleigh 1796 and 1796; E.A. No. 57 Pitney 1802 and 1807; E.A. No. 45 Somerton 1797 and 1798; E.A. No. 8 Street 1797 and 1798; E.A. No. 44 Compton Dundon 1797 and 1798; Woolavington 1797 and 1798; Stawell 1797 and 1798.

enclosed without an Act.[1] If some of the remaining parochial allotments remained as open commons, as seems the case,[2] it was not long before they were enclosed.

Finally, it is interesting to note how the diversion of the River Cary to the Dunball outfall severed the connexion between the drainage of King's Sedgemoor and that of the River Parrett thereby slightly easing flood intensity in the Parrett. This was similar to the way in which the probable diversion of the River Brue towards the Highbridge outfall had severed the connexion between the Brue and Axe valleys during the Middle Ages. Human activity had altered the drainage system of the Levels and increased the number of its divisions. Henceforth the problem of draining the Levels was going to be approached within the framework of five distinct drainage regions, each with its own outfall, namely the Southern Levels, King's Sedgemoor, the Brue valley, the Axe valley and the Northern Levels.

The Southern Levels (1770–1840)

The Southern Levels stood apart from the rest of the regions in their late reclamation, in the lack of a comprehensive drainage scheme, and in the abandonment of gravitational drainage with the adoption of steam pumping. Because of the physiographic variation within the Southern Levels there was no obvious or readily applicable solution to the problem of its flooding. Ridges and 'islands' divided it into a series of isolated moorland basins (Fig. 1) and it was difficult to combine the needs and requirements of each basin with a single plan for the draining of the whole region. The physical separation of the area into basins resulted in a divided effort to drain it, a characteristic which is reflected in the scattered nature of the area of reclamation depicted in Figure 16.

The successful draining of each basin was also hampered by the poor condition of the main river channels on which they depended ultimately for the evacuation of their surplus water. The Parrett, in particular, was shallow, winding and narrow, and consequently unable to cope with excessive amounts of water. Its condition was aggravated by the unrestricted entry of sea water which penetrated

[1] Locke Survey , f. 436.

[2] Marshall, *Rural Economy of the West of England* (2nd ed., 1805), II, 384, said that the moors 'when wholly appropriate [i.e. enclosed] and sufficiently freed from surface waters will be one of the most valuable Districts of equal extent in the whole island'.

as far upstream as Langport at spring tides. On such occasions, if the river was already full of water, over-spilling and extensive flooding in the moorland basins resulted, and gravitational drainage back into the Parrett was arrested for many weeks.[1] These adverse physical factors caused great uncertainty and dissension as to the best methods of improving the channel, and may, in part, explain the lateness of individual moorland reclamation schemes in this area (Fig. 16). Reclamation schemes did not begin until 1795, when they were almost completed elsewhere in the Levels, and did not end until 1838, when activity in the remainder of the Levels was floundering. Collinson accurately observed these regional differences in 1791 and said that the moors of the Southern Levels were not 'divided by ditches, planted on each side with willows like those about Glastonbury, but are rich flat open commons skirted with Highlands'.[2] When reclamation finally took place in the Southern Levels each project tended to be a more complete draining scheme than was the case in the other regions. The marked differences in relief meant that the area to be dealt with was always strictly limited by surrounding higher ground, and this gave a certain unity, which was often lacking elsewhere in the Levels, to the conception of each scheme. The schemes for West Sedgemoor and West Moor were outstanding in this respect. Further, the lateness of reclamation meant that advantage could be taken of the experience gained in draining other areas of the Levels.

The first drainage scheme in the Southern Levels was for North Moor which lay to the north of Athelney 'island'.[3] The basis of the scheme was a main rhyne which was excavated for 2¼ miles through the centre of the moor, and fourteen new rhynes led into it. The embankment, which excluded the River Parrett from the moor, was raised and strengthened, and a clyse built in it where the main rhyne passed through the embankment. The pattern of this scheme, with a central rhyne, collateral drains and a river embankment with a clyse,

[1] For a discussion of the flood behaviour of the Parrett, see p. 15 above. For contemporary descriptions of flooding in the Parrett valley, see J. Beaumont and H. Disney, *A New Tour thro' England performed in 1765, 6 and 7.* (1768); and Anon., *Observations made during a Tour through part of England, Scotland and Wales* (1780), 43, where it was said that 'in wet winters people have been known to come from the Parrett in boats to the very doors' of houses in Somerton, presumably sailing from an overflowing River Parrett, across a flooded King's Sedgemoor, and up the River Cary.

[2] Collinson, *History and Antiquities of the County of Somerset*, I, 84. In a print facing p. 24 of this work, there is a view across West Sedgemoor, looking from Stathe to Redhill. The moor is completely open, without a tree, and without a rhyne dividing it.

[3] E.A. No. 128, 1795 and 1798.

was typical of the rest of the moorland reclamations alongside the Parrett and its tributaries, and two years later, in 1797, almost identical schemes were begun for King's Moor alongside the River Yeo, and in each of the moors in the Tone valley.[1] Other schemes in 1818 for Week and Port Moors, and in 1819 for Wet Moor in the upper Parrett valley, were very similar.[2] Of the remaining reclamations perhaps those for West Sedgemoor in 1816, and West Moor in 1833,[3] are the most interesting and important, from the point of view both of their size and of the drainage patterns that evolved.

West Sedgemoor was described just prior to its reclamation as 'one of the most improvable tracts of waste land now remaining within the county...for in its present wet and saturated state caused by inundations and the want of a better drainage, it injures rather than benefits the stock now kept therein and is, of course, of no value to the commoners...' Consequently there was comparatively little opposition from the bulk of the commoners to draining, although some of the more prosperous farmers of the nearby surrounding uplands, who had most to lose by draining and enclosure, did attempt to obstruct proceedings.[4] But the general change of attitude paved the way for the improvement of the moor.

The pattern of rhynes laid out in the West Sedgemoor scheme did not wholly conform to the physical requirements of draining the moor, and, in this respect, the draining of the moor was very similar to that of King's Sedgemoor; for according to the percentage of the length of the existing river wall alongside the Parrett that the parochial austre tenements maintained, they received a like proportion of the newly-drained moor (see Table VIII below). First the moor was divided almost in half by the Middle rhyne, according to the respective portions of the West Sedgemoor Wall which the parishes on the north and south sides *in toto* were liable to maintain. The portions were:[5]

[1] E.A. Nos. 43 (1797 and 1806) and 39 (1797 and 1800). The upper end of the Tone valley at Creech was drained under E.A. No. 36 (1816 and 1822).

[2] E.A. Nos. 111 (1818 and 1820) and 102 (1819 and 1826).

[3] E.A. Nos. 85 (1816 and 1822) and 101 (1833 and 1838).

[4] S.R.O. Dodson and Pulman MSS. Box 1 of 94, O.B.10, 'Observations by C. Chilcott'. See also Box 2 of 94, 'Letter Book', Letter from Mr Beadon to Mr Parry (1816), which shows that the drainage of the moor almost failed because Parry, who was manorial lord of Curry Rivel, felt that the newly-drained moor would not offer such an imposing view from his intended new country house as did the undrained moor.

[5] The discrepancy, of just over 5%, between the amount of wall work done and the amount of land allotted was due to the subtraction of 5·85% from the land allotted to North Curry, and its distribution to Capland, Broadway, Hatch Beauchamp, Burton,

The regions of draining

	Wall work (%)	Land allotted (%)
North side	50·63	45·60
South side	49·87	54·40
Total	100·00	100·00

Table VIII. *Wall work and land allotted in West Sedgemoor* (1816–22)

Parishes	Feet of wall work	% wall work	Land allotted to nearest acre	% of land allotted
West Hatch	260	4·79	119	4·93
North Curry	1,850	34·08	566	28·23
Soil allotment	—	—	115	
Stoke St. Gregory	620	11·43	300	12·44
Weening Way	18	0·33	—	—
Week	30	0·55	9	0·37
Oath	48	0·89	14	0·58
Moortown	40	0·74	27	1·11
Curry Mallet	420	7·74	147	6·10
Fivehead	320	5·90	129	5·34
Swell	170	3·13	89	3·70
Capland and Broadway	280	5·16	164	6·80
Hatch Beauchamp	240	4·42	137	5·69
Burton	80	1·47	44	1·83
Beercrowcombe	250	4·61	144	5·98
Hambridge	180	3·33	87	3·62
Curry Rivel	620	11·43	248	13·28
Soil Allotment	—	—	73	
Total	5,426	100·00	2,412	100·00

Therefore the Middle rhyne, which was the main rhyne, did not follow the natural course of the drainage in the moor. (Compare the pre- and post-drainage patterns in Fig. 23.) Secondly, the size of the parochial allotments was made in proportion to the number of ropes or feet of wall assigned to each parish,[1] the boundaries between these parochial allotments being rhynes which jutted out at right angles from the Middle rhyne. On the north side of the moor, however, only

Beercrowcombe, Hambridge and Curry Rivel in particular. This mode of division was arranged by the commissioners for the drainage of the moor, but the reason for it is not clear. See S.R.O., Dodson and Pulman MSS. Box 1 of 94, 'Letter Book', Letter from Mr Beadon to Mr Tripp, 16 April 1815.

[1] Based on various documents in S.R.O. Dodson and Pulman MSS. Box 1 of 94, O.B.10, and also E.A. No. 85 (1816 and 1822).

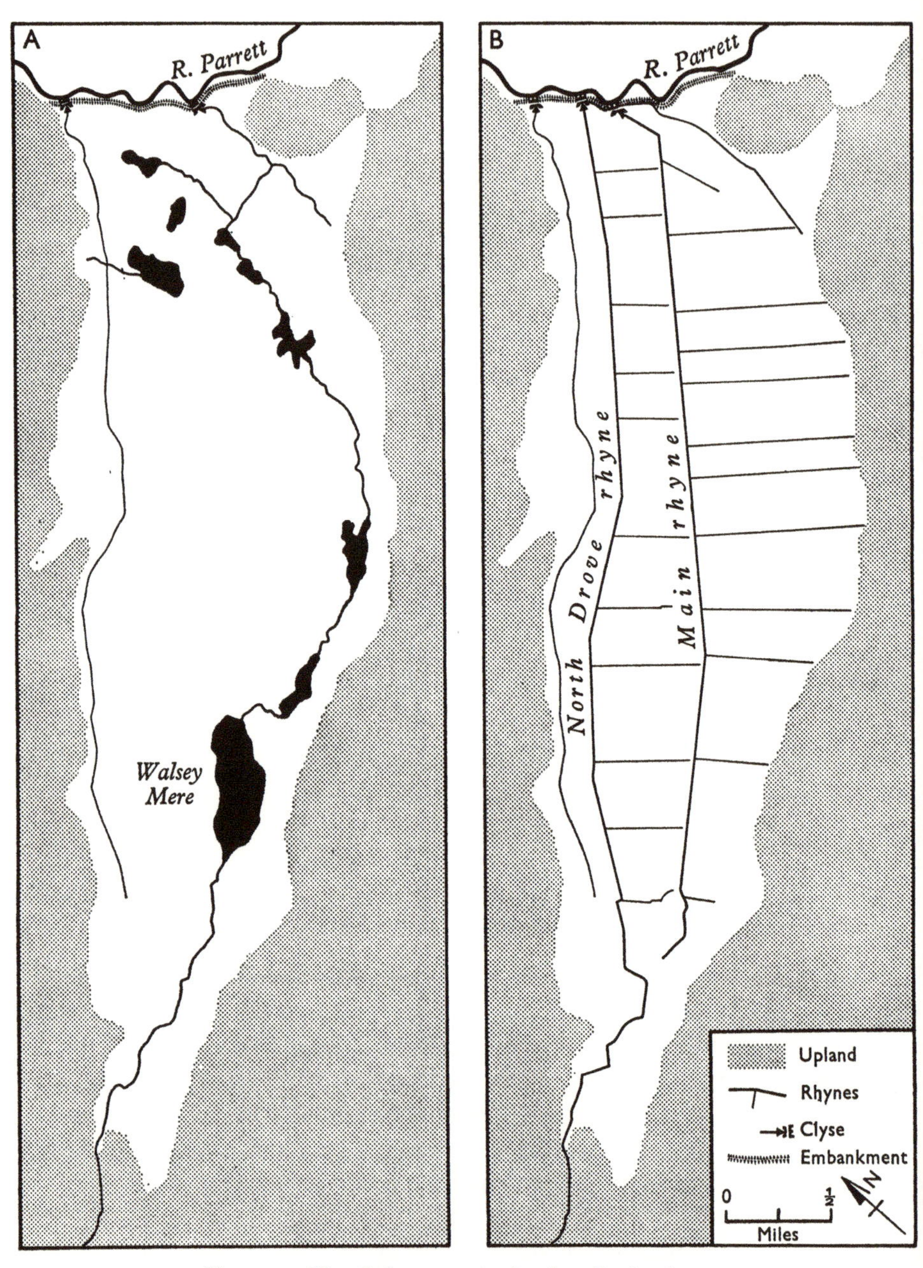

Figure 23. West Sedgemoor: A—in 1809; B—in 1822.

three parishes had an interest in about 1,400 acres, which meant only four rhynes. Extra drains had to be made within the parochial framework; therefore the North Drove rhyne was dug parallel to the Middle rhyne; it did not conform to any parochial or tithing boundary but was purely a matter of good drainage.

The draining of West Moor in 1833[1] was probably the most advanced and comprehensive scheme of moorland reclamation ever attempted in the Levels. A catchwater drain was dug to intercept the run-off from the surrounding uplands, and it had outfalls at either end which dealt merely with the run-off water. In the moor itself the usual network of rhynes was laid out and all rhynes led ultimately into a main centre rhyne, 3½ miles long, which ended in a clyse at the junction of the rivers Isle and Parrett. Irrigation was also provided for: two supply drains were connected to the catchwater drain from a weir at Slab Gate on the River Isle and a weir at Combe Bridge on the River Parrett. These could be regulated to release water throughout the moor during dry weather. (Fig. 24.)

It seems certain that, as in other regions of the Levels, little permanent improvement resulted from these reclamation schemes, for although they were adequately conceived in themselves it was assumed that their drainage would pass easily into the main rivers. But high flood levels in the main rivers due to 'freshes' or tides could last for weeks on end and effectively seal all the exits of the moorland basins by closing the outlet clyses. Neglect of the Parrett outfall and channel, in particular, vitiated whatever success was gained in draining the moorland basins, and consequently flooding was both regular and persistent in the moors. For instance the moors above Langport were regularly flooded and roads made impassable;[2] the Commissioners of Sewers could not view West Sedgemoor and the Tone Moors in 1828 'owing to the height of the water';[3] and reports during later years indicate that flooding had been widespread for a long time.[4] In an endeavour to provide a partial answer to the problem of high flood levels in the main rivers many schemes were suggested for the improvement of the Parrett outfall, all of which

[1] The draining of West Moor was finally taken over by the Parrett Navigation Company in 1836, which nearly jeopardized the scheme by proposing to make either the centre drain or the catchwater drain into a navigation canal. The company's proposal did not succeed.

[2] S.R.O. Combe MSS. 37, Estate and Household Administration Papers, 1813–19.

[3] S.R.B./*P.C.S.* Langport, 17 September 1828.

[4] For instance see Acland, footnotes on 46–8, and Clark, 108–10.

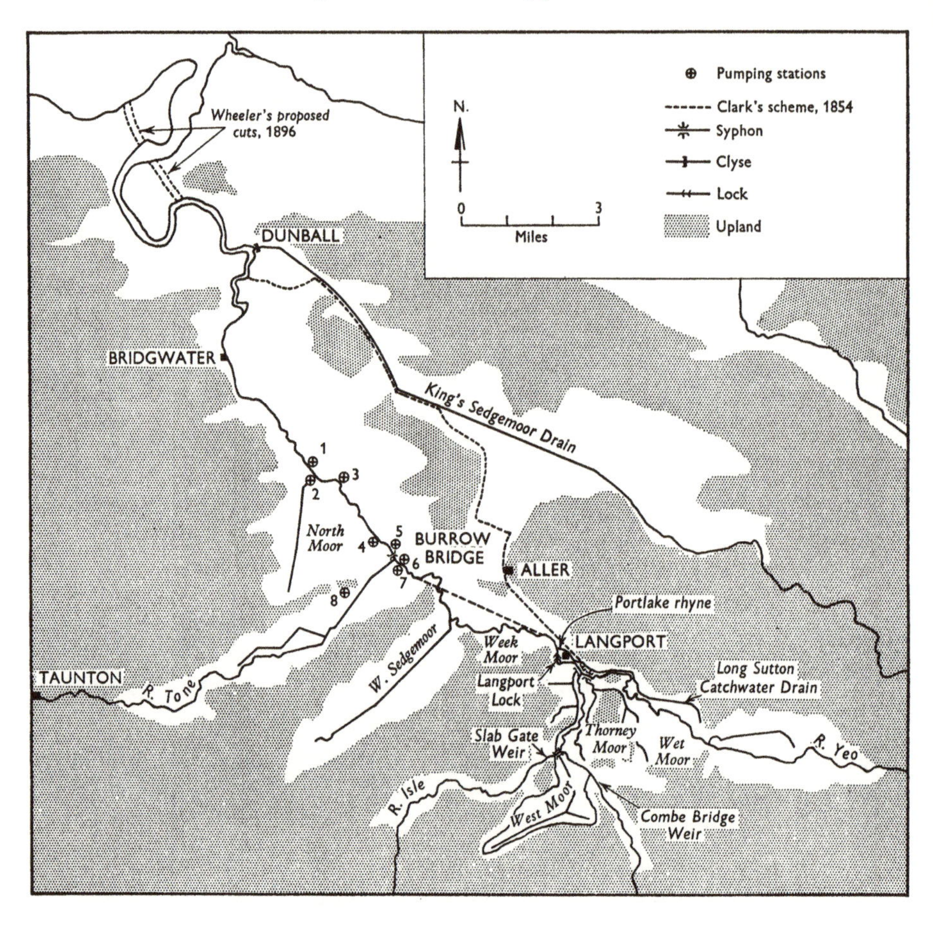

Figure 24. The Southern Levels: 1833–1900. Pumping stations: 1, Chedzoy, 1861; 2, North Moor, 1867; 3, Westonzoyland, 1830; 4, E. Saltmoor, 1837; 5, Aller Moor, 1869; 6, Southlake, 1869; 7, Stan Moor, 1864; 8, Curry Moor, 1864.

included proposals for widening and deepening the river, making new cuts, removing shoals, and erecting either training walls or sluices.[1] Although in all but Phelps's scheme of 1835 the requirements of drainage were subordinate to those of navigation, not even the

[1] (i) J. Rennie, *Report* (12 May 1814); (ii) J. Easton and R. Anstice, *Report* (13 December 1824); (iii) W. Armstrong, *Report on the Navigation of the River Tone from Taunton to Bridgwater* (1824); (iv) J. Jessop, *Prospectus of the Proposed Improvements of the Port and Harbour of Bridgwater* (23 November 1829); (v) H. H. Price, *Report on the Establishment of a Ship Canal and Docks at the Port of Bridgwater* (1835); (vi) W. Phelps, Plan for 'the effectual drainage of the moors round Langport and its vicinity', in his *Observations on the Great Marshes...*, 28–38.

navigation schemes, which might have given some relief, were proceeded with.

In the upper Parrett valley, on the other hand, some measure of main channel improvement did occur in conjunction with navigation works. In 1795 the Yeo Navigation Act[1] led to the general improvement of the banks and channels of the Yeo, and to the utilization and enlargement of the Portlake rhyne which cut across the constricting bend of the Parrett at Langport (Fig. 24). Navigation was later extended to the upper Parrett and Isle rivers in 1836[2] by the improvement of the main channels and the raising of the banks, works which were planned to give a waterway of 3 ft in depth throughout; however, in order to maintain this depth the construction of a lock at Langport was necessary. This meant higher water-levels in the three rivers above, and consequently a greater likelihood of flooding, especially if the lock was not opened in time to allow a 'fresh' to work through, as often happened.

Agitation by landowners for safeguards reached such a pitch that the Parrett Navigation Company had to make special arrangements whereby the draining of the moors above Langport Lock could be made independent of the main rivers if the need arose.[3] The Portlake rhyne was made the key to the new scheme of draining because it by-passed Langport Lock. The moors north of the River Yeo were drained by the Long Sutton Catchwater Drain which was dug to connect with the Portlake rhyne. West Moor, Thorney Moor, Muchelney Level and other lowland areas were also provided with new channels, which were carried under the Parrett by culverts to join the Long Sutton Catchwater Drain (Fig. 24).[4] In this way the whole of the moors above Langport could be drained independently of the main rivers, an obvious advantage in view of the constant high flood levels in them and the inefficient operation of the Langport Lock. The whole scheme took the form of a flood-relief system, which initially brought much benefit to the area.

[1] Yeo Navigation Act, 1795. 35, Geo. III. See also S.R.B./P.N. 8, *Report on the River Yeo Navigation from Langport to Ilchester* (Anon., 1795), also *Statement as to Navigating the Yeo, from Lord Ilchester* (1836).

[2] 6 and 7, William IV, c. 101. This Act instituted the Parrett Navigation Company. For a contemporary description of this river traffic see Brit. Mus. Add. MSS. 33691, ff. 117–18 (1826).

[3] S.R.B./P.N. 8, 'A meeting of subscribers to the measure', 27 April 1836.

[4] The Long Sutton Catchwater Drain was not able to cope with the combined overspill waters, and in 1839 an Act was procured which, among other things, provided for the enlargement of the drain to 25 ft in width.

On the other hand, in the town of Langport itself and in the moors below the town, conditions became no better and there is evidence to suggest that the works of the Parrett Navigation Company above Langport had aggravated the risk of floods by releasing more water more quickly into the lower reaches of the river. With static or even deteriorating conditions, the proprietors of the individual moors below Langport eagerly explored the possibilities of steam pumping; the engines installed in the Fens since 1820 and working with a fair degree of success probably pointed the way to what could be done in the Levels.[1]

The Southern Levels were admirably suited to pumping, for once more their physiographical division into basins, and the man-made embankments constructed in the past were of significance, this time in defining and limiting the area with which any pump had to deal. Because of the difficulty of gravitational drainage, steam pumping was a reasonable, though precarious, solution to the problem of draining. Walls were built higher and wider, and engines installed to pump water over them into an already swollen river. The pumps could not stop flooding but merely ensured the quicker evacuation of water, thereby minimizing the damage done to crops and grass. Small, independent Drainage Boards were created to administer the areas affected by pumping so that draining activity in the Southern Levels was advancing in what were literally, as well as metaphorically, watertight compartments; but there was still no comprehensive plan for the region. The future was threatened with some difficulties which were not going to be easy to solve.

In spite of these potential problems the first steam engine was installed in Westonzoyland Moor in 1830. The 1,600 acres of land which lay between Lake Wall and Burrow Wall were made into a Drainage District by Act of Parliament. A 27 h.p. scoop-wheel pump was installed on the Parrett bank, and the existing drainage channels within the moor wère reconditioned and reorientated towards the pump.[2] Unfortunately, the pump was barely effective because it was too small and therefore unable to cope with the water that accumulated in the moor; moreover, the pump outfall was placed at a point too low in the bank for it to be able to discharge when the level of the Parrett rose. In addition to these difficulties, the cost of erecting and

[1] Darby, *Draining of the Fens* (2nd ed. 1956), 220–5.
[2] E.A. No. 93 (1830 and 1843).

maintaining the pump was an enormous strain on the small area available for rating;[1] over £4,000 had been spent on erecting the pump, and £250 had to be found each year to run the engine. The rate of over 5*s.* per acre per annum levied over the moor was out of proportion to the benefit received and consequently there was little prospect of rectifying the constructional defects of the pump.[2] Seven years later, the East Saltmoor proprietors on the opposite side of the river to Westonzoyland, decided that a 10–12 h.p. engine was needed 'in order to pump or lift out such superfluous waters into the River Parrett inasmuch as the waters of the said river are frequently higher than the Level or surface of the lands in East Saltmoor...and impede and altogether prevent drainage thereof'.[3] Only a 6 h.p. engine was installed, however, and in time it too proved to be 'greatly insufficient for its purpose'.[4] In contrast to the Westonzoyland project, no drainage board was formed. (For the location of the pumps, see Fig. 24.) In 1842 a further scheme, for Stogursey, was considered by the Commissioners of Sewers but nothing eventuated, although three years later a 6 h.p. pump was erected in Southlake Moor. Like the Westonzoyland engine, it was for the same reasons unable to clear the water effectively and its running cost was a grievous expense on the adjoining lands.[5]

Because of the scarcity of evidence it is difficult to assess contemporary opinion on the steam pumping. Phelps, in 1836, thought that the larger-scale application of steam pumps to drainage would be too expensive to be practical, and this opinion was echoed by others.[6] A more important consideration which Acland foresaw, was that steam pumping alone would solve nothing; in his view it was inevitable that attention would have to be paid soon 'to relieve the present overburdened channel...either by enlarging it or by diverting the water which now causes it rapidly to overflow'.[7] One thing was certain; because there had been few wind pumps in the Levels, pumping was

[1] The rateable area did not include the lands of Sowy 'island' from which a considerable amount of water drained towards the pump.

[2] Clark, 107.

[3] S.R.B./*P.C.S.* Bridgwater, 22 February 1837. See also entries for 6 June 1837, and October 1838 for further information.

[4] Clark, 107.

[5] See S.R.B./*P.C.S.* Bridgwater, 7 July 1842, and Clark, 107–8.

[6] Phelps, *Observations on the Great Marshes...*, 24; Acland, 132; and J. Darby, 'The farming of Somerset', *Journ. Bath and West Eng. Soc.* v (1873), 167. Darby noted that some steam engines in the Levels were costing 18*s.* per acre per annum to run.

[7] Acland, 50.

a radical departure from the previous practice of gravitational draining. All in all, despite the expenditure of much time, energy and money, the drainage condition of the Southern Levels by about 1830 was little better than it had been thirty years before.

The Northern Levels (1770–1840)

The Northern Levels were a microcosm of the whole of the Somerset Levels. The early and well-settled coastal clay zone and the often flooded and largely neglected peat and lowland clay moors behind it were small-scale examples of their counterparts south of the Mendip Hills. The events in this region were almost a recapitulation of the problems and methods of draining that characterized activity throughout the Levels from 1770 to the first few decades of the nineteenth century.

The 'naturally rich' grazing lands of the coastal clay belt were drained by an intricate network of rhynes of unknown antiquity, and were protected from tidal inundation by a line of massive sea walls that filled the gaps between the natural defences of Brean Down, Worlebury Hill, Middle Hope and the Clevedon-Portishead Downs (Fig. 25). These walls and rhynes were subject to the most intricate maintenance procedures, and the minutes of the local Commissioners of Sewers for the Northern Division during the eighteenth century are a monotonous catalogue of neglected duties, views of works, fines, and maintenance work done. The impression left, after reading the minutes, is that, despite a certain amount of inevitable slackness and evasion of duty on the part of some frontagers, the commissioners were vigilant, and the work of maintenance was reasonably well done, but there is no evidence of any positive improvement of the drainage. Certainly there were no major floods after the beginning of the eighteenth century to put the rhynes and walls to a serious test.

Inland, the peat and lower clay lands were in a poor condition which was 'disgraceful to the owners', and they were subject to 'frequent inundation and sometimes in rainy seasons...covered with water for four or five successive months'.[1] The threat of flooding, as in many other parts of the Levels, was not only from rainfall and run-off, but also from the penetration of the sea up the major estuaries like that of the Congresbury Yeo, in which water at high tides flowed about five feet above the surrounding districts, seven miles inland.

[1] Billingsley, 130 and 17.

Nothing was necessary, noted Billingsley, 'but effectual draining to make...[these moors]...as good land as any in the country'. The whole district could be 'advanced by 10 to 15 shillings per acre',[1] if tidal sluices were constructed, main drains regraded and widened to

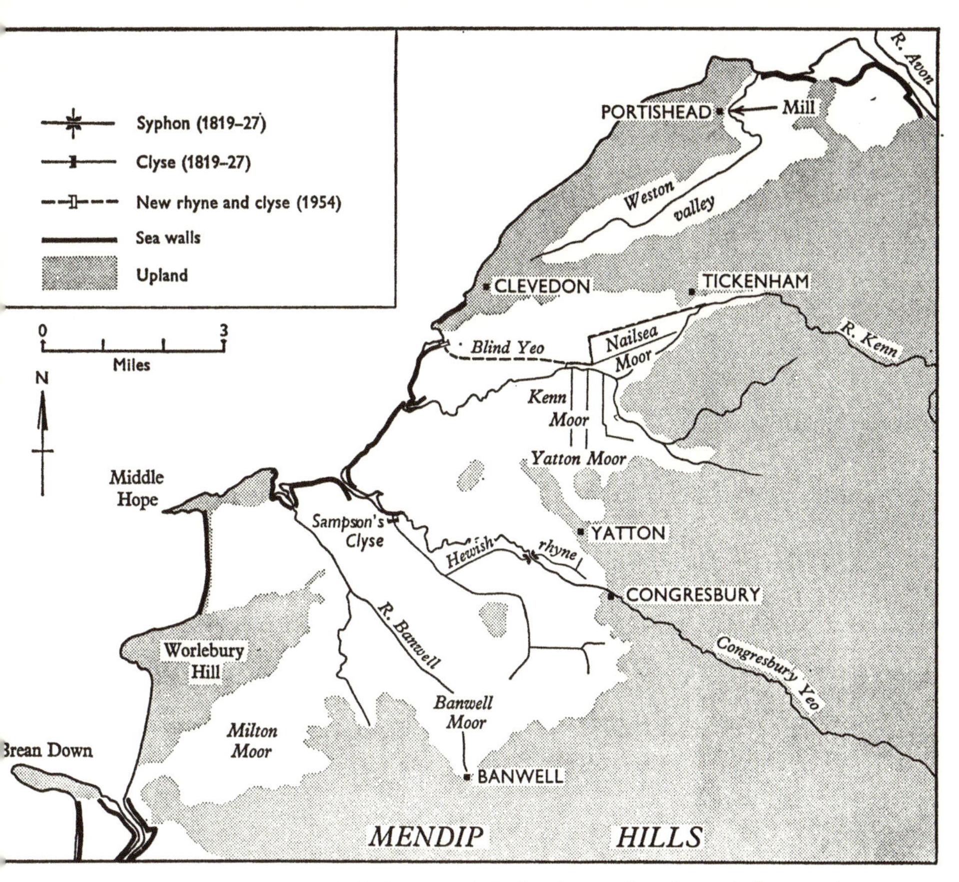

Figure 25. The Northern Levels: drainage works in the nineteenth and twentieth centuries.

increase their flow and scouring power, and lateral drains cut. By 1798, when Billingsley was writing, little had been done to enclose and drain the flooded moors in this region, except for a very early enclosure, in 1751, of 174 acres by private agreement in Yatton Moor, a small piece of 160 acres in 1793 in Milton Moor in Kewstoke, and another small portion in 1795 in Banwell Moor (Fig. 16).[2]

[1] *Ibid.* 18.

[2] E.A. Nos. 3 (1751 and 1751), 47 (1793 and 1794) and 42 (1795 and 1797) respectively.

Reclamation by Parliamentary Enclosure Act in the Northern Levels was late, as it was in the Southern Levels. It was not until the years between 1799 and 1813 that nearly 8,000 acres of the low-lying lands behind the clay belt were reclaimed and a network of drains cut across them. Activity was concentrated in the three worst-affected areas, consisting mainly of peat moors. They were the isolated Weston valley which opened out into the Avon estuary;[1] the Tickenham, Nailsea and Kenn moors, and the moors on either side of the middle section of the Congresbury Yeo. But, like their counterparts elsewhere in the Levels, these individual moorland reclamations were not a lasting success since they were not related to one another and not co-ordinated with the main outlets for drainage.[2]

The first move to integrate the drainage of the individual reclamations occurred in the Weston valley in 1810 when the local proprietors appointed Josiah Jessop as engineer to rectify the drainage (Fig. 25).[3] The crux of the drainage problem in this simple and well-defined basin lay in the presence of the Portishead mill at the mouth of the valley. The mill-owner had the right to use the main drains as millponds by admitting sea water from the high tides into them, and slowly letting out the water to drive the mill. The result for the drainage of the moor was disastrous. 'I am inclined to think', wrote Jessop in a preliminary survey, 'that there is not a perceptible discharge from the Moors, four hours out of twenty-four.'[4] After considering the idea of by-passing the mill with a new channel, Jessop decided that the only practical course of action was to destroy the mill, for the tidal millpond without a freshwater scour would soon silt up and render the mill unusable. After its removal all that was needed was the general widening, deepening and regrading of the existing rhynes, coupled with a few short cuts to link up the various parts of the existing drainage system.[5]

Negotiations over the purchase of the mill, after going on

[1] Locke described the valley as so waterlogged and susceptible to sudden flooding that it was 'dangerous to cross the moor'. Locke Survey, f. 133.

[2] E.A. No. 78, Clevedon (1799 and 1801); E.A. No. 70, Locking (1800 and 1801); E.A. No. 46, Tickenham (1801 and 1803); E.A. No. 60, Portishead and Weston-in-Gordano (1807 and 1809); Yatton and Kenn (1810 and 1815); E.A. No. 123, Weston-super-Mare (1810 and 1815); E.A. No. 24, Uphill (1813 and 1818); E.A. No. 129, Wraxhall and Nailsea (1813 and 1819); E.A. No. 120A, Hutton (1836 and 1849).

[3] S.R.B./S.W. 4, Proc. of the Comm. for the Weston Drainage, 16 July 1810, and E.A. No. 52 (1810 and 1815).

[4] S.R.B./S.W. 4, Correspondence, Jessop to Leman, 13 January 1810.

[5] S.R.B./S.W. 4, 'Specifications of work to be done', 10 September 1810.

throughout 1810, were concluded successfully, and work started on reconditioning the existing rhynes. But all was not well. Some of the local landowners, particularly those on the slightly higher and drier upland edges and at the mouth of the valley, objected to being rated for the new works. William White was called in to survey the moor and make the assessment for rating and also to make a report on the drainage condition of the moor; he was then appointed engineer in place of Jessop.[1] Within a fortnight White produced a report, which, while noting that the fall in the moor was not as great as Jessop had thought, concurred with his opinion that the destruction of the mill was essential for better drainage.[2] But the commissioners were dilatory in the prosecution of the work, and White's second report a year later spoke of uncleared rhynes in 'a deplorable state', overgrown by weeds and blocked by slips, so that 'the progressive improvement of the Lands, which there was every reason to expect would, by this time, have taken place, has been delayed'.[3] On the whole the work was done very indifferently and the main rhynes were not made deep enough, so that even when the Award was made in 1815 and the works handed over to the Commissioners of Sewers little permanent improvement resulted. During the next twenty-five years maintenance was 'not well attended to', the drains got progressively worse as slides reduced their width, and vegetation was not cleared. Flooding remained a regular feature of the lowest, western parts of the moor.[4]

At about the same time as improvement of the Weston valley drainage was being undertaken, the proprietors of the lowlands on either side of the Congresbury Yeo were becoming painfully aware of the poor drainage of their area, flooding not having been alleviated by the rhynes dug under the authority of the moorland reclamation schemes. The proprietors presented a petition to the Court of Sewers[5] and attended the court to state their case, asserting that over 2,500 acres could be improved by up to £1 an acre through better drainage.[6] But, as in the case of the Axe drainage, the local commissioners delayed making a decision because they were not sure of their power

[1] S.R.B./S.W. 4, Proc. of the Comm. for the Weston Drainage, 11 June 1811. The reason for this replacement of Jessop by White is not revealed.

[2] S.R.B./S.W. 4, White's Report, 25 June 1811.

[3] S.R.B./S.W. 4, White's Second Report, 11 June 1812.

[4] S.R.B./S.W. 4, Report by Mr J. M. Tucker, 6 March 1840.

[5] S.R.B./S.W. 4, Petition from fifty-four landholders in the North Marsh, 21 October 1810.

[6] S.R.B./S.W. 4, *P.C.S.* 1801–12, 22 May 1811.

either to make new drains or to tax the whole area for improvements. The commissioners were trying to negotiate a deal with the promoters of the Bristol and Taunton Canal which was planned to pass through the moors, with the object of ensuring that these promoters should bear some of the costs of drainage.[1]

Events dragged on interminably. In 1819 the landowners abandoned the idea of doing the improvements through the Commissioners of Sewers and obtained an Act of Parliament. John Rennie was asked to provide a report and plan which were accepted.[2] His proposals rested on the need to circumvent the high water-levels in the Congresbury Yeo, which occurred with the free entry of the tide up its embanked channels as far as Congresbury village, thereby stopping the free flow of the water from the adjacent moors. He suggested the following works:

1 A new cut from the Yeo estuary leading back into the inland moors to join with the existing rhynes made during the enclosure in 1809. This would provide a greater fall (Fig. 25);

2 A sluice, (Sampson's Clyse), should be built at the end of the new cut to restrict the sea water;

3 Another new cut (Hewish rhyne), should branch north-eastwards into Congresbury Moor, passing under the Yeo by means of a culvert;

4 Many other rhynes should be widened and deepened.

All these works were carried out, and, with the expenditure of nearly £17,000 between 1819 and 1827, the drainage of these inland moors was made independent of the Congresbury Yeo by the provision of the new cut and sluice. Rennie's scheme, in contrast to its northern counterpart in the Weston valley, had grasped boldly the principal need for the successful drainage of the Levels, namely, the provision of new cuts across the coastal clay belt to achieve maximum fall, scouring power and storage capacity during tide-lock. Success was immediate and lasting; as late as 1851 it could still be said that 'a great improvement' had been achieved[3] and that 'the benefit' had 'been very great'.[4] The success of the scheme was a standing reproach to those responsible for the neglect of the Kenn, Nailsea and Tickenham peat moors to the north, which were still flooded regularly and deeply throughout the remainder of the century.

[1] S.R.B/S.W. 4, *P.C.S.* 1801–12, 16 December 1811 and 30 June 1812.

[2] Rennie's plan has never been found, but all evidence points to the acceptance of his plan. See E.A. No. 139 (1819 and 1826).

[3] Acland, 129.

[4] Clark, 128.

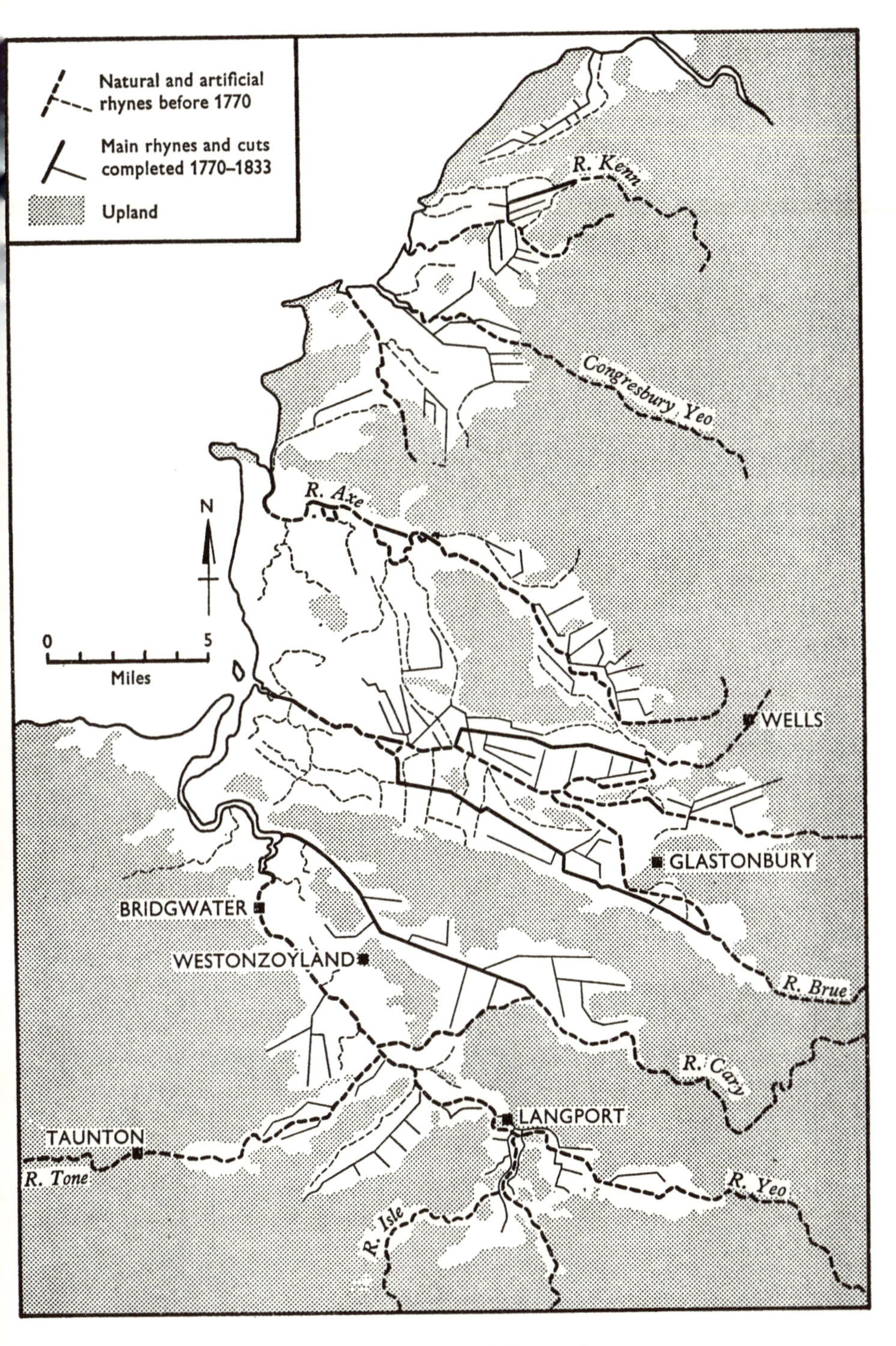

Figure 26. The principal rhynes, 1833.

Drainage conditions by 1830

To sum up, the overall position of drainage by about 1830 was disappointing and unsatisfactory. It would be true to say that the greater part of the Levels were drained in theory but not in fact, and that most areas still remained liable to severe inundation, although the new drains cleared the water off the land far more quickly than was the case formerly. The drainage system of the Brue valley was inadequate, that of King's Sedgemoor was only a little better, but some of the individual basins in the Southern Levels and portions of the Northern Levels were in a deplorable state. Only the Axe valley and the Congresbury moors, of all the peat areas, were fairly free from flooding. It was not only the actual inundation that caused trouble. Increasingly defective maintenance and administration were also beginning to sap the vitality out of the movement for better drainage. Through all this gloom, however, one fact shone forth: the blind antagonism to draining had disappeared and there was a growing realization that the problem had to be tackled seriously, even though the work so far had not produced the best results. After the stagnation of one and a half centuries, things had begun to happen. From this changed attitude had grown the only positive achievement of the age, the establishment of the pattern of drainage channels which was to be the basis of all subsequent activity (see Fig. 26).

6

THE CONSEQUENCES OF DRAINING

The distinctiveness of draining activity in the peat moors of the Levels for the half century between 1780 and 1830 was displayed in many ways. The most obvious feature was the extent and rate of activity compared with that of a century before, and, more significantly, with that of a century after. The pattern of major channels and rhynes, which for the first time intersected all the land of the Levels, is shown in Figure 26; they were a mark and a result of the enormous upsurge of activity in the draining of the region. But, besides these more obvious geographical and historical achievements, there were other elements in the landscape that marked this as a period of special note. There were the changes in the economy of the region with the extinction of the commonable wastes, the attempts to better the texture and fertility of the peat soils, the introduction of arable farming, and the improvement of the grassland pastures. There were also the cutting of the multiplicity of boundary rhynes that gave the fields their shape and size, the making of new roads and settlements, and finally the deliberate shaping of the visual scene with the planting of trees.

Although this was a period of rapid and far-reaching change, of deliberate creation rather than of slow haphazard formulation of processes and phenomena as in the past, which were the result of generations and even centuries of effort in fabricating the country in a piecemeal fashion—there was much in the new scene that had its origins in the old. Nor could it be otherwise, for it was the clay belt farmers, personified by Locke, who had pioneered the way in grassland improvement by manuring, by draining, and by their experiments with stock rearing and dairying during the middle years of the eighteenth century. When they turned their attention later to reclaiming the adjacent peat moors they transferred many of the features of the clay belt to these new lands. Therefore, even though there were some new problems to be faced in the peat moors, their solution lay in the familiar methods of improvement. The result was the creation of phenomena and processes, not markedly different in kind, but different in form.

THE AGRICULTURAL CONSEQUENCES

Perhaps the changes in the agricultural practices of the region were the most important geographical consequences of the reclamation and draining of the Levels during the late eighteenth and early nineteenth centuries. These changes occurred in a regional economy that had altered very little since the Middle Ages. With the exception of parts of the coastal clay belt, the farming emphasis was overwhelmingly pastoral in nature, with the typical marshland pursuits of fishing, fowling, reed-cutting, and peat-digging adding some variety. The piecemeal reclamation and upgrading of the land had merely intensified this pastoral predominance, and, as far as can be ascertained, this basic pattern did not change during the succeeding centuries. In 1698, Celia Fiennes passed through a 'large common' near Taunton, 'which on either hand leads a great waye goode riche land with ditches and willow trees, all for feeding cattle'.[1] Fifty years later, in 1747, Defoe found the Levels 'wholly imployed in the breeding and feeding of cattle', and of the four principal products of the county, one was the manufacture of woollen goods, and the other three were centred on the Levels; they were: '(1) Fat oxen...as large and as good as any in England. (2) Large Cheddar Cheese, the greatest and the best of the kind in England. (3) Colts in large numbers on the moors and sold unto the Northern Counties...'[2] One is left in little doubt that the situation had remained largely unchanged throughout the rest of the eighteenth and early nineteenth centuries. Locke's careful account of the rise of the clay belt graziers, Billingsley's map of the land use of the county in 1797 (Fig. 27),[3] the evidence of the 1801 agricultural returns,[4] together with the various agricultural commentaries of the period,[5] all support this view.

It was not long, however, before the changes in the draining of the

[1] *Journeys of Celia Fiennes*, ed. Morris (1949), 243.

[2] Defoe, *A tour thro' the whole island of Great Britain*, ed. Cole, 270 and 271.

[3] This map is based on one in Billingsley's book of 1794. See also G. East, 'Land utilization in England at the end of the eighteenth century', *Geog. Journ.* LXXXIX (1937), 167–72.

[4] P.R.O. H.O. 67/2, The Agricultural Returns for Somerset, 1801. As this was a survey of cultivated land, many of the Levels parishes did not make a return, and of those that did, the acreage of cropland was negligible. Comments like the following were typical; Lympsham, 'rich and adapted for pasture'; Mark, 'most of the land entirely calculated for grazing'; E. Brent, 'for the most part a Dairy parish'; and Uphill, 'chiefly meadow and pasture land'.

[5] The works of Billingsley and William Marshall in particular.

Levels led to the extinction of the commonable peat wastes, which brought about changes in some portions of the traditional economy of the region. In 1800, a commoner told Warner, with regret, that there was a time 'when these commons enabled the poor man to support his family and bring up his children. Here he could turn out his cow and pony, feed his flock of geese and keep his pig.'[1] This was

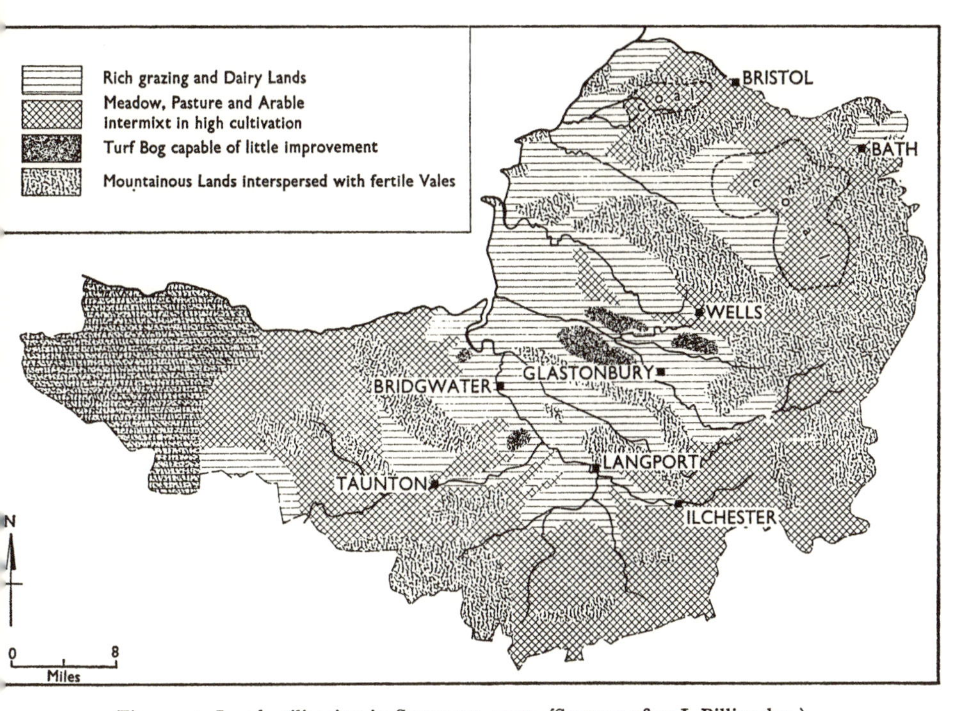

Figure 27. Land utilization in Somerset, 1797. (Source: after J. Billingsley.)

the commoner's view, though most agricultural writers could see little to commend in this pre-drainage situation. Billingsley, in particular, condemned it as wasteful because the cottagers could rarely afford to hire extra pasturage when floods occurred and they had to leave their cattle in either 'a stunted or starved condition'. Besides, there were the moral and social overtones of the commoning system, which jarred on the sensibilities of the 'improvers':

> The possession of a cow or two, with a hog, and a few geese, naturally exalts the peasant, in his own conception, above his brethren in the same rank of society. It inspires some degree of confidence in a property,

[1] Warner, *A Walk through some of the Western Counties of England*, 50.

inadequate to his support. In sauntering after his cattle, he acquires a habit of indolence. Quarter, half, and occasionally whole days are imperceptibly lost. Day-labour becomes disgusting; the aversion increases by indulgence; and at length the sale of a half-fed cow, or hog, furnishes the means of adding intemperance to idleness.[1]

In much the same vein, Locke commented on the effects of the enclosure and draining of Yarrow Moor, where the cottagers had previously been famous for their 'idleness, poverty and a species of sheep stealing that evaded the existing laws made against it', but after the draining, many of the cottages 'became farmhouses and now constitute a reputable village'.[2]

But the future was bleak for those commoners who could not establish their claims to the newly-reclaimed moors. Happily, there were alternative forms of employment, chief amongst these being the construction of the new drainage works which needed a great amount of labour. An indication of the numbers involved can be obtained from a calculation based on the total expense of excavation and wages paid in the King's Sedgemoor drainage scheme.[3] From these data a total of 54,400 'man weeks' can be deduced. Assuming that the work began in 1796 and ended in 1799 (4 years = 208 weeks), then an average labour force of about 260 men would have been needed to complete the work, and this did not include the ditching of the parochial allotments. Peat digging also increased in the raised bogs of the Brue valley as the demand for fuel grew. It seems, as indeed was forecast by Billingsley, that the draining of the Levels did not necessarily throw a lot of displaced commoners on parish relief, and this view is supported by the investigation of other sources.[4]

On the other hand, two of the other traditional occupations of the Levels, fishing and fowling, declined appreciably as the new drainage destroyed old breeding grounds. After the draining of part of Creech Moor it was said that 'the sportsmen of Taunton lament to this day that through perfect drainage of the moor, the favourite refuge of these birds [snipe] has been destroyed'.[5] By the mid-nineteenth

[1] Billingsley, 51 and 52. [2] Locke Survey, f. 526. [3] Billingsley, 57 and 196.

[4] *Ibid.* 174–5. For an investigation into this problem in Somerset, see J. F. Lawrence, 'Somerset, 1800–1835' (unpublished M.A. Thesis, University of Durham, 1938), 107. For similar conclusions elsewhere in England, see Sir John Clapham, *An Economic History of Modern Britain* (1926), I, 113–21, and J. D. Chambers, 'Enclosure and labour supply in the Industrial Revolution', *Econ. Hist. Rev.* v (1953), 322–4.

[5] J. Darby, 'An account of the reclamation and improvement of land in the Vale of Taunton Deane', *Journ. Bath and West Eng. Soc.* x (1862), 78.

century it was said that fowl shooting was 'nearly at an end', although when the moors were flooded in autumn and winter the flat-bottomed boats came out and fowling began again.[1] In an attempt to preserve a portion of the old economy, decoys were constructed to encourage the fowl. There were probably 14 decoys in the Levels, a county total which rivalled that of Yorkshire but fell well behind that for the Fen parts of Norfolk and Lincolnshire, with 26 and 39 respectively, and Essex with 29. In 1823 3 decoys were constructed in King's Sedgemoor by the Marquis of Bath, and between 1868 and 1882, when records of the catches were kept, an annual average of 1,200 duck, teal, and mallard were netted.[2] But although the yield was impressive, the number of decoys declined, and by the end of the century there were only 6 in existence; 3 in King's Sedgemoor, 2 in Sharpham Moor, and 1 in Cossington Moor. Many must have gone the way of one near Nyland, in the Axe valley, which was filled in and drained, and was said to have become 'good pasture grounds, supplying the milk for Cheddar Cheese'.[3]

Another, and quite peculiar, aspect of the old economy which also suffered from the draining, was the thousands of geese which were kept for their feathers. With the draining and consolidation of the peat moors and with the growth of better strains of grass, it was said that cows had 'superseded the feathered tribe and prove to be more profitable stock'.[4]

Methods of improving the peat moors

As parts of the old economy passed away with reclamation and draining, efforts were made to improve agricultural practices and increase the productivity of the moors. But it was found that draining alone did not ensure increased fertility of the peat soil, for it was acid and deficient in plant nutrients, as well as being friable and lacking in tenacity. It could barely support the poorest strains of grass, let alone be turned into meadowland or be cultivated. Therefore new methods were devised to improve the quality of the soil, namely, the paring

[1] Brit. Mus. Add. MSS. 33691, Sketch 35, under which is the following caption: 'Children at play in a flat bottomed boat only six feet in length near the Parrett River. These small boats are used for wild duck shooting in the winter floods'. See also E. Jeboult, *A General Account of West Somerset* (1873), 11.

[2] E. Harting, 'Wild fowl decoys in Somerset', *Downside Review*, v (1886), 218.

[3] Brit. Mus. Add. MSS. 33727, f. 660. There were at least two filled-in decoys near Cheddar, and one each at Compton Dundon, Aller, Meare Pool, and Kenn Moor.

[4] W. Phelps, *The History and Antiquities of Somersetshire*, (1839), Vol. II, Book I, 567.

and burning of the peat, the claying of the surface with the silt from the rivers and rhynes, as well as from marl pits further afield, and the warping of the moors with silt-laden water. Thus, there was a conscious effort to alter the structure and character of the soil.

Paring and burning. One method of improvement was to plough the peat in April and May when it was reasonably dry, then pare and burn the topsoil. The ashes were ploughed back into the ground and a mixture of buck wheat and Dutch clover was sown immediately, the clover being used for grazing in case the wheat failed to mature if the ground was too dry. Ideally, this treatment was followed by more intensive draining.[1] Paring and burning, however, tended to lower the surface of the peat and eventually made it more susceptible to flooding,[2] and the fine ashes were often blown away before they could be ploughed in. In any case, the benefit from burning was limited, for, as Billingsley commented, 'if followed up with successive corn crops, the strength of the land is so exhausted by the forced fertility that a rest of eight or ten years is necessary to prepare for its repetition'.[3] Used alone, this method of improvement was most successful in areas where the peat soil already had an admixture of clay, as in King's Sedgemoor, where the practice lasted for many years.

In the Brue valley, where the peat had no natural mixture of mineral constituents, paring and burning were not successful. The soil needed the addition of clay to bind it together and give it fertility.

Claying. The clay underneath the peat was anything from 7 to 25 ft down and consequently too deep to be turned up by the plough, as was done in the Fens, where, in any case, it became easier because of peat wastage. Clay and other soil could be obtained more easily from the surrounding uplands, which were not more than two miles from any point of the peat moors. Unfortunately, the poor condition of the droveways made the transporting of the clay a difficult operation.[4] Nevertheless, some experiments were made. In Glastonbury Heath, one farmer improved his lands by draining and claying until their value rose from 'one shilling per acre to thirty shillings'. His expenses

[1] Billingsley, 183–4.

[2] For Sturge's comments on this, see Acland, 129.

[3] Billingsley, 131–2.

[4] Brit. Mus. Add. MSS. 33727, f. 606. Of the Brue valley it was said 'the bogs were so soft and at times so flooded that neither wagons nor carts could have passed over the fields' (1832).

were £10 an acre for 100 to 150 cartloads of red earth laid on the peat, each cartload being brought a distance of nearly two miles. When the earth was spread over the peat the old sward was said to have been stifled naturally and new grass sprang up immediately.[1] Another farmer, in Sutton Mallett, set about removing the peat from fields in King's Sedgemoor and carting it onto the thin soil of the Polden Hills in the same parish. He then removed the upland soil by back carriage to the peat moor. When the two soils were mixed the ground produced 'a luxuriant coarse sour grass'. Locke was enthusiastic about it and said that if the whole parish could be transformed in this way and if the farmer prosecuted 'this hitherto neglected improvement with vigor... [then] a Nabob's Plum of twenty thousand pounds will be laudably secured without any East Indian crime or hazard'.[2]

The claying of the land in the Brue valley received a sharp set-back when it was realized that the clay taken from the lower Polden slopes and from other areas where the soil was derived from Lower Lias clays contained properties which caused 'souring' in milch cattle and young stock. Pastures situated on these soils contained excessive proportions of molybdenum, and they were locally known as 'teart' pastures. Although the significance of this relationship was not fully appreciated before 1855, soil was not being taken from the northern Polden slopes as readily as before; the red marls of the southern slopes of the hills were preferred,[3] although the expense of carting these marls over the hills and down into the Brue valley greatly limited their use.

A more successful method for the improvement of the peat lands was to use the silt which was deposited in the rhynes and rivers of the Levels.[4] Locke had been an early exponent of this practice, and among many examples of it was the improvement of 330 acres of peat bog to the north-west of Meare Pool between 1811 and 1842. The peat bog was first drained and allowed to consolidate; then the marsh plants disappeared, to be succeeded by a thick natural crop of fern. A canal, sixteen feet wide and with a lock, was cut from the Brue into the centre

[1] Billingsley, 184–6. With this particular example in mind, Billingsley advocated that claying accompany paring and burning where possible.

[2] Locke Survey, f. 436.

[3] Acland, 56. The close relationship between 'teart' areas and Lower Lias-derived clays was finally pointed out conclusively in 1855 by J. A. Clark, in 'On the scouring lands of Central Somerset', *Journ. Bath and West Eng. Soc.* III (1855), 52.

[4] For example, Billingsley, 186, for the improvement of seventy acres of peat bog at Glastonbury.

of the bog to enable barges to bring silt from the river. Once the silt was spread over the land the fern gave way to clover and other grasses. In its original state the land had been purchased for 18*s*. or less per acre, but it soon let for between 30*s*. and 40*s*. per acre per annum. The expense of the work was £12. 10*s*. an acre and some 33,000 tons of silt were spread on the peat.[1] So beneficial were the results that other farmers were encouraged to 'clay' their lands with silt, and the practice became widespread in the peat moors. It is interesting to note that the projected navigation and improvement schemes for the King's Sedgemoor Drain, in 1829 and 1854, provided for the transport of silt from the Parrett estuary into the peat moor.[2]

Warping. The superior quality of those pastures which were regularly flooded pointed the way to controlled flooding or warping as a method of improving the peat soils. Land in the Parrett, Tone, Isle, Yeo and Brue valleys was admirably suited to this practice because the water in these rivers was heavily charged with silt.[3] The water of the River Cary, however, had little suspended matter in it after passing over the clay area around Somerton, and was therefore of little value for warping. The River Axe was of relatively little use at all, since it issued from the Carboniferous Limestone of the Mendip Hills.

After 1780, warping became a more practical method of improving the peat than it had ever been because the water could be led over the moors by the network of drainage channels that had recently come into existence, and because the desirability and requirements of summer penning for irrigation had also become more widely recognized. Moreover, warping was less costly than carting the silt long distances from the rivers and into the moors. Yet, in spite of its many advantages, warping did not become as widespread during the early years of reclamation as was commonly expected, because it depended upon the minute control of water-levels based upon a sufficient number of inlet and outlet sluices, and it was difficult to get the co-operation of the multitude of small proprietors in the Levels in this

[1] E. Galton, 'An account of improvement of a shaking bog at Meare in Somersetshire', *Journ. Roy. Agric. Soc. Eng.* VI (1845), 182–7. See also S.R.B./*P.C.S.* Wells, 30 November 1830, for the effects of this scheme on the drainage of the surrounding moors. Similar results from other experiments in the Levels had been noted earlier by T. Davis, 'On the management of marshlands', *Letters and Papers, Bath and West Eng. Soc.* X (1805), 324–7.

[2] See pp. 219–20 below.

[3] For examples of larger-scale schemes in Glastonbury and Burtle Moor, see Clark, 119–20, and Billingsley, 186.

matter. Furthermore, the autumn and winter floods were often too violent to be controlled, and swept across the Levels irrespective of rhynes and sluices. It would have been an advantage if warping had become widespread for it would have aided the deposition of silt in the fields rather than in the river channels, thereby helping to relieve the pressure of flood waters and avoiding the unnecessary breaking of banks.

One development in the warping of the peat lands was particularly significant. Peat-cutters were allowed to dig over the surface of the peat, leave the land in 'bends' with gutters in between, and then flood the cut-out pits with thick water. It was said, that 'as soon as a thin layer of silt is deposited the natural grasses spring up and go on improving with every subsequent flood'. Land thus treated became worth from 40*s.* to 50*s.* per acre within a few years.[1] This method of improving the peat was considerably cheaper than any other, especially as the income from letting out the plot to the turf diggers had already been received. This practice was carried out increasingly during the nineteenth century, and land so reclaimed was selling for £70 an acre in 1873.[2] Unfortunately, however, this improvement was only creating a problem for the future, for the turbary was cut down 7 or 8 ft and sometimes as much as 14 and 15 ft; although it 'welled-up' a little afterwards, the general level of many thousands of acres of the Brue valley was depressed by anything from 3 to 5 ft, making it more susceptible to flooding.

With the introduction of the steam pump there was a possibility of a more minute regulation of water-levels; and the creation of Internal Drainage Districts also meant that landowners tended to co-operate more readily to ensure the success of controlled flooding. Thus warping became a regular practice in the late nineteenth century. This was particularly true of the moorland basins of the Southern Levels which lent themselves readily to limited flooding. Notices of the intended flooding of the moors were served almost annually.[3]

The introduction of arable farming

The attempt to improve the peat moors, after their initial draining, had been undertaken with the intention of either upgrading the

[1] Acland, 56. For other earlier examples see Billingsley, 186.

[2] J. Darby, 'The farming of Somerset', *Journ. Bath and West. Eng. Soc.* v, 155–6.

[3] For examples of controlled flooding in Westonzoyland Moor, Curry Moor, and King's Sedgemoor, see S.R.B./W.S. 1; C.M. 1 and 2; and K.S. 1.

pasture land or preparing the newly-drained ground for arable farming. But cultivation was eventually doomed to failure because of uncertain drainage conditions and defective husbandry. Billingsley thought that the 'egregious slovens' of the marsh were making too much money, too easily, from the fattening of cattle to worry about the cultivation of the moors; 'where nature is most bountiful her gifts are the least prized', he observed.[1]

Nevertheless, local experiments in cultivation were made in more favourable areas during the initial years of draining, and good to above average returns were to be had from the newly-drained and broken-up soil of the Levels for those few who made the effort.[2] Some of the yields were exceptional; for example, on the clay lands of Mark parish ten successive crops or wheat yielding about 50 bushels per acre, without manure or fallow, were known; in one place nineteen successive crops had been grown without any obvious deterioration by 1797, and four years later the crop was said to be 'as prolific as was never known before'.[3] Most of the new cultivation, however, was confined to King's Sedgemoor where good crops (for example 40–5 bushels of wheat and 80 bushels of oats per acre) were taken after the first few years of drainage. But such bounty could not last for ever; constant ploughing and burning tended to reduce the level of the moor, which became so 'wet and soft as to be incapable of cultivation and remained in a neglected state for years'.[4] By 1830, the greater part of the moor was entirely exhausted by continual cropping and in a deplorable state. 'The chickweed', wrote Acland, 'considered comparatively harmless, is here quite a plague, rising up to eight inches thick in a fortnight...There seems to be no remedy but the constant use of horse hoe and...a better system of cropping.' Constant and uninterrupted cropping was the most serious defect of the farming technique. Changes had already been made in the cropping technique in an attempt to improve conditions. The land was sown with clover, but this crop could not be grazed by cattle because of the waterlogged state of the ground; instead it was mown continually, and thus nothing in the form of manure was returned to the soil. Attempts to improve the fertility of the soil by claying were restricted to the edges

[1] Billingsley, 187 and 188.

[2] For example, of new cultivation in the Levels at Locking and Yatton in 1801, see P.R.O., H.O. 67/2.

[3] Billingsley, 187 and 177, and P.R.O., H.O. 67/2.

[4] Acland, 131.

of the moor owing to the inadequacy and softness of the roads for carting; elsewhere the spreading of ditch scourings over the peat became a prevalent practice. A partial answer to the agronomic problem lay in a suitable rotation, and swedes, rape, turnips, mangolds and potatoes were cultivated alternately with cereal crops and 'fed-off' by sheep, which were light enough to stand on the boggy land. The roots had to be cultivated by spade because horses and ploughs sank into the peat. Nevertheless, as late as 1850, the old system of paring and burning and of taking several corn crops in succession, relieved only by an occasional fallow, was still stubbornly pursued in the cultivated parts of the moor.[1] This system did not end until about 1850 when the upland farmers complained that, although they could grow 'as much straw on Sedgemoor as ever, there was a deficiency in the yield of grain to what it used to be'.[2] In places on the other moors alongside the River Parrett the soil was similarly pared and burnt, and rape or some other green crop was sown, after which wheat and oats were taken alternately for five or six years. This rotation was repeated monotonously.[3]

The peat soils of the Brue valley were more difficult to cultivate, and only those areas that were clayed or warped received attention. Because of the already considerable expenditure entailed in these processes, crops were less likely to be jeopardized by bad farming.[4] The more sterile heath centres of the raised bogs were not cultivated but deliberately planted with firs, birch, alder and other acid-resisting plants.[5]

Perhaps the most successful crops in the newly-reclaimed moors were roots and withies. The roots were well adapted to the more favourable clay grounds, and, by 1850, their use was rivalling the hitherto unchallenged value of the pastures of the coastal grazing lands; it was thought that 'the degenerate days of the mangold-wurzel and swede turnip have made great changes in the marsh aristocracy'.[6] Withies were not introduced on a large scale until 1825 when large portions of West Sedgemoor were planted. Later, West Moor and Curry Moor became the main centres of cultivation, and the withy

[1] Acland, 56-7.

[2] Darby, 'The farming of Somerset', *Journ. Bath and West Eng. Soc.* v, 145.

[3] Acland, 57.

[4] For an example of successful arable cultivation in the Brue valley, see Acland, 56–7, and Locke Survey, f. 101.

[5] Phelps, 'On the formation of peat bogs...', *S.A.N.H.S.* iv, 104.

[6] Acland, 52. See also pp. 126–7 above.

remained strongly localized in the Southern Levels.[1] The plant was admirably adapted to ground that flooded regularly as it was planted in autumn and survived inundation during the winter.

The comparative ease of pastoral farming in the Levels put cultivation at a disadvantage, but it is also certain that the limited area cultivated, and the unwillingness to take pains over arable husbandry, were largely due to the indifferent success of the attempts to improve the drainage. Whereas the new rhynes evacuated the water more quickly than before and helped to dry out the land, floods were still uncontrolled and swept over the Levels with regularity between November and March of nearly every year. Floods were widespread and frequent. In October 1799, Marshall saw barley being harvested from boats in King's Sedgemoor.[2] Farmers knew that 'sometimes extraordinary crops are grown—sometimes the whole is lost by floods', and that they took the 'risk of having the whole of their crops destroyed by a high flood in July'.[3] In 1860, it was reported to the Commissioners of Sewers that 'the late rainy season has told with melancholy effect on farmers in the district [i.e. the Parrett valley]. We have seen corn cut and uncut standing in water, we have seen probably two-thirds, certainly half, of the hay crop entirely destroyed as fodder and fit only for litter'.[4]

In addition, it was asserted that defective drainage was coupled, from 1810 onwards, with what amounted to a deterioration of the position because extensive pipe draining in the uplands had increased enormously and had accelerated the run-off into the main streams. This is something which cannot be measured, and, indeed, the effect of pipe draining in increasing flooding is not proved by what evidence there is, but the stress that was laid on it as a cause of deterioration during the nineteenth century leaves one in little doubt that there was at least some cause for complaint.[5]

[1] For more specific details of its introduction into the Levels, see C. V. Dawe and J. E. Blundel, *An Economic Survey of the Somerset Willow Growing Industry*, Univ. of Bristol, Bull. No. 9 (1932).

[2] Marshall, *Rural Economy of the West of England*, 271.

[3] Acland, 57 and 47.

[4] A. Whitehead, *The Improvement of the Somerset Level* (1860).

[5] Clark, 109; Acland, footnotes on 47 and 130; and Grantham, *Report on the Floods in Somersetshire, 1872–73*, 5. The most modern information which tends to contradict the nineteenth-century assertions is contained in the *Report of the Land Drainage Legislative Sub-Committee of the Central Advisory Water Committee*, 8, and Appendix IV, 79–85 (H.M.S.O. 1951). For an earlier investigation and discussion which hinted at these contrary views, see J. B. Denton, 'On the discharge from under-ground drainage and its effect on the

Added to the unreliability of the drainage system were the uncertainties of the weather; lack of sunshine and the greater frequency of wet weather had always made cereal cultivation difficult in the west of England, and it was not surprising that these uncertainties had tended towards the exploitation of the moors so that as much as possible was taken out of them, without any investment to ensure their future fertility. Whereas Billingsley, Marshall, Acland and Sturge had all expressed enthusiasm and optimism on the prospect of cultivating the moors, Clark, in 1853, expressed serious doubts about their value as arable land. By 1873, J. Darby felt that either the moors should be returned to permanent pasture (as was happening at the time) or that the drainage should be radically improved if arable farming was to succeed. National trends towards a conversion from arable to pastoral farming strengthened the former view. Indeed, by that time (1873) Darby observed that only 'a few fields may still be seen that are under cultivation'.[1] Everything seemed to point towards the same conclusion that the physical environment of the region was not suitable for arable farming. The experiment in cultivation was at an end. It had been limited in extent and temporary in character.

Improvements in the grasslands

The experiments with moorland improvement and with cultivation had resulted in only a thin, patchy veneer of economic change, and the Levels, both during and after the great surge of drainage activity, remained almost entirely covered with grass. It was the grassland that benefited most from the new drains; the quality of the grass improved, and the amount of land for grazing was extended, both in area, when the new ground was reclaimed, and in the length of the grazing season, with the quicker evacuation of water in the periodically flooded lands. The value of the moors rose, and Locke, who was a keen observer, noted the following increases in the rent per acre of some of his own grounds:

Quality of land	1755	1796
Best	20*s.*—25*s.*	60*s.*—70*s.*
Medium	12*s.* 6*d.*—17*s.* 6*d.*	45*s.*—55*s.*
Poorest	2*s.* 6*d.*—10*s.*	30*s.*—40*s.*

arterial channels and outfalls of the country', *Minutes of Proc. Instit. Civil Engineers*, XXI (1861–2), 48–81, and the discussion afterwards, 82–130.

[1] Darby, 'The farming of Somerset', *Journ. Bath and West Eng. Soc.* V, 154 and 147.

One farm, which let for 10*s*. per acre in 1747, added an average of 5*s*. to its value every seven years until in 1803 it had become worth £250 per annum, instead of £50 as in 1747.[1] These figures do not wholly reflect the rise in values due to improvement alone and must include the general rise in prices that occurred during the period. Nevertheless, Locke gives many other examples which support his calculations of waste land in the moors which was improved by draining. For example: 900 acres of Blackford Moor were reclaimed and became worth 70*s*. an acre, and Lyng Moor likewise advanced to 60*s*. per acre; King's Sedgemoor, 'a boggy flat of insignificant value', rose from 2*s*. 6*d*. per acre to 34*s*.; and 3,200 acres in Godney Moor rose from 2*s*. 6*d*. to over 60*s*. per acre with draining. Again, there was Church Allerton Moor, which was 'now let at three pounds per acre, but in its flooded, uninclosed state was of little or no value'; and 1,000 acres of peat bog in Meare became worth 40*s*. 6*d*. per acre with reclamation.[2]

The question naturally arises, how did these values compare with those on the uplands during the same period? In the parishes of Chilton, Cossington, Puriton, and Woolavington, along the Polden Hills, and Loxton in the Northern Levels, poor arable land varied from 4*s*. to 10*s*. per acre. Normally, dry portions on the moorland edge were worth 40*s*. to 50*s*. per acre, but, if flooded, fell to 30*s*. Newly-drained lands in these parishes, however, could command from 60*s*. to 80*s*. per acre and, it was stated, even 100*s*. in some cases. At Wedmore 'island' flood-free arable upland was worth 20*s*. per acre, but the newly-drained moor immediately below was worth double this amount.[3] One may suspect that Locke's figures were somewhat exaggerated in order to make the contrast more marked, for he was never modest in his claims; however, Arthur Young, writing in the same year as Locke, found newly-drained pasture lands in the vicinity of King's Sedgemoor letting for between 40*s*. and 60*s*. an acre, which seems to prove Locke's statements.[4] Billingsley's

[1] Locke, 'Historical account of the marsh-lands of the County of Somerset', *Letters and Papers, Bath and West Eng. Soc.* VIII, 64–5.

[2] Locke Survey, ff. 74, 270, 386, 210, 426 and 300 respectively. Locke also noted improvements at Glastonbury, f. 209; Kennard Moor, f. 262; Mark, ff. 289–90; Middle Meads, f. 303; Street, f. 43; and the possibility of reclamation in Edington, f. 187; and Drayton, f. 178.

[3] *Ibid.* ff. 127, 154, 367, 517, 274–5 and 473.

[4] A. Young, 'A farming tour of the South and West of England', *Annals of Agriculture*, XXX (1798), 314.

figures for King's Sedgemoor show a rise from 5*s*. to 30*s*. with draining, and these figures almost agree with Locke's.[1] Again, in 1798, Billingsley noted that 3,000 acres of peat at Wedmore were 'previously rendered unproductive by inundation and the consequences, six or seven months of the year...which land is now let with liberal allowance of profit to the occupier from 30 to 50 shillings per acre'.[2]

It is unfortunate that there are no tax returns or assessments which would enable one to compare rent per acre throughout the Levels with that in the surrounding uplands between the period before drainage and enclosure in the late eighteenth century and the period after drainage and enclosure in the early nineteenth century. But it is possible to arrive at some idea of relative values in 1815 by using the assessment to the Property Tax of that year,[3] which provides a check on some of the statements of Locke, Billingsley and others (Fig. 28). The parish totals for this assessment are the total of the annual rent for the lands, buildings and other forms of real property, including mineral workings, within each parish. Therefore precautions must be taken in its use; urban parishes and parishes with large mineral workings must be excluded; but in the agricultural parishes, the part of the assessment concerned with buildings can be assumed to cancel out.[4]

Looking at the country as a whole, we find that the average rent of over 30*s*. per acre was amongst the highest in Britain and reflected the prosperity of farming at this time, this average rental, or more, being found only in Middlesex, Lancashire, Worcester, Warwick, Leicester and Northamptonshire. But the county average obscures the local variations. Giving evidence on Somerset farming before the Royal Commission on Agriculture many years later, Mr W. C. Little said, 'Somerset is a county which has every variety; it is almost an epitome of England',[5] and this view is plainly reflected in parochial rents shown in Figure 28, which have a distribution that is complex and varied in the extreme. But if one concentrates on the

[1] Billingsley, 198.

[2] J. Billingsley, 'On the uselessness of commons to the poor', *Annals of Agriculture*, XXXI (1798), 32.

[3] *B.P.P.* XI (1830–1), 394–402.

[4] For a discussion on the use of the Property Tax, see D. B. Grigg, 'Changing regional values during the Agricultural Revolution in South Lincolnshire', *Trans. Instit. Brit. Geog.* XXX (1962), 91–103.

[5] 'Report on Devon, Cornwall, Dorset, and Somerset', Royal Commission on Agriculture, *B.P.P.* XIV (1882), 423.

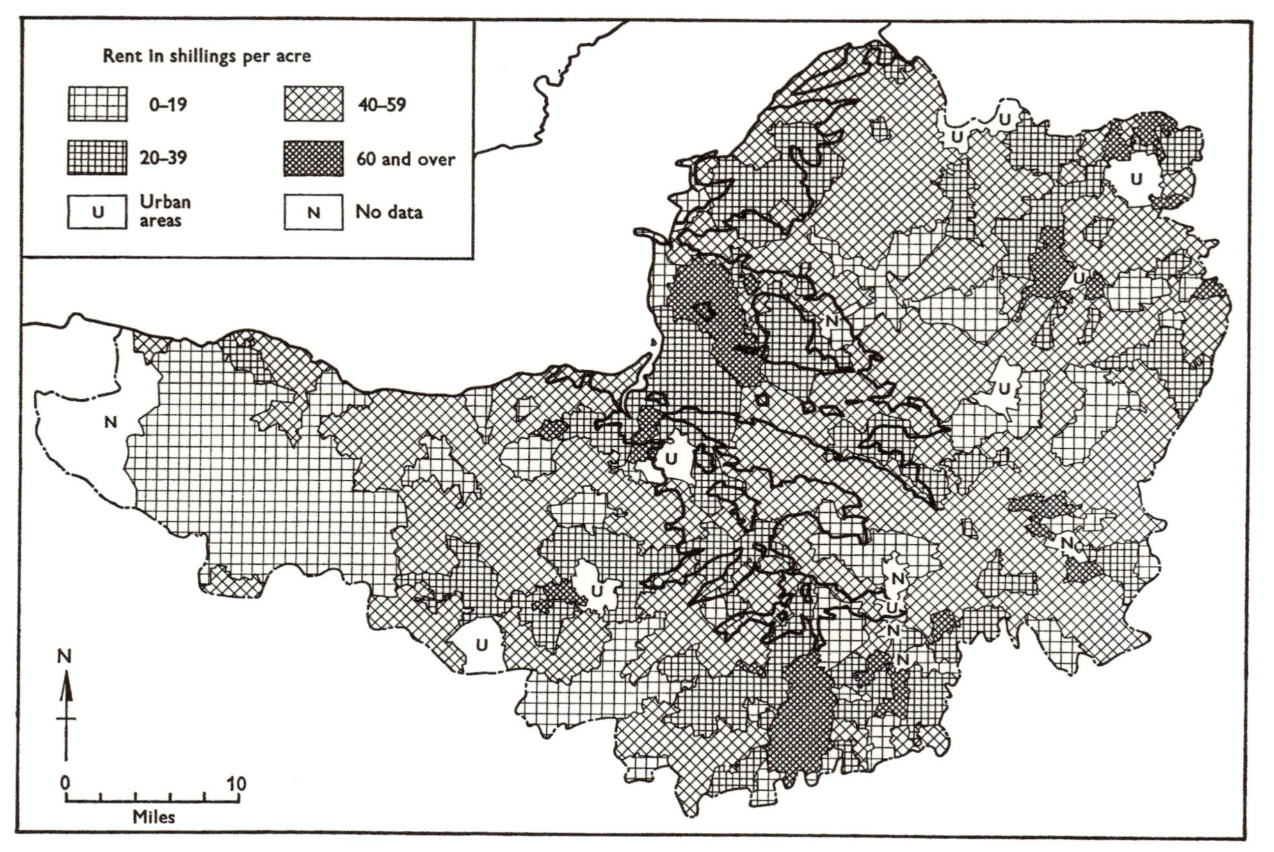

Figure 28. Rents in Somerset, 1815.

area of immediate interest, it is apparent that the coastal clay belt had rents that were over 50*s.* per acre (E. Brent, Mark and Brent Knoll all being over 70*s.* per acre), rents which were equalled only in the mixed farming zone between South Petherton and Crewkerne, on the southern border of the county. Even where the rental on the clay belt dropped to below 50*s.* per acre, as in the Northern Levels and around Bridgwater, it was still at the high level of over 40*s.*, which was equivalent to the rental of the newly-drained inland moors of the northern side of the Brue valley, of the Tone valley, West Sedgemoor, and parts of the upper Parrett valley. In the Levels, only the southern side of the Brue valley, the eastern end of King's Sedgemoor, and the Gordano valley had rentals below 30*s.* per acre, and this was as much a reflection of the inclusion within these areas of parts of the low-rental upland zones of High Ham and the Polden Hills as of the poor drainage of the moors. With this evidence there is every reason to believe Locke's assertions that drainage had caused a marked increase in the value of the peat moors.

With the extension and improvement of the moorland pastures, the regional differences in pastoral activity in the clay and alluvial lands of the Levels were carried into adjacent portions of the peat moors. Most of the Axe valley and parts of the north-eastern end of the Brue valley became a predominantly dairying and cheese area after the example of Cheddar. The remainder of the Brue valley was largely given over to grazing and fattening because of the influence of the clay belt graziers of Huntspill and Burnham. The Southern Levels and King's Sedgemoor were also mainly summer fattening grounds, although in some areas there was a definite emphasis on hay, which was cut and sold to upland farms.

The extension, through draining, of dairying and summer fattening strengthened the traditional aspects of the economy of the Levels, particularly the fattening of cattle, and the lowlands became even more important as the centre of a complex arrangement that extended throughout the west of England for the rearing, fattening, and marketing of stock. The Levels' grazier had a marked preference, almost to the exclusion of all other breeds, for the red Devon and Somerset cattle. These he bought in the fairs of northern and eastern Devon mainly during early spring, and in other fairs in the remainder of Somerset at intervals throughout the rest of the year (Fig. 29). His second stock were the black Welsh stores, particularly bullocks, which were transported by ship across the Bristol Channel to such places as Uphill, Bridgwater and the north Devonshire ports of Bideford, Barnstaple, Ilfracombe and Lynton. Other cattle were driven across the lower Severn on the Beachly-Aust banks at low tide and depastured in the Northern Levels. The cattle were fattened on

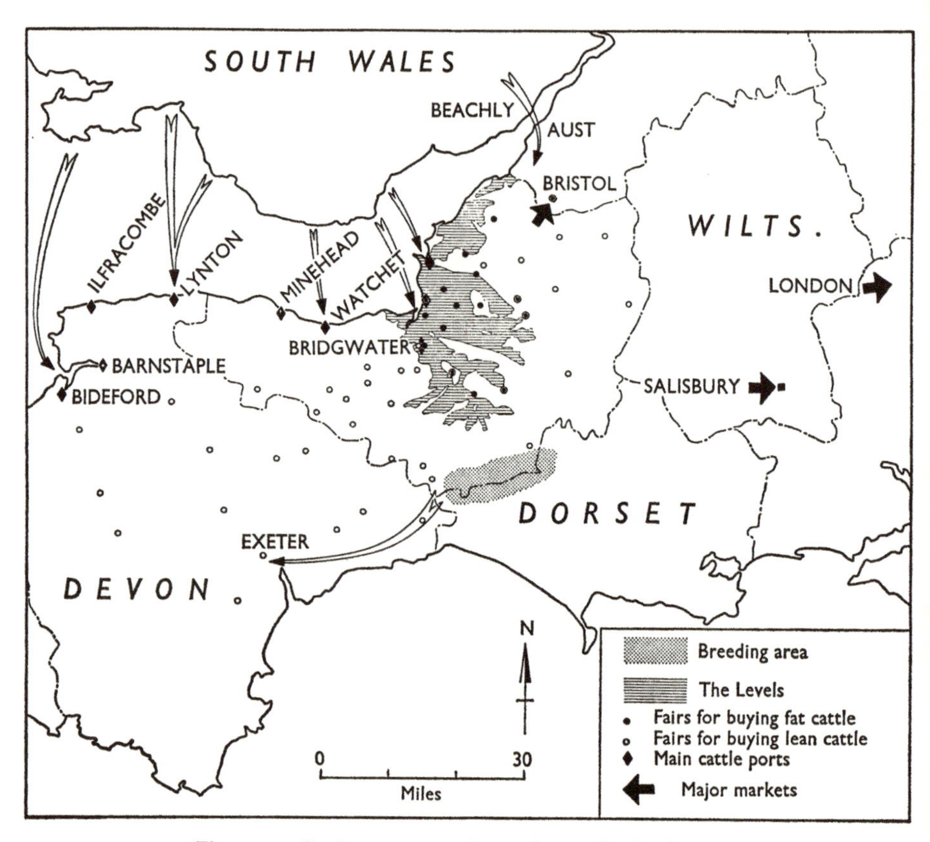

Figure 29. Cattle movements in south-west England, *c.* 1800.

the clay belt pastures and the newly-drained moors, and sold in the fairs of the Levels, at least one fair being held every month. They were then driven to the three main markets of Bristol, Salisbury and Smithfield, the journey to Smithfield taking nine days.

The supply and rearing of young weaning calves for the Devon breeders, particularly those near Exeter, was localized in the upper Axe valley astride the south Somerset-north Dorset county border, although this was not the only source of young cattle, which were also bought regularly in the Stolford and Sherbourne fairs.[1] It was a system that rivalled that of the Leicestershire graziers in its complexity and extent, and it was the basis of the economy of the Levels in the early nineteenth century.

THE CHANGING COUNTRYSIDE

But the effects of draining did not end with changes in the economy, agricultural practices and land improvement; they were particularly manifest in the landscape itself, in the pattern of rhynes and fields, the new roads, and the increased density of settlement.

Rhynes and fields

Apart from the more important drainage channels each field was surrounded by rhynes. These were the finest and ultimate part of the drainage network, but they were also boundary ditches. Open ditches were preferred to hedges because they were quicker and easier to establish. In addition, there was a strong prejudice on the part of the graziers to hedges and trees, as they thought that the hedges harboured flies which teased the cattle and hindered the progress of fattening, and that the trees disturbed the free circulation of the air. But an even more telling argument for rhynes was that they were cheaper than any other form of boundary. The price of fencing a rope (20 ft) at the end of the eighteenth century, in Somerset, was described as being, for mortar walls, 11*s*. 6*d*., for dry walls, 8*s*. 3*d*., for quickset hedges, 4*s*. 9*d*. and for rhynes, 2*s*. or less. Added advantages of the rhynes were that water for summer drinking could be held in them by penning, and that the network of channels provided a means for the minute control of flooding, for pest control and for warping the land, provided, of course, that co-operation between neighbouring landholders was first achieved. Thus cheapness, coupled with a duality of function, left little room for alternative forms of division, and a dense network of rhynes was soon to cover the moors.[2]

This pattern of rhynes and ditches controlled the size of the fields.

[1] Billingsley, 238–58; Young, *A Farmer's Tour through the East of England*, IV, 7; *idem*, 'A farming tour of the South and West of England', *Annals of Agriculture*, XXX; and Warner, *A Walk through some of the Western Counties of England*.

[2] Billingsley, 175–6, 78–83, and 57. See also J. Billingsley, 'On the best method of inclosing, dividing, and cultivating waste lands', *Letters and Papers, Bath and West Eng. Soc.* XI (1809), 30. Most of the boundary rhynes were 8 ft wide at the top, 3½ ft wide at the bottom, and 5 ft deep.

On the older settled lands of the clay belt there was a preference for large fields, about 20 acres or more, and in Pawlett Hams and in the Huntspill Moors many fields were 40 acres in size (Fig. 30B).

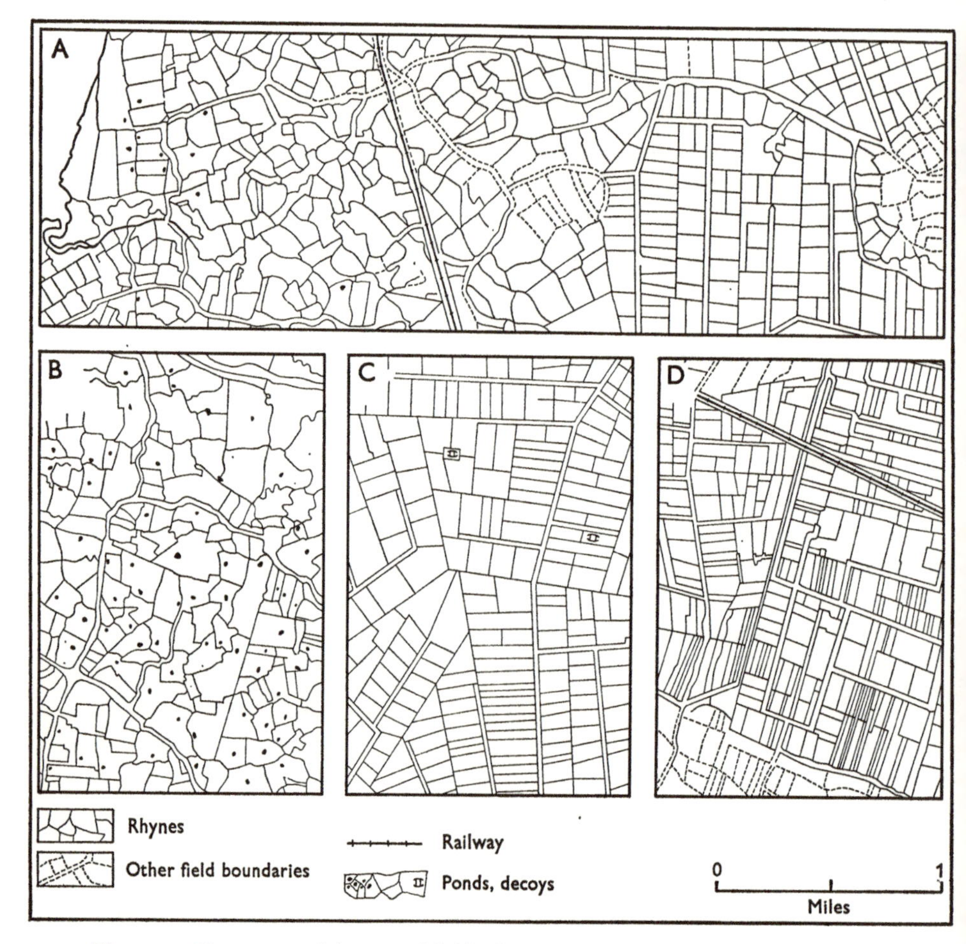

Figure 30. The pattern of rhynes and fields: A—a section across the Northern Levels from near Kingston Seymour to Kenn and Nailsea moors; B—on the coastal clay belt, in Berrow parish; C—the eastern end of King's Sedgemoor, about half a mile south of Walton; D—the southern Brue valley, Catcott and Edington heaths.

Graziers felt that a larger field would fatten stock sooner than two small fields half the size. Whatever the truth of this, and there were some strong contrary views both in Somerset and in other grazing districts of Britain,[1] the fact remains that the fields were generally

[1] See W. G. Hoskins, *The Making of the English Landscape* (1955), 145.

larger in these older settled districts than in the areas drained after 1770. The same was true of the older reclamations around the edges of the moors and alongside the main river channels; for example, on the fringes of the Brue valley, near Meare Pool, and on the moors near Glastonbury. There were two exceptions to this; in the Northern Levels the majority of fields were about 10 acres or less, and few were greater than 15 acres, and in the area of medieval reclamation between Sowy 'island' and the Parrett river, plots of 3, 4 and 5 acres were general. This may be a reflection of the piecemeal reclamation in small blocks and of the multiplicity of rights in medieval times.

In contrast to the fields of the older settled areas, the new fields of the eighteenth- and nineteenth-century enclosures and reclamations were not as large, the average field rarely exceeding 15 acres, and most commonly being 10 acres but often as little as 5 acres. The small size of these fields resulted from the limited area of common grazing that was left to be divided between the austre holders who were able to establish successfully their claims before the enclosure commissioners. In the case of King's Sedgemoor and the later reclamations of West Sedgemoor and West Moor, the number of claims was rigorously, even harshly, controlled and limited, and, as a result, the field sizes were larger, the average being between 10 and 15 acres with some as large as 20 or 25 acres (Fig. 30C); but in the Brue valley and the Northern Levels the fields were smaller. In addition to the division of the commons into fields there was the complication of turbary rights, which boosted enormously the number of legitimate claims to a part of the moors. Minute, elongated plots, stretching into the peat bog centres, are a distinctive facet of the field and drainage patterns of these areas (Fig. 30D).

There were also striking differences in the shapes of the fields in the new and of those in the old reclaimed areas of the Levels. The determining factor in the clay belt alignments was the disposition of the natural drainage channels in the clay, which tended to be winding and circuitous in nature. Linking divisions between these natural channels were straight; nevertheless, the general result was to produce irregularly shaped fields (Fig. 30A). In the newly-reclaimed areas, however, divisions were uncompromisingly rectilinear and straight, giving rise to a regular and monotonous scene. R. Warner, who travelled through the Brue valley in 1800, commented on the new landscape, 'no variety can occur in one uniform level, no intimacy in a ground divided into regular quadrilateral inclosures and moors intersected by rectilinear canals'.[1] Despite local variations in size and shape, however, the over-riding impression was one of uniformity throughout the Levels.

Roads

As with the rhynes and field boundaries so with the roads; an irregular pattern of tracks indicates older settlement, and straight lines and right angles readily identify the newly-reclaimed areas. These differences show up clearly in Figure 31, which depicts the pattern of roads in the central part of the Levels in 1782 with the additions up to 1822.[2] All in all, there were few roads in the Levels before 1782, with the exception of the clay belt where a distinctive radial pattern of roads emanated from Brent Knoll and dominated the otherwise irregular pattern of the coastal lands. The clay belt was linked in the north and south to the peripheral uplands, but there were limited contacts inland across the seasonally flooded moors. The only way to overcome the obstacle of the marsh was to construct causeways which linked the islands. For example, Wells was linked to the coast by a causeway across the Panborough-Bleadney gap to Wedmore island and from there by another causeway from Mark to Watchfield. Glastonbury was similarly linked to Taunton, by Street causeway to the Polden Hills, by Greylake's Fosse to Sowy island, by the Burrow Wall and Tone embankment to Athelney island, and by the Baltmoor Wall to Lyng peninsula. Meare and Westhay islands were joined to Glastonbury by a track along the top of the River Brue embankment. Although often of very great antiquity and very impressive, these roads were merely cross-country roads linking distant towns and had little to do with the utilization of the moors.

The new pattern of roads was more dense and distinctive. The roads were often built on the refuse thrown up from two new rhynes dug on either side, and they struck out across the moors in straight lines. Usually one main track penetrated either the centre, or at least the driest part, of the new parochial allotment in the commons, and the

[1] *A Walk through some of the Western Counties of England*, 59.

[2] Based on Day and Master's map 'The County of Somerset' (1782) and W. Greenwood's 'Map of the County of Somerset' (1822). One very interesting detail of the Day and Master's map that has not been shown is the type of boundary, if any, that ran alongside the roads. The boundaries depict areas of enclosure very clearly.

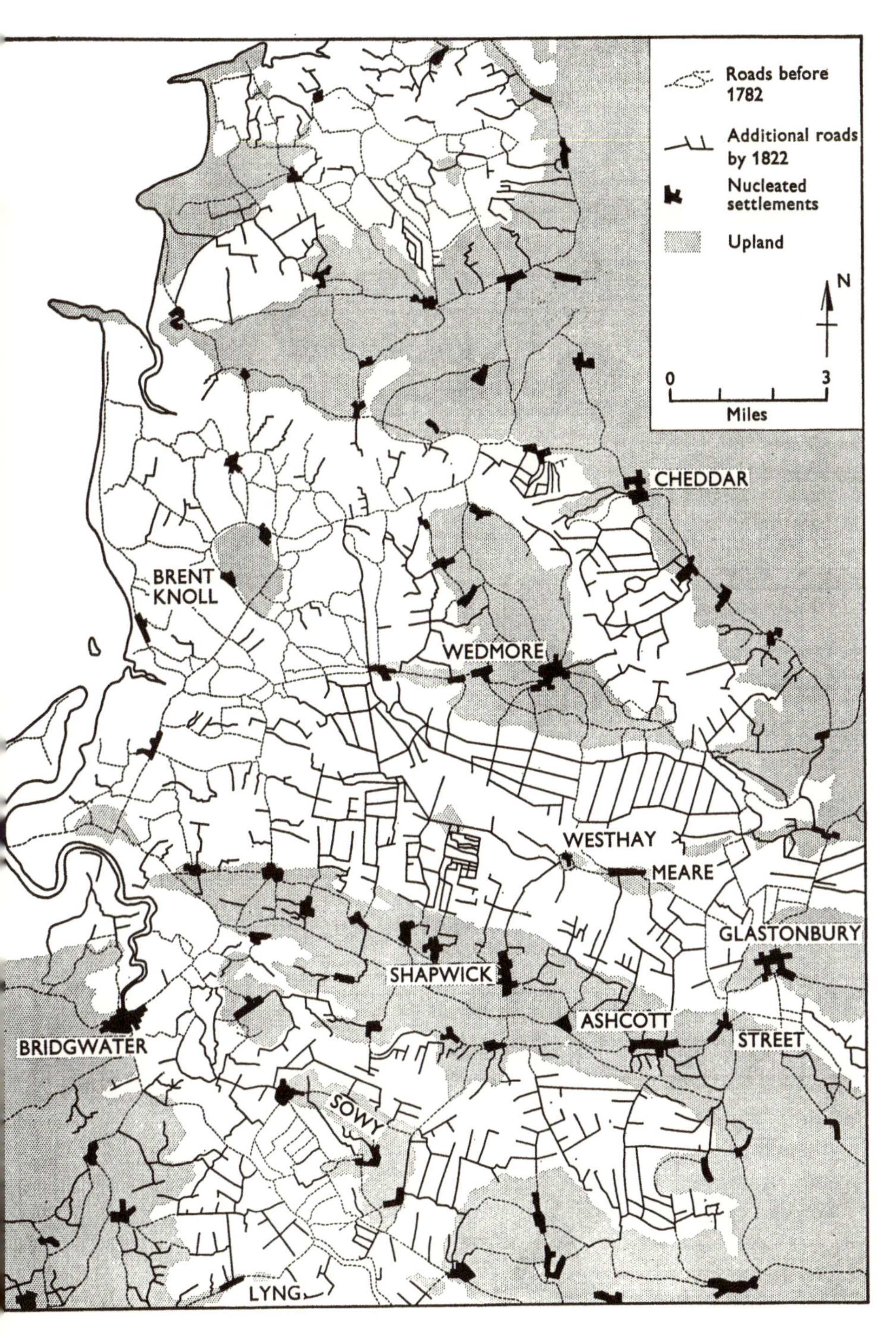

Figure 31. Roads in the Levels: 1782–1822. Only the main roads in the uplands, prior to 1782, are shown, but all those in the Levels are shown. (Sources: based on Day and Master's map 'The County of Somerset' (1782) and W. Greenwood's 'Map of the County of Somerset' (1822).)

minor roads branched off at regular intervals towards the parish boundaries or the worst-drained areas, rarely linking up with the roads of adjoining parishes. This stickleback-like pattern of minor roads jutting out into the moors is particularly evident in the southern part of the Brue valley and in the Axe valley, and similar isolated 'no through' roads penetrate the once common lands in the Northern Levels. It is significant that most of these roads were, and still are, called droveways and were merely elongated strips of grassland used for moving stock; only a few were surfaced and metalled. However dense this new pattern was, compared with that of the period before reclamation took place, the fact remains that the number of roads was not sufficient to provide direct access to every field, a problem that is still with the Levels farmer of today. The paucity of roads was largely due to the costs involved in enclosure. In comparison with an upland enclosure, a lowland enclosure usually entailed greater administrative and legal costs, both in procuring an Act of Parliament and in producing an Award. Surveyors' costs were also greater owing to the multiplicity of claims to be worked out, and there was the added expense of bridges, culverts and sluices. Despite the much cheaper divisions by rhynes rather than by hedges, the overall cost of the lowland enclosure was greater than its upland counterpart, and economies were obtained only by reducing the length of the roads.[1]

In contrast to the minor access roads in the moors, some of the main roads were well surfaced, particularly when they became turnpikes at the end of the eighteenth century; for example, the whole length of the Bristol–Bridgwater road on the clay belt. The turnpike era also brought about the creation of new links in the Levels, such as the north-south roads across the Brue valley from Wedmore, through Westhay to Shapwick, and from Meare to Ashcott. With the creation of a new network of roads, wheeled carriages and carts became more common. The old market cross and a stone conduit in Glastonbury, the situation of which had been 'inconvenient since the introduction of wheeled carriages in the marsh district', were demolished and removed.[2] But the most important change of all was the way in which these new lines of communication broke down the isolation of the 'island' communities, just as had happened fifty years before to the

[1] Billingsley, 57 and 58.

[2] Phelps, *History and Antiquities of Somersetshire* (1839), Vol. II, Book I, 496–7 and 560.

coastal villages.[1] Better access meant not only better returns in farming, but better facilities and amenities.[2]

Settlement

'This country', wrote Billingsley about the Levels, 'has been heretofore much neglected, being destitute of gentlemen's houses, probably on account of the stagnant waters and unwholesome air.'[3] But it was not only the noticeable lack of large country houses and estates that differentiated the settlement pattern of the Levels from that of the surrounding uplands, but a lack of any settlements at all. The recurrent flooding in autumn and winter put house-building on the moors out of the question, and the constant references to the unhealthiness of the damp miasmal situations on the 'islands', and on the upland edges leave one in no doubt as to its detrimental effect. Billingsley averred that 'agues and low fevers from the humidity of the air, impregnated with the exhalations from the stagnant waters of the marsh, prevail very generally during the vernal and autumnal seasons', and were only subdued by summer drought and winter frosts; and that he had 'observed' the 'happy effects' of reclamation in the Brue valley on the health and comfort of the adjoining villages.[4]

Whatever the truth of these assertions, and there is much evidence to support them, the absence of settlement in the Levels is clearly displayed in Day and Master's map of 1782 (Fig. 32). The plot of the number of 'farms and cottages' on this map can be checked against the estimates of the number in Locke's 'Survey' (*c.* 1789) and in Collinson's *History and Antiquities of the County of Somerset* (1791). For some settlements Collinson's figures are a little higher than those on the map, but on the whole they show a remarkable degree of conformity. The situation in 1822, after which date most of the reclamation had taken place, is shown in Figure 33, and is based on Greenwood's map. Everywhere there appears to be a greater density

[1] One example of this was in Meare where Phelps noted that 'fifty years ago the only approach to the village was by a road passable only for a horse from Glastonbury, the other sides being open moors and bogs, impossible except in the midst of summer. By drainage they have become more consolidated, and under the enclosures which have taken place, roadways were marked out to communicate with Wedmore on the north, and Ashwick and Shapwick on the south...', *History and Antiquities of Somersetshire* (1839), Vol. II, Book I, 566–7.

[2] *Idem, Observations on the Great Marshes and Turbaries of the County of Somerset*, 20.

[3] Billingsley, 167.

[4] Billingsley, 54–5; Locke Survey, f. 130; and Phelps, *History and Antiquities of Somersetshire* (1836), Vol. I, Book II, 49.

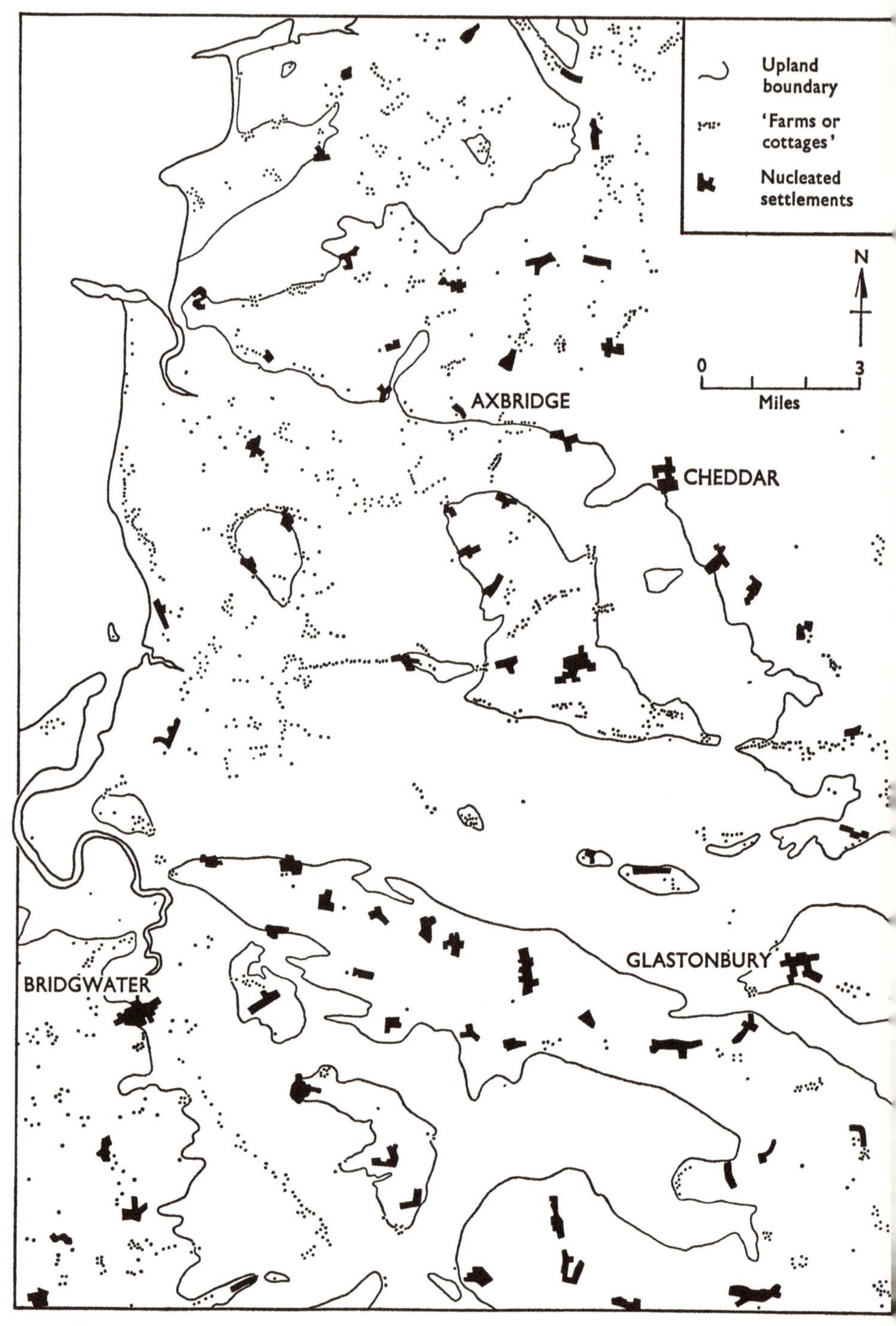

Figure 32. Settlements in the Levels, 1782. (Source: based on Day and Master's map 'The County of Somerset' (1782).)

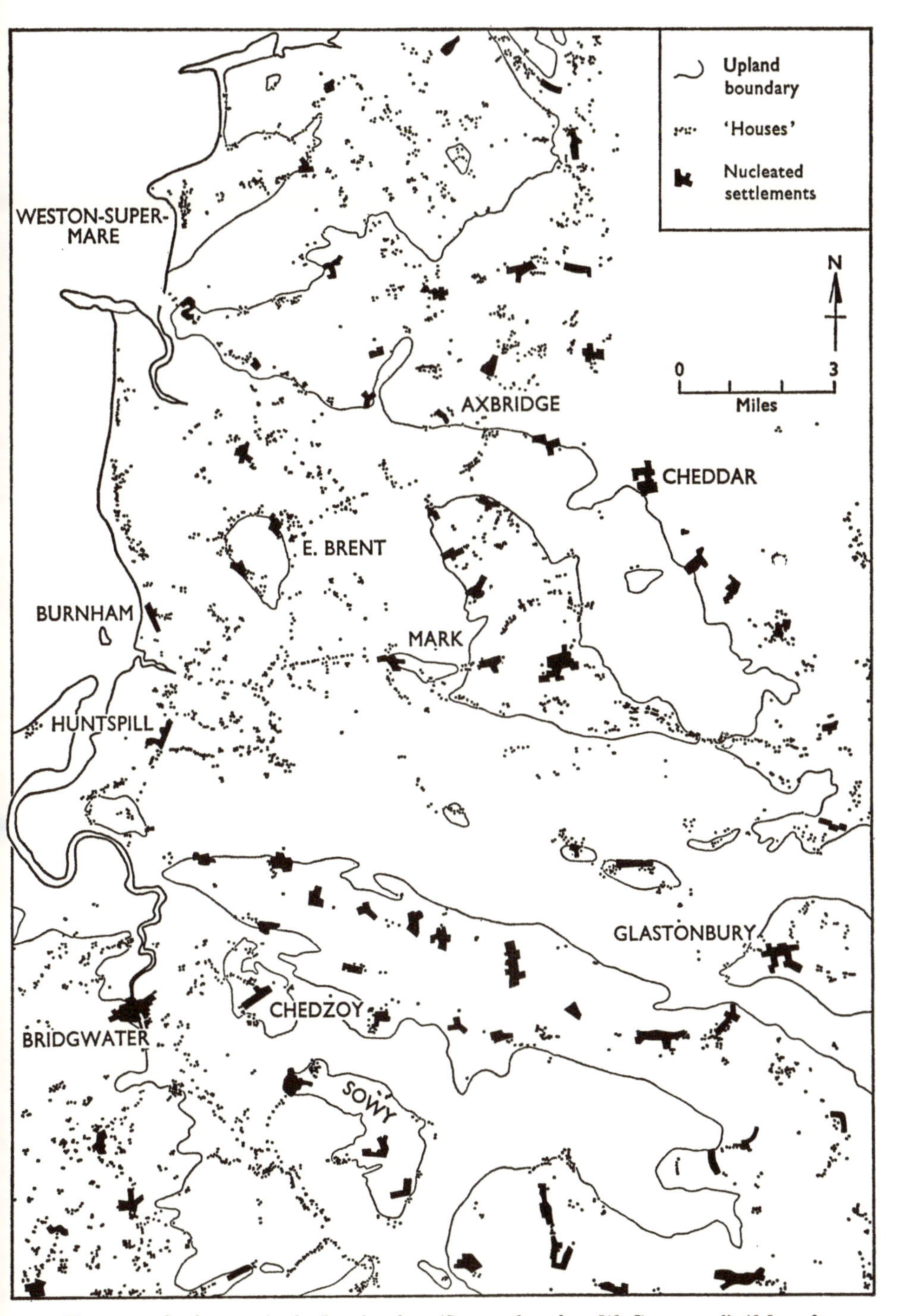

Figure 33. Settlements in the Levels, 1822. (Source: based on W. Greenwood's 'Map of the County of Somerset' (1822).)

of settlement which is so widespread and general as to defy detailed comment. This greater density may be no more than a reflection of the different techniques of compilation of information and cartographic representation employed in the two source maps. But particularly noteworthy was the dispersal of houses to new areas of settlement, north and east of Huntspill, south of Mark 'island' in Mark Moor, in the north Brue valley moors, on the northern edge of Sowy 'island', in North Moor, in the moors south and west of Chedzoy 'island', along Lake Wall, and particularly along the Parrett banks from Bridgwater to the Tone confluence. In only a few cases did settlement move into the middle of the moors; because of the recurrent threat of floods it clung mostly to the flood-free edges of the uplands and 'islands', which were barely more than two miles away in any direction. This tendency for the farms to be in the uplands was still common much later; in 1851 Acland commented that 'nothing could be worse than the arrangement of farms' in King's Sedgemoor, though some new ones were then being built on its edges.[1] Documentary evidence of the changes between 1782 and 1822 is very sparse. Locke noted that 14 new houses were built in East Brent after 100 acres of land were enclosed, and that 10 houses were built on 17 acres of land enclosed in Burnham, 'with the expectation of many more'.[2] But a glance at the maps tells the story better than any words can.

[1] Acland, 56.

[2] Locke Survey, ff. 81 and 103.

7
THE LOST YEARS

In order to achieve the control of flooding in the Levels, the drainage authorities, after about 1830, had to adopt a threefold plan of action. First, they had to maintain the existing system efficiently; secondly, they had to extend and improve the existing, but sometimes imperfectly conceived and constructed, drainage schemes; and thirdly, they had to plan and implement new works, particularly in the main rivers, that would bring about a better control of flooding. Yet these aims were never fully realized because of local administrative and financial difficulties. These difficulties form a unifying theme through the record of draining in the later nineteenth and early twentieth centuries, and, until they were solved successfully, work on the scale needed for true flood control in the Levels could never be carried out. In this way, the administrative and financial problems were as important as the physical and technical problems of draining, and they were certainly no less difficult to overcome. As a consequence, the remaining years of the nineteenth century were lost years for draining. The maintenance of existing works was neglected, few new works were constructed, and the drainage of some areas of the Levels deteriorated.

The ever-pressing administrative and financial problems certainly loomed large in the story of the draining of the Levels in the nineteenth and early twentieth centuries, but their importance should not obscure the physical and technical difficulties that also had to be faced. Insufficient emphasis had been laid on the need for maintaining the main river channels and their tidal outfalls. The need was not new; it had long been recognized,[1] but little had been done, except to confine troublesome sections of the rivers with massive walls and embankments, the construction of which always seemed easier than to widen existing channels, or to excavate new ones. Drainage work during the seventeenth and eighteenth centuries had barely taken into account the state of the main rivers, and the failure of the moorland

[1] See pp. 53 and 113 above.

reclamations of 1770–1830 to achieve perfect drainage, finally demonstrated that draining was being tackled the wrong way round. The Axe, Brue and Congresbury comprehensive Drainage Acts were a concession to this view, but their achievement was temporary and limited, and the high levels of the rivers during flood time still remained.

Some of the problems, it must be admitted, were only imperfectly understood, particularly those relating to the River Parrett. Deposition of sediment reduced the capacity of the channel to pass off water quickly, but it was not known whether the sediment was of marine or freshwater origin, and the remedial works necessary were therefore uncertain and open to much dispute, particularly the question of building a clyse. In addition to the immense natural difficulties and the varying opinions on how to overcome them, there was a conflict of interests, the interests of navigation at Bridgwater, the interests of the 'Bath' brick manufacturers,[1] and the conflict between the coastal clay belt farmers and those of the inland peat moors. The developing technical equipment of the nineteenth century enabled some changes to be made; steam pumping offered a temporary respite to the inland moors but solved nothing, for the high level of the main rivers during periods of prolonged rainfall and high tides still remained.

ADMINISTRATIVE AND FINANCIAL PROBLEMS

The administrative and financial difficulties arose largely from the inefficiency of the Court of Sewers which was the administrative body for the drainage. All new works completed under the Enclosure and comprehensive Drainage Acts had been handed over to the court for maintenance. A thorough examination of the proceedings of the Commissioners of the Court throughout the nineteenth century reveals that the needs and problems of the drainage system had long outgrown their powers, which had been acquired gradually since the Middle Ages. They could do little more than authorize the repair and maintenance of existing works, and levy rates to cover the cost. New works could not be constructed readily without recourse to Parliament, and without the prior consent of the landowners representing three-quarters of the value of the land in question. To achieve this latter requirement was almost impossible. Another cause

[1] See p. 216 below, and n. 3.

of ineffectiveness was that the commissioners consisted mainly of landowners or people of property, in fact, those most likely to be reluctant to undertake new works for fear of rating themselves.

All the commissioners could do was to appoint juries of landowners to 'view' and supervise the work of maintenance, and to report offenders to the court. But, if Acland's lengthy account of the procedure of the court is true, their methods of dealing with offenders had become little more than a farce:

The defaulter demurs to the right of the demand on his labour, or tries to shift it on to his neighbour. If the court quashes his plea, after a certain amount of vain struggling, he is summarily convicted and called upon to submit himself to the mercy of the court. If he is humble, he is at once fined 2*s.* 6*d.*, or some other sum, according to the will of the court. If he recalcitrates, he is threatened with the awful consequences of summoning a jury of the county (the legal process authorized by the Act), which are as mysterious as the unknown results contemplated in the event of the Speaker of the House of Commons putting on his hat. In fact the legal jury is hardly ever summoned, and the man is joked or coaxed into paying his fine. Like other ancient institutions, the Court of Sewers is said (by those who feel no inconvenience from it) to work well, and always ends with a good dinner.[1]

Indeed, their procedure had become cumbersome, their threats innocuous, and their administration tardy and rather rusty from disuse.

It is true that the multiplicity of liabilities, numbering literally thousands, which had come into being with the reclamation of the moorland areas and the completion of the comprehensive drainage works, between 1770 and 1810, made supervision more difficult. In these circumstances it is not surprising that many defaulters escaped penalty and that the court was inefficient. Moreover, the system of rating was not good, each maintenance liability was rated individually and confusion was commonplace. In addition to the inherent difficulties of the Sewer Court were the worsening economic conditions after the end of the Napoleonic Wars. The effects of these conditions can be seen in the Brue valley, where the commissioners decided to collect only one-third of the current drainage rate in 1817 and in 1829; the position became so complicated that in 1841 they decided to waive all debts and start afresh.[2] Supervision and maintenance became so bad that it was stated in the Wells Divisional Court that 'doubts, confusion and uncertainty have induced many commissioners

[1] Acland, 45–6. [2] S.R.B./*P.C.S.* Wells, 16 December 1829.

to withdraw themselves from the Court', and it was suggested that the sewers laws be thoroughly revised and improved, prior to bringing a bill before Parliament for their amendment.[1] Even Phelps, who was a conscientious member of the court, had to admit that 'many irregularities had crept into their practices after the lapse of years, and inconveniences arise from the want of further powers to carry into effect improvements and order new works'.[2] As a consequence, the court, which should have met at intervals of six months or less, sometimes did not meet for years, and maintenance suffered accordingly.[3]

Eventually, an Act was passed in 1833 giving the commissioners power to construct and maintain new works, but these were confined to maritime works and works connected with navigable rivers, a restriction that excluded many important rhynes.[4] Added to this limitation was the requirement, once more, that the prior consent of the owners of three-quarters of the land affected was needed for any new works. As a result, the concession was rendered almost ineffective and no improvements were made. The commissioners could also purchase land for the maintenance and improvement of existing works but not for new works, despite the power they had to carry out the new works.

A factor which undoubtedly contributed towards the inefficiency of the commissioners was the anomalous basis of their territorial jurisdiction, which was geographically unsound. The places in which the individual juries met are depicted in Figure 34, in which the limits of the jurisdiction of the various divisions of the commissioners are also shown.[5] Except for the boundaries along the watersheds of the Polden and Mendip Hills, these administrative limits did not conform to the physical limits of the five major drainage areas of the Levels.[6] For instance, a part of the western end of the Axbridge Division drained into the Wells and Glastonbury Division, and a portion of the coastal clay belt at Huntspill, which naturally drained towards the River

[1] S.R.B./*P.C.S.* Wells, 13 April 1831.

[2] Phelps, *History and Antiquities of Somersetshire* (1836), Vol. I, Book II, 56.

[3] For example, the Wells Court did not sit from the beginning of 1839 to 1843, and the Langport Court from June 1854 to January 1856.

[4] 3 and 4 William IV, c. 22. See also Report of Royal Commission on Land Drainage, *B.P.P.* x (1927), (Cmd. 2993), 1057 f. for a review of this legislation.

[5] Information for this map is culled from the Proceedings of the Divisions of the Court of Sewers, deposited in the S.R.B.

[6] These are the Axe valley, the Brue valley, the Cary valley and King's Sedgemoor, the Southern Levels, and the Northern Levels.

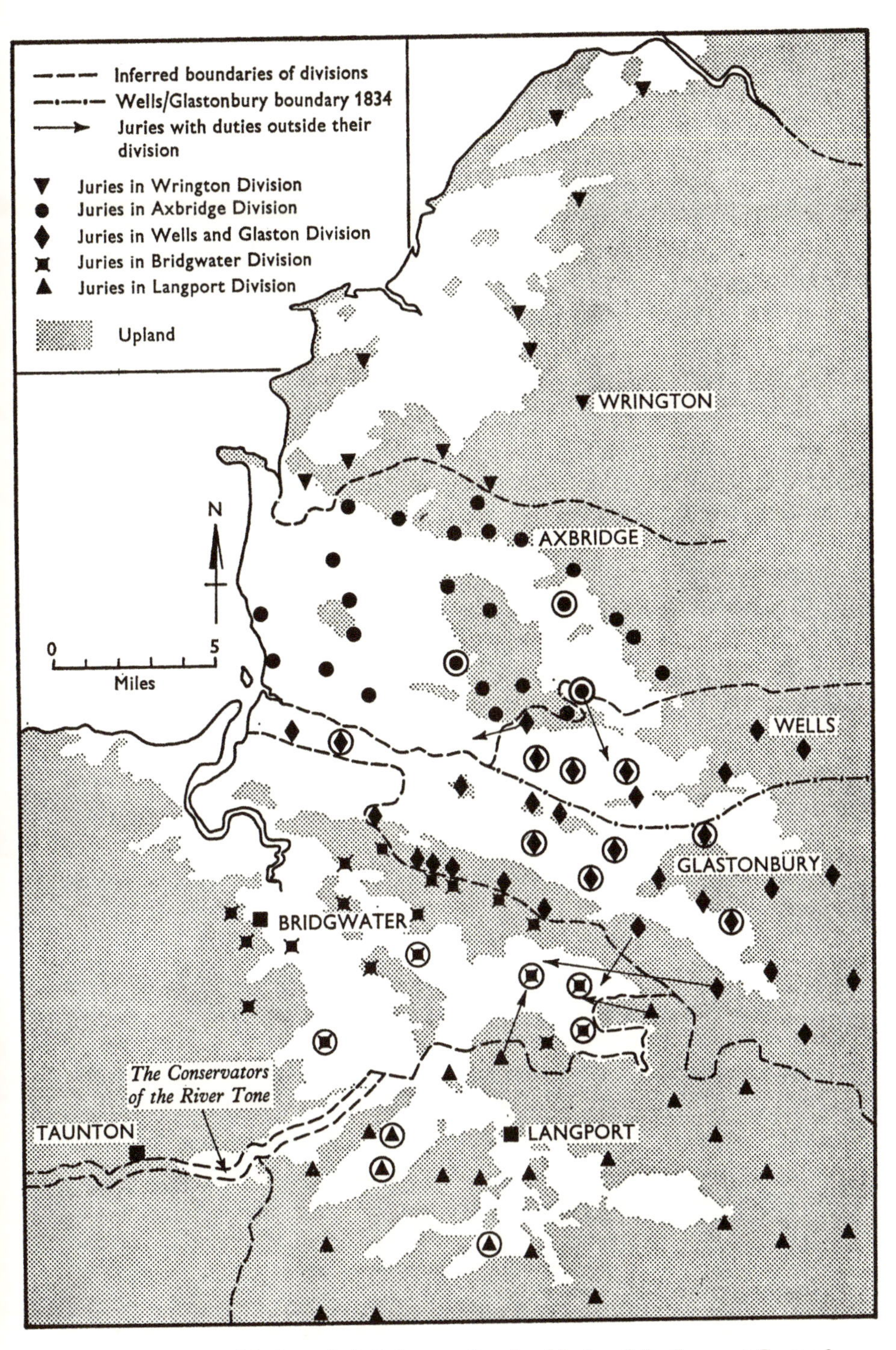

Figure 34. Areas of drainage jurisdiction: to show local juries of the Somerset Court of Sewers in existence from 1789 to the mid-nineteenth century. Encircled symbols show juries created 1770–1830. (Source: S.R.B./*P.C.S.*)

Brue, was included within the jurisdiction of the Bridgwater Division. The separation of the Langport and Bridgwater Divisions was illogical and detrimental to good administration, because the headwaters of the Parrett and Cary were severed from their outfalls. Consequently, neither King's Sedgemoor nor the Cary valley was an administrative entity as a drainage area, nor was the River Parrett. The new juries created to administer maintenance in the newly-reclaimed moors did nothing to amend these anomalies, for they were created within the framework of the old divisions. The administrative position was further aggravated in 1834 when, in an effort to increase their efficiency in the Midland (Brue) Division, the commissioners agreed to divide responsibility between two courts, one at Wells as before, and the new one at Glastonbury with responsibility for the River Brue and areas to the south. Although the commissioners from one court did occasionally attend the proceedings of the other the division resulted in a dire lack of communication over common drainage problems.[1] Some areas were not supervised at all. The River Tone, for example, was outside the jurisdiction of the Sewer Courts and was maintained by the Conservators of the River Tone, whose interest in the maintenance of the river was purely navigational. Even this interest slowly languished after the construction of the Bridgwater–Taunton Canal in 1825, and the condition of the river deteriorated.[2] Thus many of the limits of jurisdiction cut across the natural divisions of the drainage, and, as co-operation between the various authorities was almost non-existent, the drainage suffered.[3]

The limited financial basis on which the court had to operate during the early nineteenth century was equally unsatisfactory, as can be gauged from Figure 35. In this map the number of acres in a parish which were considered to receive benefit from drainage works (and consequently were assessed for drainage rates) is expressed as a percentage of the total area of the parish. The rated lands were then known as Sewerable Lands.[4] It can be seen that large areas within the

[1] S.R.B./*P.C.S.* Wells, 8 August 1827 for preliminary moves and pleas, and 18 August 1834 for details of the subdivision of duties.

[2] J. Savage, *The History of Taunton* (1822), 383–400.

[3] In 1834, in an effort to increase efficiency, the Midland Division (Wells and Glastonbury) abandoned the jury system and appointed dyke-reeves instead, but with very little effect on improving the administration.

[4] The returns of Sewerable Lands are to be found in the Proceedings of the Commissioners of Sewers, and also in separate schedules in S.R.B. Record Room, South Case 4.

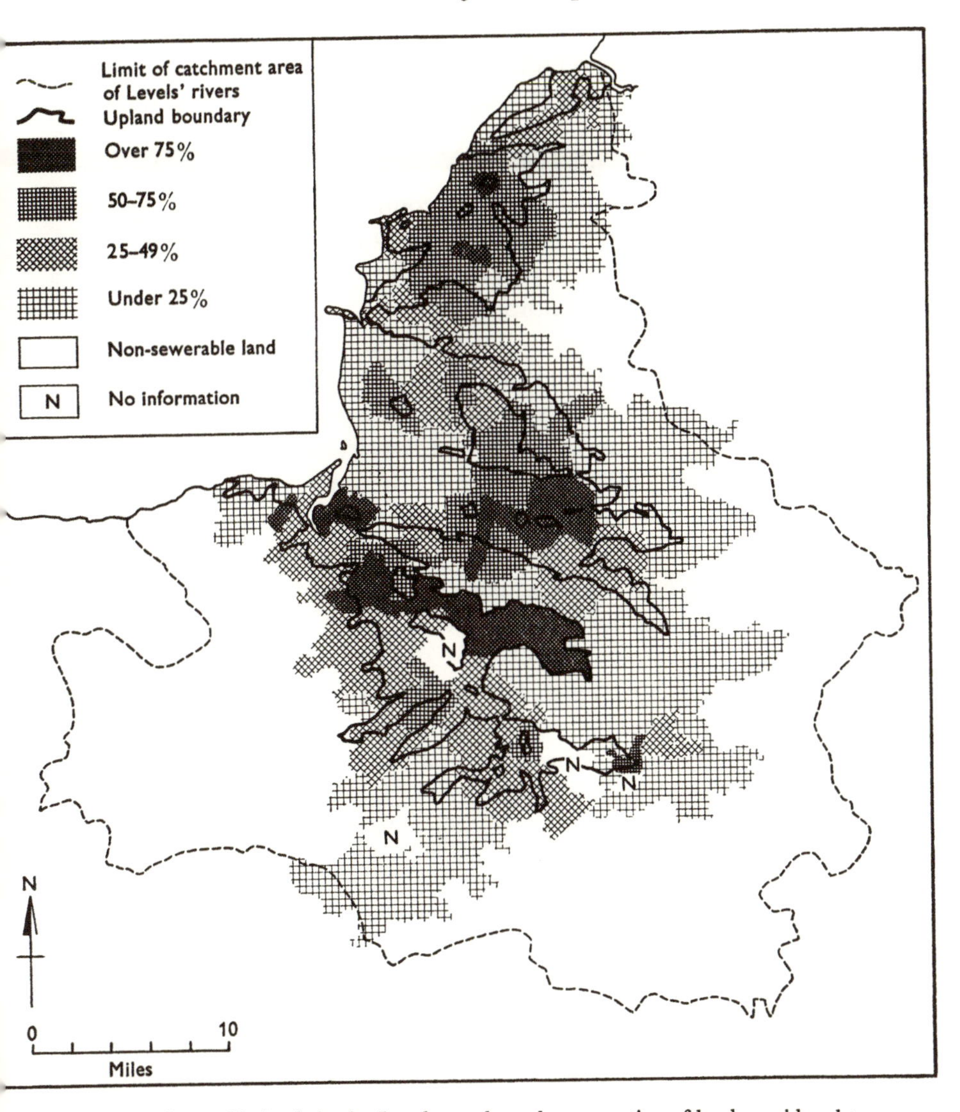

Figure 35. Sewerable lands in the Levels: to show the proportion of land considered to avoid damage or to derive benefit from the maintenance of drainage works during the early nineteenth century, and so rated. Calculated on the basis of parishes.

catchments of the main rivers contributed nothing towards the upkeep of drainage works from which they undoubtedly benefited, and in most of the upland parishes as little as 10% of the land, or even less, was rated. The situation was better in the inland moors, although only in King's Sedgemoor and parts of the eastern end of the

Brue valley was the assessment realistic. The absurd position of many of the lowland parishes is indicated by the extreme example of Lympsham, where only 124 out of 1,940 acres, or about 6·5 % of the land, were considered to derive benefit from the sea walls and therefore obliged to contribute towards their upkeep, when in fact, the whole of the parish and the surrounding country was protected by these walls. Coastal lowland in the Northern Levels was more highly rated and the overall position in this region was far more realistic than in the areas south of the Mendip Hills.

Clearly, there were two well-defined objectives towards which all subsequent drainage legislation had to move; first, the establishment of one authority over each or all catchment areas, with adequate powers to supervise and co-ordinate the drainage activity and construct new works; secondly, a more widespread and uniform basis of rating over the whole of a catchment area to raise money for improvements. Until the river catchment was regarded as the basic hydrographical unit for all activity there could be no hope of success in draining and reclaiming the Levels.

The Land Drainage Act of 1861 did little to meet these two requirements. It did, however, among other things, provide for: (*a*) the creation of Internal Drainage Districts outside the jurisdiction of the Court of Sewers, with authority to manage their own drainage affairs, make new drainage works, and rate the land within their boundary; and (*b*), a revision of the laws affecting the Commissioners of Sewers, giving them greater powers of rating and maintenance.[1] But by the very wording of its preamble the Act excluded control of both major and secondary catchment areas.[2] Consequently, the boundaries of the new Internal Drainage Districts were ill-adapted to the needs of draining, and an even greater administrative division than before came into existence. (Compare Fig. 34 with Fig. 36.) More serious was the fact that the Internal Districts had no specific obligation to repair the main channels that passed through their area, and the Commissioners of Sewers had no power to rate the districts to maintain these channels, which were left to deteriorate.[3]

[1] For further details, see R. B. Grantham, 'The Land Drainage Act, 1861...', *Journ. Bath and West Eng. Soc.* XIII (1865), 119–32.

[2] Internal Drainage Boards could only be established for 'any bog, moor or other area of land that requires a combined system of drainage, warping or irrigation'.

[3] For details, see R. B. Grantham, 'Arterial drainage and outfalls', *Journ. Bath and West Eng. Soc.* XV (1867), 252.

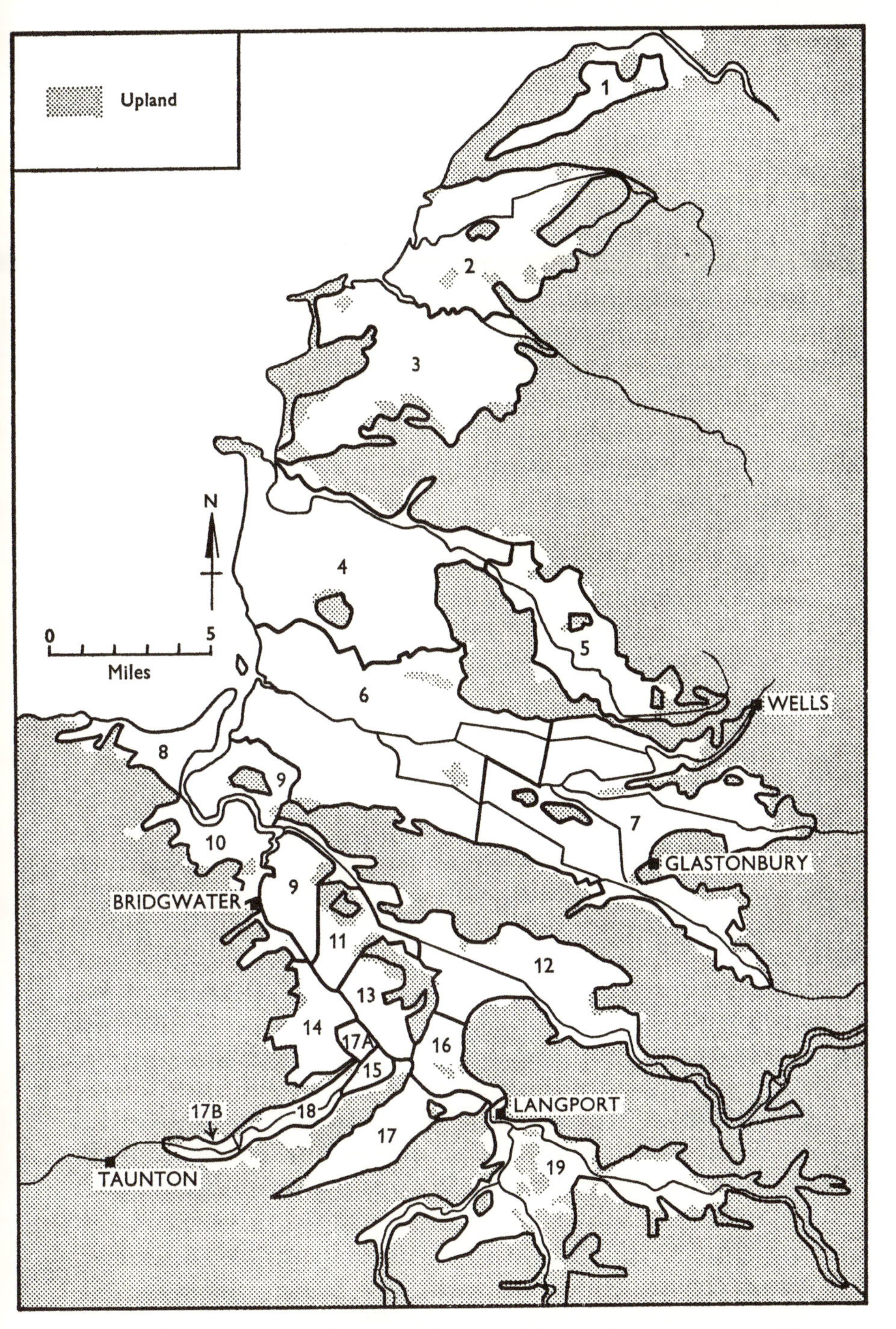

Figure 36. Internal Drainage Districts: 1, Gordano Valley; 2, North Somerset; 3, West Mendip; 4, Lower Axe; 5, Upper Axe; 6, Lower Brue; 7, Upper Brue; 8, Stockland; 9, Bridgwater and Pawlett; 10, Cannington and Wembdon; 11, Chedzoy; 12, King's Sedgemoor and River Cary; 13, Othery, Middlezoy and Westonzoyland; 14, North Moor; 15, Stan Moor; 16, Aller Moor; 17, West Sedgemoor; 17A, West Sedgemoor (E. Saltmoor); 17B, West Sedgemoor (Upper Curry Moor); 18, Curry Moor; 19, Langport. Boards 1–3, north of the Mendip Hills, were not created until after 1930. (Source: S.R.B., 'Map of Boards and principal Drainage Works' (1951).)

Five new districts were created in the Southern Levels for Aller Moor, Stan Moor, Curry Moor, North Moor and the Chedzoy Moors. King's Sedgemoor was taken away from the general supervision of the Bridgwater and Langport Courts of Sewers and placed under a special sewer commission of its own. Once more the physical separation of the moorland basins in the Southern Levels favoured a divided drainage effort (Fig. 36). Inevitably, lack of control over the major rivers, which were still under the jurisdiction of the sewer commissioners, led to worsening conditions. It was said later that the efforts of the Internal District Boards had been 'poorly rewarded for the simple reason that they were perfectly impotent to carry out the works that were necessary to relieve their lands',[1] that is to say, main river works.

All these changes had taken place south of the Mendip Hills whereas the Northern Levels were by-passed by events, and the Commissioners of Sewers, with their traditional methods of administration and maintenance, remained in operation until 1930.

The extensive flooding of 1872–3 (Fig. 37) brought about Grantham's report on the state of the drainage of the Levels. He found the drainage administration totally inadequate and wrote:

The difficulties are not altogether of an engineering character, but are also (indeed chiefly) legislative, the administrative authority having from the length of time and change of circumstance become totally incompetent to meet the present requirements arising from the advancing improvement of the agriculture. The Commissioners of Sewers cannot perform their duties, as indeed, they have themselves admitted.[2]

He recommended that the administration be remodelled and put on a more comprehensive scale, and that *ratione tenurae* be abolished as a basis for maintenance. More severe floods during 1876, and 1877,[3] brought matters to a head and the Somerset Drainage Act was passed in 1877.[4] With it, the old sewers divisions south of the Mendip Hills were finally abolished and a further eleven new Internal Drainage Districts were created, making a total of nineteen in all throughout

[1] G. D. Warry, *Ubi Voluntas Via Fit: The Somerset Drainage Act, its Principles and Practice* (1880), 7–8.

[2] Grantham, *Report on the Floods in Somersetshire in 1872–73*, 20.

[3] The rainfall of the latter half of the 1870s and early 1880s was heavy and consistently well above average. See *Climatological Atlas of the British Isles* (1952), 53. For the 1877 flood, see E. T. MacDermot, *The History of the Great Western Railway* (1931), II, 344.

[4] For details of the working of the Act, see W. R. Poole, *The Somerset Drainage Act, 1877* (1881).

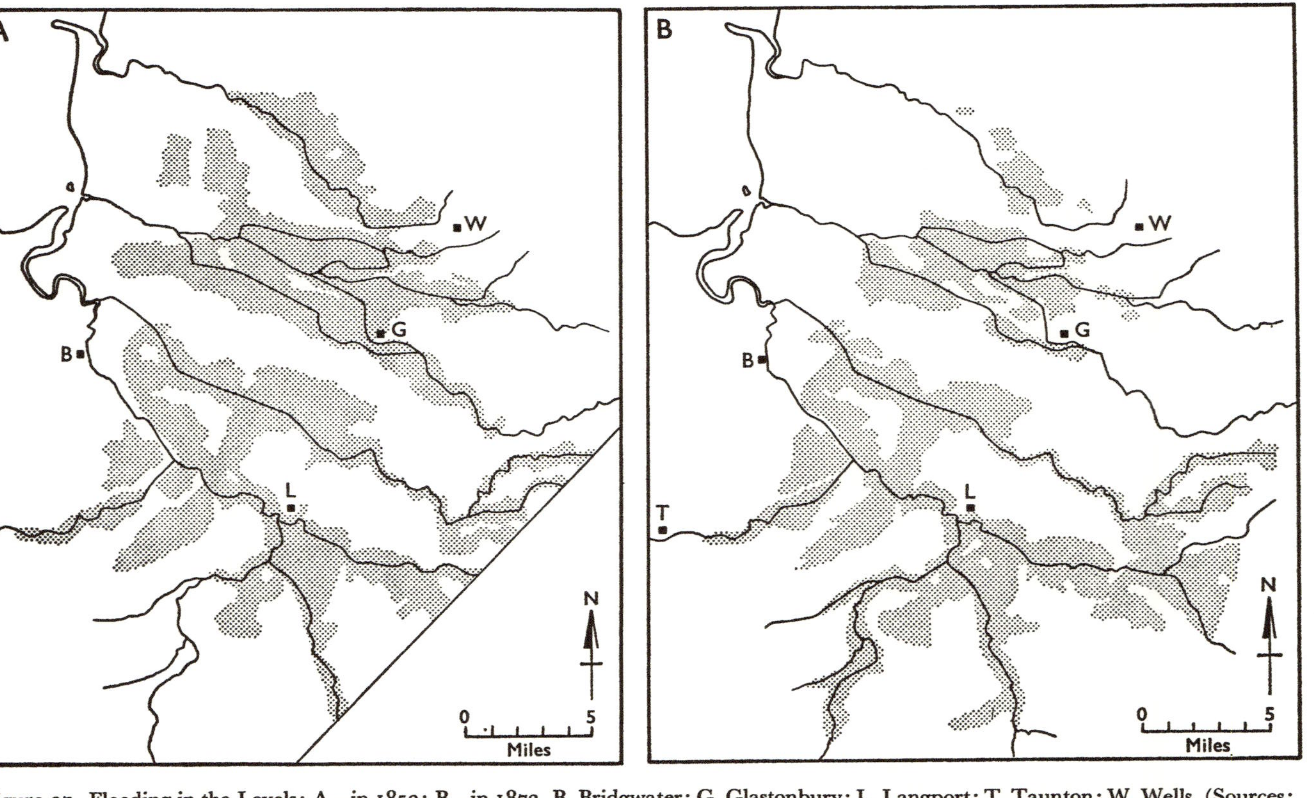

Figure 37. Flooding in the Levels: A—in 1853; B—in 1873. B, Bridgwater; G, Glastonbury; L, Langport; T, Taunton; W, Wells. (Sources: 1853—after a map showing 'Borders of moors subject to floods', in J. A. Clark, 'On the Bridgwater and other Levels of Somersetshire', *Journ. Bath and West Eng. Soc.* II (1854), 99; 1873—after a map showing 'Flooded areas' in R. B. Grantham, *Report on the Floods in Somersetshire in 1872–73* (1873).)

the Levels.[1] (See Fig. 36.) Each district now had control over its own works, including the main rhynes and seawalls, and they set about reorganizing and rationalizing the mass of maintenance and rating procedures that had evolved haphazardly during the preceding 600 years. This freed the new body, known as the Somerset Drainage Commissioners, from routine maintenance worries, and they were able to concentrate on the supervision of the drainage as a whole, and on seeking long-term remedies to the flood problem. Yet, as old problems receded new ones arose. Foremost amongst these was the opposition of the Internal Boards to the new commission and its policy. This separation of responsibility was a major cause in the failure of the age to concert and execute a comprehensive programme of improvement.

From the standpoint of financing drainage works the new Act was far in advance of existing drainage legislation in Somerset, or, for that matter, in other lowland areas in Britain. First, liability *ratione tenurae* was now commuted to a rent charge on the land, based on the Poor Assessment. Secondly, each Internal Drainage District was able to recoup the cost of local works by rates within the whole of its particular area. These rates and their collection varied according to the needs of maintenance and pumping, and in some cases, as in the Upper and Lower Axe Districts, they were abandoned completely because no work was carried out. Thirdly, the Somerset Drainage Commissioners could levy rates on land which 'received benefit',[2] but, according to the time-honoured interpretation of this phrase, the areas away from the main rivers and upland areas were not included.[3] Only 121,400 acres, or one-fifth of the total area of the catchments of rivers Parrett, Tone, Cary, Brue, Axe, Yeo and Isle, were considered to receive benefit. Because of this, the Commissioners could never hope to raise sufficient money to remedy the major defects of the drainage.

[1] One more was created, Southlake, which was later amalgamated with the Othery and Middlezoy Board.

[2] According to J. C. A. Roseveare in 'Land Drainage in England and Wales' (1932), 2–3, the boundary of land considered to receive benefit was 'two feet above flood level'. For a discussion on the agricultural desirability of designating all lands eight feet above flood level as receiving benefit, see B. A. Keen, 'Land drainage: The area of benefit', *Agriculture, Journ. Ministry of Agriculture*, XLIII (1936), 521.

[3] A Select Committee of the House of Lords, appointed in 1877 to enquire into the formation of Drainage Authorities, had advocated the rating of uplands as a means of raising money towards the cost of the maintenance of works from which the upland areas benefited, but this plea went unheeded. Appendix III to the Report of the Royal Commission on Land Drainage in England and Wales, *B.P.P.* x (1927) (Cmd. 2993), 1100–05.

These restrictions lasted in Somerset until about 1925, when both local and national agitation[1] finally culminated in the Land Drainage Act of 1930. With this Act the two main problems that had bedevilled efficient drainage administration were finally tackled; one authority had control over the whole catchment area of the Levels; and the basis of rating was extended so that a precept could be placed on the upland areas. In spite of these administrative improvements, it still took a long time before any start was made on new drainage works. With this background of the ever-pressing administrative and financial problems in mind, we are better able to appreciate the technical problems of draining during the nineteenth and early twentieth centuries in all the regions of the Levels—in the Southern Levels and King's Sedgemoor, in the Brue valley, in the Axe valley, and in the Northern Levels.

THE REGIONS AND THEIR PROBLEMS

The Southern Levels, 1840–1900

The problem of draining the Southern Levels overshadowed all other undertakings during the nineteenth century. The individual reclamations carried out between 1793 and 1830, and the experiments in steam pumping, had done little to stop flooding in the individual basins. For example, in 1850, West Sedgemoor was 'still in a very bad state', and could be grazed only in the driest of seasons.[2] Three years later it was said that the Chedzoy moors, the Sowy moors, Aller Moor, North Moor, the Tone moors, in fact the whole of the Parrett valley below Langport, were flooded almost annually, and that the position was little better in the upper Parrett valley, despite the relief channels that had been constructed by the Parrett Navigation Company.[3] Exceptionally heavy rainfall in 1852 turned the Levels into a sea and they were 'completely under water', rendering

[1] For instance, the Report from the Joint Select Committee on the Ouse Drainage Bill, *B.P.P.* VI (1927), 145, Sessional Paper 113; The Report of the Joint Committee on the Doncaster Area Drainage Bills, *B.P.P.* IV (1928–9), 723, Sessional Paper 88; and finally the Report of the Royal Commission on Land Drainage in England and Wales, *B.P.P.* X (1927) (Cmd. 2993), 1047.

[2] Acland, 50. See also *Taunton Courier*, 5 February 1840, for flooding around Athelney.

[3] For details of the conditions in these moors and the effects of the flooding on the local inhabitants, see E. Chadwick (comp.), 'Sanitary effects of land drainage'. Report of Poor Law Commissioners on an Inquiry into the Sanitary Condition of the Labouring Population of Great Britain. *Journ. Roy. Agricult. Soc.* IV (1843), 157–9. See also Clark, 109.

all traffic impossible. The meadows around Taunton were flooded, and at Langport and upon Sedgemoor the waters accumulated in great depth to the 'heavy loss of the graziers and farmers of that rich pasturing district'.[1] The next year proved to be no better (Fig. 37).

The general failure of the reclamation and pumping schemes to improve flood conditions was the result of the unsatisfactory state of the main river channels on which depended the evacuation of all surplus water.[2] These channels had received scant attention during preceding years and their condition had been allowed to deteriorate. The defects of the channel of the River Parrett were the worst, and they may be catalogued as follows:

1 The constricted, winding, and ungraded condition of the channel between Burrow Bridge and Langport;

2 The unimpeded entry of the tide which caused siltation in the river and reduced its capacity to carry off flood water;

3 The shoals and constrictions in the channels of the upper Parrett, and also in its tributaries, Tone, Isle and Yeo.

In an endeavour to alleviate the first of these defects various proposals were put forward to divert the waters of the Parrett away from the narrow section of the river between Burrow Bridge and Langport. One noteworthy plan was to construct an overflow weir about a mile below Langport, which would divert the excess flood waters of the Parrett into a new channel that was to pass via Aller Moor and King's Sedgemoor to the Dunball outfall. The plan had considerable merit and was vigorously advocated by Clark as the best solution to the problem, with the minor amendment that the weir be constructed above, rather than below, Langport (see Fig. 24 for Clark's scheme).[3] Yet neither these proposals, nor others in 1853 to widen the Parrett between Burrow Bridge and Langport, came to fruition,[4] largely, one feels, because of the difficulty that the two divisions of the Court of Sewers, above and below Langport, had in agreeing over matters of responsibility and rating. The defects persisted and after severe flooding and damage in 1860 Arthur Whitehead proposed that the general improvement of all the main channels be undertaken. He

[1] J. Algernon Clarke, 'Trunk drainage', *Journ. Roy. Agricult. Soc.* xv (1854), 6.

[2] The local farmers knew that 'from the inefficiency of the outlets, the water...will probably remain until March or even April, stagnating upon the meadow land'. See also R. Smith, 'On bringing moorland into cultivation', *Journ. Bath and West Eng. Soc.* v (1857), 111.

[3] Acland, 49–50, and Clark, 111–14.

[4] S.R.B./*P.C.S.* Langport, 2 February 1853.

estimated that the cost of improvement would barely exceed the damage done by flooding in any single year;[1] yet his report went unheeded as the Commissioners of Sewers were powerless to execute new works. The only alternative way to implement his suggestions was to obtain an Act of Parliament, but this procedure was too costly to commend itself to local landowners.

With little immediate prospect of improved conditions in the main river, opinion swung back to steam pumping as a means of improving the drainage of the moorland basins. Two factors had helped to bring about this revival of interest in pumping. First, pumping was greatly encouraged by the invention of Appold's centrifugal pump in 1851, which was much more effective in lifting water than was the scoop wheel. Secondly, the new Land Drainage Act of 1861 permitted the formation of Internal Drainage Districts, which were most suitable for the limited area of low ground in the moorland basins.[2] As each new Internal District was created a steam pump was erected to drain the area. Thus, in 1861, a 30 h.p. engine was installed to drain the Chedzoy Moors, and Westonzoyland's old scoop-wheel pump was replaced by a new Appold-type pump in the same year. Three years later Curry Moor had a 45 h.p. engine, and Stan Moor a 30 h.p. engine. In 1867, North Moor had a new engine; in 1869, Aller Moor had a 30 h.p. pump installed and the old Southlake pump was replaced by a more efficient one.[3] (For the location of the pumping stations see Fig. 24.)

Initially the pumping schemes worked well and the improvement they produced satisfied the majority of the local farmers. This immediate success was great enough to induce landowners in six other areas in the upper Parrett, Isle and Yeo valleys to attempt to install steam engines and pumps.[4] However, the inhabitants of Langport

[1] Whitehead, *The Improvement of the Somerset Level.* See also the *Bridgwater Times* and *Somerset County Gazette*, 17 August 1859, for references to severe damage by storms and overflowing of the Parrett, in the vicinity of the Sowy Moors. The same newspapers also carried accounts of a series of excessively high tides and storm damage by flooding on 5 and 12 October, and 2 November 1859 and 1 February and 19 September 1860.

[2] See page 206 above.

[3] For general information and specifications of these pumps, see the *Second Annual Report of the S.R.B.* (1951–2), 21–3. For detailed contemporary accounts, see S.R.B./*P.C.S.* Bridgwater, 9 January 1867, for North Moor; 28 September 1864, for Curry Moor and a projected scheme for King's Sedgemoor; and 27 July 1869, for Chedzoy.

[4] They were Muchelney, West Moor, and four other areas which are not known but which must have been Thorney Moor, King's Moor, Wetmoor, and Perry and Drayton Moors. See S.R.B./*P.C.S.* Langport, 10 May 1864 and 4 November 1868. Figure 31 shows the location of these areas.

strongly opposed the plan because of the 'inconvenience already suffered by the town in times of flood and its probable increase by such pumping operation'.[1] Their objection was given added weight by the worsening flood conditions which resulted from the neglect of the flood-relief system that had been constructed in the upper Parrett valley by the Parrett Navigation Company;[2] because the Langport Lock was not operating, the river above the lock had become silted and narrow, and the channels of the relief system were decayed and almost useless. Eventually, the formation of the proposed six districts was not permitted, but, in order to provide for a better form of drainage supervision, one large district to cover the whole of the upper Parrett area was created; however, no steam pumps were installed (Fig. 36). The outcome of pumping operations in the lower Parrett showed that the fears of the Langport inhabitants had been justified, for what seemed to be increasingly high water-levels in the main rivers caused serious seepage through the river banks. Indeed, it was desirable purposely to flood the moors at the height of the flood threat in order to obtain an equal pressure on both sides of the banks and so keep them intact. While pumping gave brief respite and improved the individual moorland basins, the uncontrolled and indiscriminate pumping by each district ultimately contributed to a deterioration of the general drainage by overburdening the already swollen waters of the Parrett.[3]

Foreseeing the consequences of this retrograde trend, local proprietors and farmers called for a report in 1869 on the best means of improving the main river channel. Bazalgette and Whitehead surveyed the area, and they recommended many new cuts through bends in the rivers above Langport, the removal of Langport Lock, the straightening and widening of the Parrett below Langport, and a sluice to exclude the tide 1½ miles above Bridgwater. The estimated cost of the works was £118,000, exclusive of the necessary purchase of the Parrett Navigation Company, legal costs and land compensation.[4] The expense would have been so great and the results so uncertain that the landowners, the Internal Districts, and the

[1] Grantham, *Report on the Floods in Somersetshire in 1872–73*, 9.

[2] For some instances of the neglect of drainage works by the company and fines by the Commissioners of Sewers, see S.R.B./*P.C.S.* Langport, 8 January 1862.

[3] Darby, 'The farming of Somerset', *Journ. Bath and West Eng. Soc.* v, 142, for comment on this.

[4] J. W. Bazalgette and A. Whitehead, *Report on the Yeo, Parrett and Isle Drainage* (1869).

Commissioners of Sewers could not agree on the implementation of even a part of the recommendations and consequently they did nothing.

The great floods of the winters of 1872 and 1873 pointed to the deplorable state of the drainage, and Grantham was called in as an independent inspector by local landowners in the Brue valley, many of these being themselves Commissioners of Sewers. His report was a melancholy testimony to decaying works and inefficient administration, for he found the main rivers and moorland rhynes in the same dismal state as Bazalgette and Whitehead had found them four years earlier. The extent of and severity of flooding during 1872–3 is best gauged by looking at Figure 37, above. The water did not begin to recede for nearly two months, all communication was severed except by boats, and, with the onset of warmer weather, conditions became appalling. 'Ague [malaria] set in early in the spring', wrote Grantham, 'and is now very prevalent on the verge of the moors among the poorer families who are badly fed and clothed, whilst even many persons in a better position in life have also been affected with it.' Everyone waited with trepidation for the outbreak of cholera and typhoid fever.[1] Grantham made no particular recommendations concerning the engineering problems of the Southern Levels as these were obvious enough and were centred around the condition of the Parrett channel, but he did recommend a thorough revision of the sewer laws, which had become confused and inefficient, and hindered the 'neither very formidable nor expensive' drainage works that were needed. He recommended that the sewer laws should be consolidated and replaced by a new Act based on the 1861 Drainage Act,[2] but the enclosure commissioners and government departments, who were asked to comment on his report, thought that this was not far-reaching enough and that a private Bill was preferable as it would take less time to get through Parliament than would the revision of the existing laws. Whatever the views on the steps necessary to

[1] Grantham, *Report on the Floods in Somersetshire in 1872–73*, 20. See also Darby, 'The farming of Somerset', *Journ. Bath and West Eng. Soc.* v, 152. Writing one year after the floods he estimated that the pastures of the Levels were barely yielding 1/6th of their potential value, and thousands of acres lay 'like a ship, waterlogged, incapable of being converted to the natural services for which they are so pre-eminently suited'. It is interesting to note that not one reference to the floods appears in any of the proceedings of the Commissioners of Sewers. In fact, by this time, the entries had dwindled to a mere formality, recording the date and place the commissioners sat, and a few fines.

[2] Grantham, *Report on the Floods*, 22.

reorganize the administration of the Levels drainage, the feeling that something had to be done immediately was underlined by further severe floods in 1876 and 1877, which once more covered the Levels. As was foreseen by the enclosure commissioners, it was impossible to extend the provisions of the 1861 Act to fit the requirements of the Levels, and therefore completely new legislation was needed.

The Somerset Drainage Act was passed in 1877 and high hopes were held for the future of draining the Levels. Legislation had not come too soon, for the draining of the Southern Levels had been steadily getting worse.[1] The Othery moors were said to be 'in a very bad state'; Aller, Stan, and West moors were 'exceedingly wet and ill-drained, as well as liable to ruinous inundation'; and North Curry moor in the Tone valley was 'getting rapidly into a worse condition'.[2] Some moors above Langport were 'in a wretched state', having only 'the coarsest herbage and unfit for stock...except for a few weeks at the height of summer, being a perfect swamp'. Following further extensive floods in the winter of 1879–80, liver-rot decimated sheep flocks and they were 'swept off *en masse* as by a plague'.[3] Two years later it was said that 'floods and foot-rot have almost ruined the farmer, if not quite'.[4]

It was unfortunate that from the time the new commissioners (appointed under the Somerset Drainage Act) assumed new duties, Somerset, with the rest of the country, entered upon a period of prolonged and unrelieved agricultural depression accentuated by a succession of unusually wet years. The predominantly pastoral economy of the Levels saved this region from some of the early shocks which arose from the importation of cheap grain in large quantities, and which affected the wheat-growing areas of the country first. After 1885, however, refrigeration and the importation of frozen meat from overseas caused a fall in fat-cattle prices, and agriculture in the Levels was severely affected.[5] A vicious circle arose whereby the land was neglected and allowed to deteriorate, with the result that it could not

[1] See p. 206 above, n. 3.

[2] Darby, 'The farming of Somerset', *Journ. Bath and West Eng. Soc.* v, 142, 145, and 146 respectively.

[3] *Idem*, 'Stream and flood in relation to agriculture', *Journ. Bath and West Eng. Soc.* XII (1880), 152 and 154.

[4] W. C. Little, 'Report on Devon, Cornwall, Dorset and Somerset', Royal Commission on Agriculture, *B.P.P.* XIV (1882), 423, Qu. 67098.

[5] *Ibid.* 35. For example, the annual rent of first-class grassland in Pawlett Hams fell by 45·5% between 1871 and 1881.

bear a high drainage rate; then without adequate funds nothing could be done to improve the drainage.

Bearing in mind these adverse economic conditions, it is creditable that the commissioners spent nearly £25,000, between 1883 and 1885, on widening, deepening and generally improving sections of the Yeo, Isle and Parrett.[1] It is difficult to ascertain what was done in the main channels as few records of the new works appear to have been kept, but the indications are that no more was done with the money than to maintain the *status quo*. A further £4,000 was spent on an embankment around Langport to protect the town from the floods which inundated it almost annually,[2] and various experiments were made to clear mud in the main rivers by dredges.[3]

These works, however, were mere palliatives and, without the backing of regular maintenance, they quickly became ineffective. Moreover, no serious effort had yet been made to rectify the defects of the section of the River Parrett that lay between Burrow Bridge and Langport. It was on remedying these defects that the improved drainage of much of the Southern Levels, and the prevention of flooding in King's Sedgemoor depended. The crucial nature of this section of the Parrett was made evident by the breaking of the Aller Moor embankment in the autumn of 1891, when over 10,000 acres of land were flooded in Aller Moor, North Moor and King's Sedgemoor; and the water did not drain away for over five months.[4] Proposals by F. Lowrey in 1893 to enlarge and deepen this section of the Parrett went unheeded.[5] Another flood, which was said to have been even more serious than the previous one, occurred during the following winter. The waters overtopped the new Langport Wall and rose to the upper windows of the houses in the town. Langport people went to cut the Aller Bank in order to relieve the pressure of water above the town, but they were met and turned back by parties of Aller

[1] S.R.B./D.C. 30. It seems fairly clear from a pamphlet entitled *Report on the best scheme for the prevention of the floods at Taunton* by 'Hydraulic Engineer' (pseudonym) in 1889, that little or no improvement had been undertaken in the River Tone.

[2] S.R.B./D.C. 39. See also Grantham, *Report on the Floods*, 6, for a description of Langport during a flood.

[3] For a summary of these experiments, see E. L. Kelting, 'The dredging plant of the Somersetshire Drainage Commissioners and the Somerset Rivers Catchment Board', S.R.B., *Fourth Annual Report* (1954–5), 15–18.

[4] S.R.B./D.C. 27–9. The Aller Bank also broke in 1872, 1891, 1894, 1924 and 1929, and was a weak spot in the flood protection works.

[5] Lowrey, *Report on the River Parrett Floods Prevention* (1893).

people who were standing guard over their own embankment which had been cut three years before in similar circumstances.[1]

With little prospect of any improvement being made in the Burrow Bridge–Langport section of the river, attention temporarily shifted to the lower end of the Parrett with the publication of W. H. Wheeler's plans to improve the river outfall and the entrance to Bridgwater Port.[2] He proposed the excavation of cuts across the loops of the Parrett and Pawlett Hams and across Hawkhurst Farm in order to shorten the course of the river and increase its scouring power (Fig. 24). As Wheeler had had considerable experience in outfall improvement works in the Fens his plan was looked upon as promising, yet it was not accepted. A close examination of current newspaper comments, letters and articles makes it clear that, besides objections to the cost of the scheme (between £100,000 and £110,000), it was firmly and finally rejected by the slime-batch owners.[3] The owners of the lucrative 'Bath' brick industry naturally opposed any scheme which would alter the tidal regime and the pattern of deposition in the river. It is relevant to note that in the past they had been able to influence successfully the proposed location of clyses in both Bazalgette and Whitehead's, and Lowrey's plans for the improvement of the river.

The century ended with one more attempt to find a solution to the problem of flooding in the Southern Levels. In 1898, William Lunn was requested by the commissioners to prepare a report on the improvement of the Parrett and the Tone. He proposed to grade the bed of the Parrett uniformly between Bridgwater and Langport at 12 in. per mile, widen the river between the existing banks wherever possible, and also remove the locks of the now abandoned Parrett

[1] S.R.B./D.C. 13–15, and the *Somerset County Gazette*, 15 December 1896, and the *Langport and Somerton Herald*, 17 November 1894.

[2] W. H. Wheeler, *Report on the River Parrett proposed Improvements* (1896). His proposals were basically the same as those of Phelps in 1836. See p. 158 above.

[3] *Bridgwater Mercury*, 29 August, 19 September, 17 October, 21 November and 12 December 1896. The slime-batches were platforms built in the river by the dumping of brick rubbish. The deposit of mud or slime left on them after the ebb tides was used for the manufacture of 'Bath', or polishing, knife-cleaning and scouring bricks. The batches were situated mainly in the stretch of the river where tidal flow was arrested by river flow, and deposition was at its greatest. Here the alumina and silica particles, which were the main ingredients of the 'Bath' bricks, were sorted out from the rest of the silt. The proportion of sand increased rapidly below Bridgwater, and the material was too fine, after about a mile above the town. Production of 'Bath' bricks reached about 8 million annually in the middle years of the century. *V.C.H.*, II, 353.

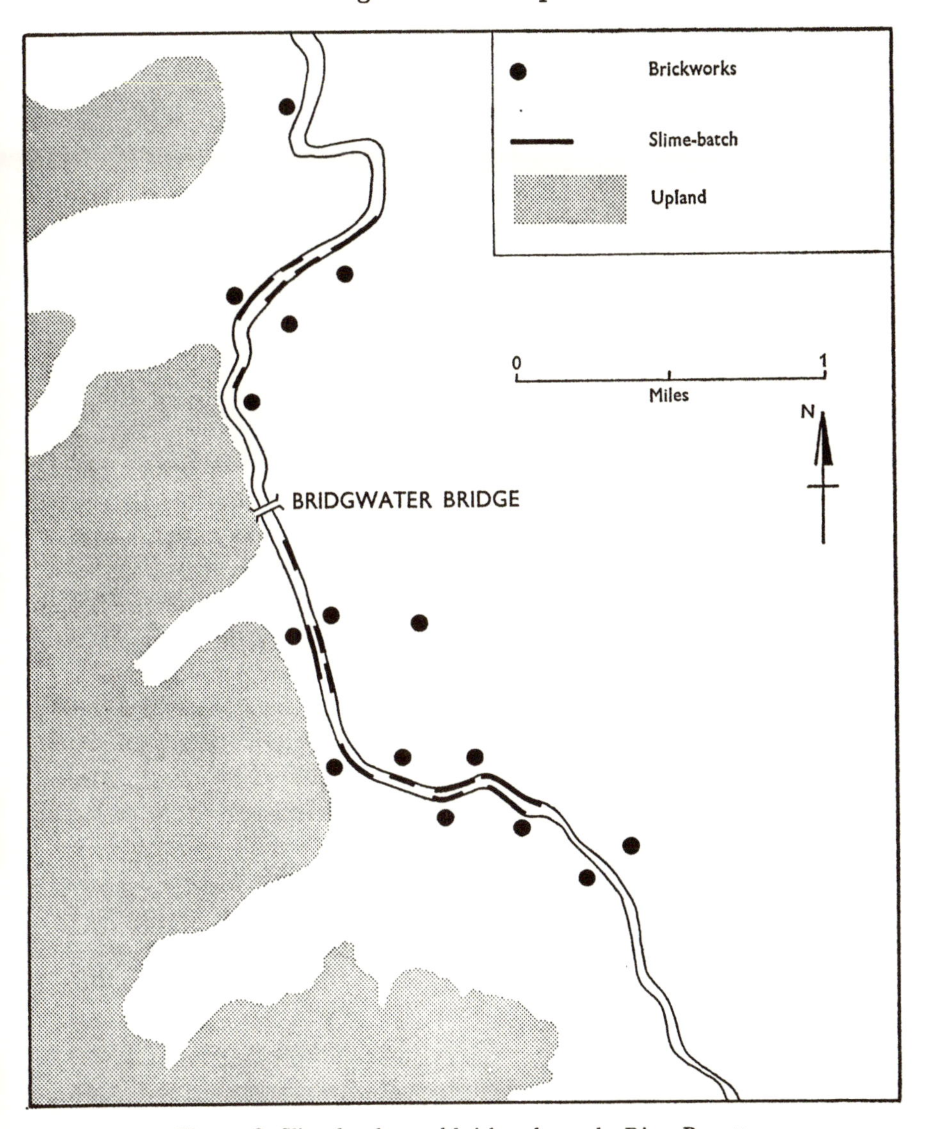

Figure 38. Slime-batches and brickworks on the River Parrett.

Navigation Scheme together with as many of the slime-batches as possible.[1] The number of slime-batches had increased during the closing years of the century; there were 25 of them in the Parrett in 1896 with a total length of 2,650 yards and they seriously obstructed

[1] W. Lunn, *Report on the River Parrett Floods Prevention* (1898).

the unimpeded flow of water in the river (Fig. 38).[1] In calling for the report it seemed that the commissioners acknowledged that something should be done, but, despite the promise of improvement that the implementation of even a part of Lunn's plan would have given, it became one more scheme to criticize and reject.

King's Sedgemoor, 1810–1900

After the special commission for the draining of King's Sedgemoor handed over its powers to the Commissioners of Sewers in 1796,[2] even the most routine maintenance of drainage works was neglected. With such sparse attention being paid to the efficiency of the existing system, there was little prospect of the rectification of the three major defects in the Main Drain, which had impaired the success of the original scheme.[3]

As early as 1811, the position had deteriorated far enough for the Commissioners of Sewers to accept eagerly an offer by the promoters of the Bristol–Taunton Canal to build a new, larger and lower clyse at Dunball. The Canal Company proposed to incorporate the clyse in an aqueduct over the Main Drain. However, the canal project was eventually abandoned and no improvement made to the clyse.[4] Conditions became worse as slips of the banks and extrusions of peat in the channel bottom blocked the Main Drain. In 1816 Robert Anstice was asked to inspect and report on the state of the clyse, the Main Drain and major subsidiary drains.[5] Typically enough, time dragged on, and the survey was not carried out until another two years had elapsed, when clyse and drains were found to be in a bad state of repair.[6] The commissioners also asked for a report on the cheapest way of widening the bridges across the drain, but nothing eventuated.[7] Apathetically, the commissioners waited another ten years before they made a new order to view and repair the Main

[1] Figure 38 is based upon information found in 'A Schedule of Slime Batches' (1896), S.R.B./D.C. 39.

[2] S.R.B./*P.C.S.* Bridgwater, 6 October 1795 for details of the transfer of responsibility.

[3] For these defects, see pp. 149–50 above.

[4] S.R.B./*P.C.S.* Bridgwater, 27 December 1810; and 21 January 1811. It seems probable that the offer was made by the Canal Company in an attempt to influence the Court of Sewers, who opposed the intended canal on the grounds that it would be detrimental to the drainage of the Levels. The subsequent history of this venture can be traced in E. C. R. Hadfield's *The Canals of Southern England* (1955), 197 ff.

[5] S.R.B./*P.C.S* Bridgwater, 15 July 1816.

[6] *Ibid.* 4 April and 24 July 1818.

[7] *Ibid.* 4 September 1818, and 23 April 1819.

Drain, but by this time the position had deteriorated so far, and the ineffectiveness of the commissioners had become so apparent, that the local landowners felt it necessary to employ Josiah Easton to inspect the Main Drain. He found the outworks dilapidated and the bed and sides of the drain in a deplorable state, and he estimated that an expenditure of £10,000 was needed to prevent the drain from becoming completely useless.[1] The cost of redeeming the drain staggered the landowners, and, fortunately for them, the need for a decision on the recommendations of the survey was delayed by the publication by Easton in 1829 of a scheme to improve the King's Sedgemoor Drain and make it navigable from Dunball to Yeovil.[2] In recommending it to the local landowners he pointed out that

> without the vast expenditure and the obtaining of an Act to invest the Commissioners of Sewers with further and sufficient powers, the present state of the drain is so insufficient and decayed that the whole benefit originally effected by that great work is likely to be in a great degree lost and the valuable district sink again into its former swampy and unproductive state.

Rarely can the needs of drainage and navigation be reconciled, and this scheme was no exception since the penning of water for navigation purposes would have impeded the free drainage of the moor. In addition to this, the proposal to pass some of the Yeo waters into the Cary to keep up its level was not likely to commend itself to the moorland proprietors, who had been endeavouring to reduce the level of water in the Main Drain. Consequently, the scheme fell into abeyance and nothing was done. The commissioners again issued an order to scour the Main Drain, and a few hundred pounds were spent on clearing it.[3] They were back where they started.

In 1842, in an attempt to break the impasse, a Mr Glyn was asked by the commissioners to investigate the best means of improving the drainage of the moor and he suggested two alternatives. The first was obvious enough; it entailed lowering the clyse sill by four feet and improving the clay belt portion of the cut from Dunball to Crandon bridge. The other remedy was more revolutionary and quite unnecessary; it consisted in turning the lower part of the existing cut into a reservoir into which water might be pumped over a barrier by

[1] S.R.B./*P.C.S.* Bridgwater, 4 July 1823.

[2] J. Easton, *Prospectus of a Plan for rendering Navigable the King's Sedgemoor and other Principal Drains*... (1829).

[3] S.R.B. Sedgemoor Draining Expenditor Accounts, 1825–36.

steam engine, the reservoir clearing itself at ebb tide.[1] Which of these two remedies was favoured is not known, but the commissioners, uncertain of their power to carry out new works, sought legal opinion and once again nothing was done. This sequence of events was becoming common in all parts of the Levels. Two years later, the clyse doors at Dunball were said to be 'in a very decayed and unsound state' and in immediate danger of collapsing.[2] It seems highly probable that this danger at last goaded the commissioners into action and the clyse sill was lowered by 20 in., and the doors replaced by new ones.[3] Nevertheless, in 1850, the condition of the moor was still 'very deficient' and 'far from satisfactory'; the sides of the cut still slipped in, the bottom of the drain 'welled up' and the moor was flooded almost annually (see Fig. 37 for the extent of flooding in 1853).[4]

With the passing of the Land Drainage Act of 1861 there was promise of better things. The moor was taken from the jurisdiction of the Bridgwater and Langport Divisions of the General Sewers Court and placed under a new special commission for King's Sedgemoor alone.[5] It had wider powers and a wider area of supervision, which included the upper Cary valley. The new administration did not result, however, in a new attack on the problem. The nearest the commission got to improvement was to consider and draw up a plan for a new outfall channel, leading from where the Crandon railway bridge crossed the drain to Horsey Pill in the Parrett estuary. The new commissioners found themselves in exactly the same position as the old sewers commissioners; their authority was limited and the basis of rating was restricted. It would have been better if they had been constituted as an internal district, as were the sewers commissioners for the moorland basins of the Southern Levels. In 1873 Grantham noted that 'since they were constituted...they have not effected any material improvement and although they have raised about 3,000 l. even proper repairs have been imperfectly effected...For years past the floods have inundated many thousand acres, and have remained

[1] S.R.B./*P.C.S.* Bridgwater, 21 July 1842 and 10 July 1843. It is possible that this was Joseph Glyn, who was a great proponent of pumping and who wrote 'Draining by steam power', *Trans. Royal Soc. Arts*, LI (1837).

[2] S.R.B./*P.C.S.* Bridgwater, 26 July, 8 August 1843 and 3 June 1845.

[3] Acland, 49.

[4] *Ibid.* footnote on p. 48.

[5] The Chedzoy area of the moor, which naturally drained towards the Dunball outfall, was made into a new Internal District, and its drainage was reorientated towards a new pumping station on the Parrett bank (Fig. 24).

on the land for months owing to the want of a sufficient outfall.'[1] In the same year, J. Darby thought that improved drainage would immediately add one-third to the low value of the moor, which was then about £30 an acre.[2]

Worse was to follow. The clyse doors had become completely rotten and in 1876 were carried away by a high tide.[3] Their collapse endangered a nearby road bridge, which was eventually demolished, and until such time as tidal doors could be fitted under the arches of a new bridge the tide surged up the drain and into the moor. The cost of £16,000 for erecting new gates and other works depleted the resources of the Board, and plunged the area so deeply into debt that even with the passing of the Somerset Drainage Act in 1877 no further improvements were undertaken in King's Sedgemoor. From then on, maintenance barely kept pace with decay; the old Cary river was useless as a drainage channel by 1886,[4] the bottom of the Main Drain continually 'welled up', and the sides of the clay belt cut slipped in. The flood condition of the Moor was very much bound up with the state of the Parrett channel, and the vulnerability of the moor to complete inundation was emphasized by the serious breaches of the Aller Bank in 1872, 1891, 1894, and again in 1924 and 1929.

The Brue valley, 1810–1900

After the works carried out under the Brue Drainage Act were completed some improvement in the state of the moors was apparent, and 'meadows and pastures succeeded the morass'. The prospects for the future seemed even better and it was confidently predicted that 'traces of the bog will be scarcely seen, a green surface having succeeded the russet appearance of the turbary'.[5] But the result proved not to be as great as was hoped for and it was a bitter disappointment that the Brue drainage scheme failed to produce a lasting improvement in the valley. With the passing of the period of agricultural expansion and high returns there was little opportunity to remedy the scheme's defects and shortcomings.

Conditions in the valley deteriorated as the maintenance of existing works was increasingly neglected. The Proceedings of the Wells and

[1] Grantham, *Report on the Floods*, 11.
[2] Darby, 'The farming of Somerset', *Journ. Bath and West Eng. Soc.* v, 154.
[3] S.R.B./D.C. 39, Dunball Clyse Papers.
[4] S.R.B./D.C. 54, 'Report of the Committee on the Old Cary River' (1886).
[5] Phelps, *History and Antiquities of Somersetshire*, Vol. I, Book II (1836), 50.

Glastonbury Divisions of the Court of Sewers abound with references to neglected duties by reluctant, apathetic and sometimes bewildered frontagers, and with feeble court orders for their rectification.[1] The Commissioners of Sewers neglected their duties as badly as did the individual proprietors. Highbridge Clyse was already in a bad state of repair by 1826; two years later it was feared that unless something was done the gates would soon collapse. Happily they were replaced by new ones and disaster was averted.[2]

The commissioners realized that the deficiencies of the original Brue scheme could be overcome only by the construction of new works; the channel of the Brue needed widening and deepening, particularly through the coastal clay belt, but the North and South Drains remained a problem to which there was no agreed solution, other than pumping. The commissioners considered procuring an Act for improving the drainage,[3] but this expensive course of action was not pursued since the opportunity arose of combining the improvement of the drainage system of the southern portion of the valley with the construction of the Glastonbury Canal, which had begun in 1829.[4] For over twenty years this scheme diverted the energy of the commissioners away from more worthwhile and lasting drainage improvements.

In order to make the entrance to the canal independent of the existing outfall and clyse, the old and now silted-up estuary of the Brue was cleaned out, and a bridge was built across the estuary with two pairs of gates, which were to exclude the tide, pen back the fresh water and act as lock gates. (See Fig. 19 for the new outfall and Fig. 20 for the line of the canal.) The line of navigation followed the Brue up to Cripps's bridge, where an abandoned meander of the river was excavated and used; the canal then followed the channel of the South Drain, branched off with a new channel at the eastern end of Shapwick Heath, and finally crossed the Brue by an aqueduct before it reached Glastonbury. The promoters of the scheme had assumed that the canal was perfectly level, and that the penning of water at Highbridge would give enough depth for navigation up to Glastonbury.

[1] See also S.R.O. Carver MSS. 3, Brue Drainage Papers: Committee order for 1 August 1814, for the removal of obstructions in the Brue.

[2] S.R.B./*P.C.S.* Wells, 19 June 1826, and 16 June 1828. See also S.R.B./*P.B.C.* 5 October 1857.

[3] S.R.B./*P.C.S.* Wells, 29 October 1830, and 5 January 1831.

[4] The original scheme for navigation in the Brue valley was along the River Brue itself, but fortunately the scheme was considered impracticable.

This was not so, for there was a significant rise from the edge of Shapwick Heath to the higher clay lands around Glastonbury. Therefore, excavation of the canal was stopped. Surveys were carried out by Richard Hammett and John Beauchamp, but the results of the levels taken were so contradictory that another survey was carried out by John Rennie, who recommended that a lock be built at Maze Wall and that new and higher banks be erected along the sides of the canal.[1] These recommendations were carried out and the measures proved successful for navigation.

Initially the canal received nothing but praise; the new entrance at Highbridge provided an additional outlet for the flood waters which had been penned in the peat moors behind the clay belt, and it was claimed that they passed off the moors in half to one-third of the time that they had previously.[2] But it soon became obvious that the penning of the water at such a high level in the canal above Maze Lock meant that the peat lands in the eastern end of the Brue valley were not draining into the canal as was hoped for, but were absorbing water and were becoming increasingly waterlogged. Eventually, the peat became so saturated that it was said to have risen many feet, and navigation in the canal was completely disrupted. In the western and lower parts of the canal, the gates of Highbridge Clyse were kept closed in order to maintain the required depth for the navigation, which meant that the water in the rhynes throughout the Brue valley was penned above the normal level. This caused even more serious waterlogging and flooding than before, and the drainage condition of the valley deteriorated rapidly.

To make matters worse, the proprietors on the clay belt lands insisted that the commissioners shut the new lock gates in order to pen water for stock purposes, and that they cut slacker-holes in the Highbridge Clyse doors to allow the excess water to escape. But the clyse-keeper kept the clyse doors shut permanently, even in flood time, and the water trickled away through the slacker-holes as best it could. Because of the lack of scouring the lock gates of the navigation outfall at Highbridge became sealed with five feet of mud on either side and were soon in 'a very bad and dangerous state'.[3] Thus both the canal

[1] I am indebted to Mr D. R. May for information on these early surveys. R. Hammett's *Report on the Glastonbury Canal* (1826) is in Taunton Museum, Pamphlets, 55 (ii). For Rennie's survey and the results see Clark, 122–3, also S.R.B./*P.B.C.* 24 July 1830.

[2] Phelps, *History and Antiquities of Somersetshire*, Vol. I, Book II, 60.

[3] S.R.B./*P.C.S.* Glastonbury, 2 October 1866.

and Highbridge Clyse gates remained closed during flood time and the water in the Brue valley lay impounded in the peat moors for want of any outfall.[1] Eventually conditions became so bad that navigation on the canal almost ceased; in any case its economic life was near its end, for in 1848 the Bristol and Exeter Railway Company purchased the canal,[2] and eventually a railway line was laid to replace it. Soon after, the lock was removed and once more the needs of drainage had priority over those of navigation. Much time and energy had gone into making the canal serve two purposes, to no purpose, and the canal, which had been one of the main reasons for the deterioration of the drainage condition of the valley, was used no more.

Meanwhile, the routine maintenance of works continued to be neglected elsewhere in the valley. Extrusions of hard peat blocked the channels, some of which broke off and worked its way downstream to clog the Highbridge outfall,[3] the accumulation becoming so great at one period that the water of the river was diverted into the canal outfall, thereby causing excessive silting at the clyse outfall.[4] Once more, the gates of Highbridge Clyse were allowed to deteriorate, and repeated pleas were made for their repair in 1851, 1857 and 1864.[5]

Some improvement was clearly necessary and a special committee of the Commissioners of Sewers was formed to make investigations into the best means of improving the drainage channels, which, in their opinion, were 'insufficient for the proper draining of the Levels'.[6] A survey was made of the rhynes and drainage works, and a report given to the commisioners by G. W. Hemans, whom they had appointed consulting surveyor.[7] Besides the generally neglected state of all drainage works and the problem of the Glastonbury Canal, Hemans singled out the insufficiency of the Brue outfall as the main defect in the drainage system because it prevented the effective discharge of the pent-up flood waters which lay in the peat moors behind.

[1] S.R.B./*P.C.S.* Wells, 10 January 1850. It was the verdict of the Sewers Court that 'the damming up of the river in wet seasons must necessarily entail serious damage to the low-lying lands within the reach of the flood waters of the Brue'.

[2] The Canal Company was bought out by the Bristol and Exeter Railway Company (11 and 12 Victoria. c. 28). See also S.R.B./*P.C.S.* Wells, 9 October 1850, and Whitehead, *The Improvement of the Somerset Level* for further details of the neglect of the canal.

[3] S.R.B./*P.C.S.* Brue, 8 June 1842.

[4] *Ibid.* 10 March 1876.

[5] *Ibid.* 5 October 1851; 5 October 1857 and 20 January 1864.

[6] *Ibid.* 20 June 1864.

[7] G. W. Hemans, 'Report on the State of the Drainage' in S.R.B./*P.C.S.* Brue, 28 September 1867.

Hemans's main proposal was to make a new cut north of the canal outfall and entrance, thereby giving the Brue an additional outlet. But the commissioners were deterred from making a decision by the cost of the scheme and by their fears that the discharge from the new outfall would cause considerable wash and damage to the existing outfall. Hemans then submitted alternative proposals to enlarge the existing outfall, but these too were rejected. Once more the commissioners were back where they started and the neglect of major drainage works was allowed to continue.[1] By 1869, Highbridge Clyse gates were stuck in a half-open position by excessive mud accumulation in the estuary, and tidal floods were averted only by the operation of the fresh-water gates further upstream;[2] by 1876, it was feared that the clyse gates would collapse as had those at Dunball.[3]

The floods of 1872–3 were widespread and disastrous for the valley (Fig. 37). In Grantham's report on conditions, he drew attention to the inadequacy of the maintenance and administration of the drainage work and said: 'The whole system is imperfect. Indeed so much so, that it is only surprising how much better the work is done than might be expected.'[4] In particular he noticed the antagonism that existed between the clay belt farmers, who wanted to conserve water for irrigation and fencing, and the peat land farmers, who wanted deeper outfalls and a quicker evacuation of surplus water. This conflict of opinion and aim had been one of the main reasons for the chaotic state of affairs in the valley, for there had never been any agreement between the upper and lower portions of the valley either on the supervision of the existing system of drainage, or on the planning of new works. But even the prospect of continously deteriorating lands did not overcome these internecine differences. In J. Darby's view, the problems of the Brue valley were largely financial, and engineering difficulties were 'nothing'. 'Someone must take the risk of failure, and it is not much to be wondered at, that no one will, where the preliminary cost is large and the risk of failure great.'[5]

The passing of the 1877 Somerset Drainage Act brought little

[1] S.R.B./*P.C.S.* Brue, 12 December 1867. During the next year, landowners in Cossington, Chilton, Edington and Catcott petitioned for an additional outlet to relieve the flood waters that lay on their land.

[2] *Ibid.* 20 March 1869.

[3] *Ibid.* 20 October 1876.

[4] Grantham, *Report on the Floods*, 12–14.

[5] Darby, 'The farming of Somerset', *Journ. Bath and West Eng. Soc.* v, 156–7.

relief to the area.[1] Between 1883 and 1889, £23,350 was spent on rebuilding Highbridge Clyse, widening and deepening small portions of the river, widening some narrow bridges and strengthening portions of the Huntspill sea wall.[2] But these works were acts of palliation; the basic defect of the insufficiency of the outfall through the clay belt remained, and the North and South Drains still could not empty their water into the Brue when the river was in flood.

The Axe valley, 1810–1900

The 1802 comprehensive Drainage Act for the Axe valley had proved more successful than its counterparts in the Brue valley and King's Sedgemoor. This success was due to the less difficult physical problems of draining the valley,[3] and to the efficiency of the local division of the Commissioners of Sewers, who, in contrast to the commissioners in other divisions in the Levels, exercised a stricter control of maintenance and a more rigorous enforcement of fines. The commissioners also showed considerable initiative in persuading the Bristol and Exeter Railway Company to attach tidal doors to a bridge that it was building across the Axe at Batch, one mile below the existing clyse at Hobb's Boat (Fig. 20).[4] Thus, a new clyse was provided, giving an additional fall and increased reservoir capacity during periods of tide-lock. Conditions in the lower river improved immediately, and in 1854 it still could be said that the drainage system of the valley was 'equal or superior to that of any other district'.[5]

In an attempt to eliminate the last vestiges of flooding in the valley a special report was prepared by Hippisley in 1867. He found that approximately 1,273 acres of land were regularly flooded in the upper end of the valley, and that this was caused by the impossibility of gravitational discharge from the moors when the River Axe was running at a high level. He therefore proposed the regrading and improvement of the bed of the Axe above Clewer (Fig. 20). Unfortunately, however, the proposals were rejected by the commis-

[1] See S.R.B./*P.C.S.* Brue, 10 March 1876 for an account of flooding during the previous winter.

[2] S.R.B./D.C. 48 and 49.

[3] See p. 141 above.

[4] The commissioners thought the bridge was detrimental to navigation and drainage, and, if they had wished, could have impeded or prevented its construction. Finally they brought pressure to bear on the Railway Company to attach and to maintain the new doors. S.R.B./*P.C.S.* Axbridge, 12 February 1836.

[5] Clark, 127. See also Acland, footnote on p. 46 for confirmation of this view.

sioners on the grounds that the main river channel was clear enough, and that the flooding was not serious.[1]

After the great floods of 1872–3 (Fig. 37), Grantham recommended nothing more than the widening and grading of the channel, to make the drainage perfect.[2] Further flooding in the peat moors in the upper end of the valley in 1875 brought forth further recommendations to grade and widen the river, also tentative suggestions on pumping in the low-lying Westbury, Rodney Stoke, and Draycott moors.[3] But nothing was done, and, with the continuation of routine maintenance, conditions became no worse. Because of this, the new Somerset Drainage Commissioners of 1877 paid no attention to the area.

The Northern Levels, 1840–1900

The Northern Levels had always been different from the other lowland areas of Somerset, but the reclamation of the individual moors, and the comprehensive schemes in the early years of the nineteenth century had brought the region very much in line with the lowland areas further south. As the century progressed, the divergence of interests and developments became apparent again and the Northern Levels remained aloof from all the administrative and financial innovations brought about by the creation of the Internal Drainage Boards in 1861, and by the work of the Somerset Drainage Commissioners in 1877, as well as from the practical experiments in steam pumping. Admittedly the need for change in the Northern Levels was less urgent than in its lowland counterparts to the south; the problems of draining were neither so severe, nor so extensive. No floods comparable to those of 1872–3 and 1891 in the rest of the Levels had incommoded these northern lowlands so completely that solutions to the problem were sought in new institutions and new administrative machinery. Consequently the Wrington Division of the Commission of Sewers remained the supreme authority in the Northern Levels for nearly a hundred years more and was replaced only in 1934 with three Internal Drainage Boards, constituted under the authority of the new Somerset Catchment Board of 1930 (Fig. 36).

Although the Northern Levels had a lesser problem to deal with, the region did seem to be insulated by the Mendip Hills from the

[1] Hippisley, 'Report on the River Axe from Cradle Bridge, Weare to Westbury Moor', S.R.B./*P.C.S.* Axbridge, 15 January 1867.

[2] Grantham, *Report on the Floods*, 14. [3] S.R.B./*P.C.S.* Axbridge, 29 January 1875.

debate, plans, controversies and experiments that were being carried on elsewhere in lowland Somerset, and, for that matter, throughout the rest of lowland England. In particular, the retention, for most works, of the system of maintenance *ratione tenurae* which had been all but abandoned elsewhere in the Levels by the end of the century, was a marked feature of this isolation. The Proceedings of the Wrington Division and of the various local juries, make virtually no mention of either events or changes in neighbouring lowlands.

Throughout the nineteenth century, it would seem that the powers of the Commissioners of Sewers were just about equal to the needs of the drainage. Some improvements were made in their organization to increase their efficiency in 1847 and 1855,[1] but the net result was merely the maintenance of the existing position, rather than the improvement of the drainage works. It looked as though some changes were imminent in 1853 when the commissioners asked Mr Penistone to prepare a report on the draining of the badly-flooded lands around the River Kenn, but the cost of making a new outfall channel through the moors and the clay belt must have deterred the commissioners from action, for nothing more is heard of the scheme.[2]

If it could be said that the Proceedings of the Commissioners in the Northern Levels in the eighteenth century were 'a monotonous catalogue of neglected duties, views of works, fines, and maintenance work done,' then the same was even truer of the nineteenth century when the enormous mass of detail on routine work and day-to-day duties overwhelms the reader and defies broad analysis. All one can say is that attention to sea walls and sluices occupied most of the time because the danger of inundation was largely maritime in origin, and that by comparison the rhynes were neglected. After examining the Proceedings of the Commissioners one is left with the impression of arbitrary working, and of a lack of urgency in their actions and in those of their officers, the dyke-reeves. The following lengthy report in the *Weston Mercury* of 14 October 1899, of the activities of the Banwell dyke-reeves is typical of many in these years.

The autumn meeting of this body was held on Tuesday last. The foreman, Mr. Henry Bisdee, with several other members of the jury, met at Ebdon's Bow at 11.30 a.m. and proceeded forthwith to inspect the sea wall, which

[1] S.R.B./*P.C.S.* Yatton Jury, 28 June 1847, and S.R.B./*P.C.S.* Wrington, III, 10 October 1855.

[2] S.R.B./*P.C.S.* Wrington, III, 4 May 1853.

they found somewhat out of repair owing, to a great extent, to the recent prolonged drought, causing large cracks therein. Subsequently the dyke-reeves met at the Ship Hotel where the usual business in connection with the several views was transacted. The Clerk was requested to give notice to the various occupiers of the sea wall to have the same properly repaired. The Commissioners of Sewers having placed a temporary bay in the main river between Ebdon's Bow and New Bow, a considerable portion of the Sewers works within the view had not been done at the usual time, owing to the great quantity of water in the several watercourses. It was therefore decided to extend the time until Saturday next, 21st. At 4 o'clock the Jury, including several visitors, sat down to dinner, which reflected the greatest credit on the landlord Mr. Kendall, after which the usual loyal and complimentary toasts were proposed and suitably responded to. The remainder of the evening was devoted to music and singing, which everyone thoroughly enjoyed.

It seemed as though Acland's judgement on the court was substantially correct: 'Like other ancient institutions the Court of Sewers is said (by those who feel no inconvenience from it) to work well and always end with a good dinner.'[1]

Whilst there is no evidence that drainage conditions became any better, there is none to show that they became worse. The Wrington Commissioners slumbered on until 1930 when national concern at drainage conditions led to the reorganization of drainage authorities throughout the country, which finally brought about their demise and replacement by three new Internal Drainage Boards, those for the Gordano valley, North Somerset, and West Mendip (Fig. 36.)

REPORTS, ARGUMENTS AND DECLINE

The year 1900 marks no special stage in the draining of the Levels, but it is a convenient date at which to pause, for from this time on, the lack of progress in the draining of the five regions presents a unifying theme of inactivity. From an examination of the annual reports of the Somerset Drainage Commissioners for the years 1900 to 1914 it is clear that only a minimum of maintenance was carried out;[2] a further report by Lunn on the River Parrett and its outfall was called for in 1912 and then ignored.[3] Drainage conditions slowly became worse throughout the whole of the Levels and agricultural

[1] Acland, 45–6, quoted on p. 199 above.
[2] S.R.B./D.C. 31.
[3] W. Lunn, *Flood Prevention and Obstructions in the River* (1912).

activity declined.[1] The annual flooding of at least all the peat lands was an assured fact, and it is probably safe to say that the drainage situation in 1900 was no better, and possibly worse in some localities, than it was a hundred years before.

With the outbreak of the First World War, and the national drive towards higher food production, the problem of the farmers in the Levels became acute. With every justification, they maintained that the inadequate drainage hindered any possibility of increasing output, a view which was more than substantiated by another serious and widespread flood in 1917. Eventually, the Somerset Agricultural Executive Committee by-passed the Drainage Commissioners and took matters into its own hands by asking for an independent enquiry into the state of the drainage. C. H. J. Clayton reported in 1918; he was highly critical of all aspects of the maintenance and administration of the drainage. In particular, he found all the main rivers in a deplorable state of maintenance and strongly urged that some work should be done on them immediately, for 'the longer it is delayed the more difficult and costly it will become'. The main rivers were the key to efficient drainage.[2] William Lunn, engineer to the drainage commissioners, immediately drew up plans for the comprehensive improvement of all the main rivers of the Levels by widening, deepening and regrading the beds, at a total cost of nearly £329,000.[3] The scheme was immediately rejected and a new report on the state of the drainage was called for. Obviously something was very wrong with the administration of the drainage. Basically the defects were four in number:

1 The unbelievable obstinacy of the Internal District Boards of the coastal clay belt. They had rejected all plans because they objected to being rated for improvements that would primarily benefit the inland peat moors and upland areas. These were the Boards of the Lower Axe, Lower Brue, Stockland, Bridgwater and Pawlett, and Cannington and Wembdon Districts (see Fig. 36). The case of the Lower Axe was a typical one. The clay belt farmers dammed up the

[1] For a general comment and survey of drainage conditions in England throughout these years, see H. Spence-Sales and J. Bland, *England's Water Problem* (1939), 189, in particular.

[2] C. H. J. Clayton, *Flooding in the Drainage Areas of the Rivers Parrett, Tone, Cary, Brue and Axe* (1918).

[3] W. Lunn, *Flooding in Districts: Report and Estimates...for Improvements in the Rivers Parrett, Tone, Brue, Axe, King's Sedgemoor Drain and Cary River* (1919).

water of the Axe at the old Bleadon clyse with clumsy baulks of timber in order to obtain summer water in their lands. If a sudden or prolonged amount of rain fell and the troublesome baulks were not removed in time (which was usually the case) extensive flooding took place in the peat lands of the Upper Axe Districts. Yet, the Lower Axe Board consistently blocked the expenditure of £6,700 for the improvement of the clyse and the River Axe generally, on the grounds that their area would not benefit. It was merely another example of the conflicting interests of the clay belt and peat land proprietors, which had so often prevented progress in the past;

2 The Somerset Drainage Commission was not disposed to overrule these rejections, because it was composed of representatives of the Internal Boards. What was even more significant, was the fact that the five clay belt boards had a preponderant voice on the commission, returning eleven out of a total of eighteen members, so that they could easily overrule proposals for improvements that they felt did not benefit them directly;

3 A comprehensive view of drainage requirements was greatly impaired because the Somerset Drainage Commissioners had, in many instances, delegated responsibility for main rivers to the Internal Boards, which did not always have an engineer to advise them. The controller of the drainage, or expenditor as he was called, was at best a solicitor or business man, but more often a local farmer with no technical experience, and, what was worse, with a narrow and limited attitude to the problems and requirements of draining;

4 The financial basis on which the commissioners had to work was pitiful; their resources were not equal to the expenses involved in the execution of the most essential works. The problem of finance was partly due to the need to repay £60,000 borrowed for long-needed improvements and major maintenance work carried out between 1882 and 1894. The repayment of this money over thirty years put an intolerable burden on the meagre resources of the commissioners by taking over half (£2,300) of their average annual expenditure of £4,500 between 1890 and 1918. Therefore, there was very little money left, and about £1,500, or less, was spent on maintenance, and this included the upkeep of all the main rhynes, rivers, clyses, and sea walls throughout the Levels. Even when the loans were at last repaid in 1920 and 1925 the commissioners were immediately plunged into

debt again by taking up new loans of £8,000 for necessary maintenance.[1] Thus the commissioners were bound to an impecunious existence by their own constitution and by the intransigence of the Internal Boards, who backed up their unwillingness to approve of new drainage measures with a refusal to collect any more special precepts issued on them by the commissioners. They argued, and, it must be admitted, with some justification, that the land had deteriorated so badly that it could not stand any increased tax.

These defects were not amended and in 1919 a new report by S. Preston, of the Board of Agriculture, was asked for by the County Drainage Committee.[2] The report was damning and the criticism could not have been put in stronger terms. Preston said that the administration and maintenance were thoroughly inefficient and he called for a stop to the endless stream of reports and recommendations about defects, the remedies for which were well known and recognized, and asked that some money be spent on improving the main rivers instead.

By this time, every part of the Levels was subject to regular flooding;[3] the area of land under water and injured by flooding in 1917 and 1919 was considered to be comparable to Grantham's calculation of approximately 70,000 acres of land flooded during the exceptionally wet winter of 1872–3. West Moor, North Moor, Aller Moor, West Sedgemoor, King's Sedgemoor (Fig. 24), and large parts of the Brue valley carried only 'poor rough grass, rushes, and sedges, and...flood regularly or are permanently water-logged'. Even the floodable area of the Axe valley was said to be 'tending to increase in recent years', and the results of defective drainage were quite as marked there as elsewhere in the Levels, 'some fields being entirely over-run with rushes and rendered practically worthless from the farming point of view'. The relative superiority of the moorland grazing pastures was lost to those of the uplands, which were now worth from 20*s*. to 25*s*. an acre more. Arable farming was all but at an end and the old cultivated fields in the moors were 'given up to

[1] An examination of the commissioners' accounts for the whole of their existence from 1881 to 1930 reveals no appreciable change in the general picture outlined above. See S.R.B./D.C. 13–15.

[2] S. Preston, *Notes...on Reports re Floods in the Districts of the Parrett, Tone, Cary, Brue and Axe* (1919).

[3] This and the following paragraphs are based largely upon J. G. Stewart, *Report to the Board of Agriculture on the Drainage of the Valleys of the Parrett, Tone, Cary, Brue and Axe* (1919).

poor, rushy sedge'. Small patches of cultivation still lingered on in the eastern end of King's Sedgemoor and in the Brue valley but cropping 'was risky on account of water-logging'. As Stewart concluded in his survey of agriculture in the Levels:

At present uncertainty as to results hangs over the farming of the lowlands like a blight. A farmer may buy stores in the spring, keep them, as is the practice, for some weeks on the high ground, and then find that he is unable to summer them on the water-soaked marshes or, should a summer or autumn flood occur, as was the case last year, he may find himself with insufficient grass to finish his stores and must needs dispose of them at a loss.

The same drawbacks affected the hay crop which was often carried away by a flood. The limitations imposed on dairying were perhaps even more serious, for there was little incentive to improve a herd because stock frequently had to be reduced during the winter. Consequently the supply of milk was 'mainly regulated by the state of the marshes'. In most years the Levels were bereft of cattle from November to May inclusive; indeed they were still 'summer grounds', and there was little to distinguish the landscape of the Levels in 1919 from that of 1819. The experiments in changing the economy, begun in the late eighteenth century, were at an end.

Despite the strong recommendation by Stewart that Lunn's plan of 1919 be put into operation and that an expenditure of £390,000 would be 'amply and fully repaid' by greater farming production, still nothing was done. The Chairman of the Commissioners, R. N. Grenville, could point to no positive achievements when he spoke on 'Somerset Drainage' to the Somerset Archaeological and Natural History Society in 1926.[1] The annual reports of the commissioners from 1921 to 1927 show that the only works constructed were new gates for Highbridge Clyse and a penning sluice at Bleadon, on the Axe.[2] Some maintenance to internal and arterial drains was carried out under the auspices of government-financed unemployment relief schemes, the most significant thing about them being, perhaps, if one takes the long view, the provision for the first time of central government funds for drainage maintenance in the Levels.

In 1929 the commissioners ended their half century of administration with almost as great a flood as had brought them into existence in

[1] R. N. Grenville, 'Somerset drainage', *S.A.N.H.S.*, LXXII (1926), 1–12.
[2] S.R.B./D.C. 32.

1877. The Tone bank burst at Athelney and the whole of the Southern Levels and King's Sedgemoor were flooded from 11 November 1929 to February of the next year and many hundreds of persons were rendered homeless.[1] Speaking at a conference to discuss yet another plan to improve the drainage system, one speaker was reported as saying that:

> It was not uncommon for certain villages and hamlets to be isolated during flood periods. That day the village of Kingsbury Episcopi was practically cut off from the outer world. Besides the loss of ordinary social amenities there was a definite menace to health. It was a pathetic sight to see families driven from the ground floor and have to take refuge in the upper rooms, while he had known cases where children had been unable to get to school for weeks at a time.[2]

The situation had now become critical. With the passing of the Land Drainage Act in 1930 it was hoped that the fitful activity and even downright lassitude which had characterized affairs during the preceding century would at last come to an end and that a start would be made on main river works to combat flooding. The Act removed the two great impediments to successful draining, for the tangle of authorities working with antiquated powers and inadequate resources was swept away. The new administrative authority was the Somerset Catchment Board and it had exclusive control over the main rivers, including those of the Northern Levels, where the Court of Sewers was finally disbanded. The old formula of benefit being abolished, the board could levy rates over the whole of the catchment area of the main rivers, by issuing general precepts in proportion to the rateable value of property in the area of county councils included within the catchment areas, and could issue special precepts on the Internal Drainage Boards, the distinction between upland and lowland still being maintained.

It was regrettable, however, that the promise of improvement held out by these long-awaited changes was largely nullified by the chaotic

[1] For details of the extent and effect of this disastrous flood, see T. Stuart-Menteath, *Somerset* (1938), 23–4, being part 86 of *The Land of Britain*, ed. by L. D. Stamp. See also the the *Bridgwater Mercury* and *The Times* of the appropriate dates.

[2] Report on a conference held at Bridgwater of the representatives of various public bodies to discuss the Report on the Drainage by W. H. Haile, Engineer to the Drainage Commissioners, entitled 'Great Floods Prevention Scheme', *Bridgwater Mercury*, 4 December 1929. The isolation caused by flooding was so great that it was said that 'In some of the most backward districts...in the once island villages of Sedgemoor, the social life of the peasant is not unlike that of his predecessor a thousand years before'. *V.C.H.* (1906), I, 336.

economic conditions of the period immediately following, and, although regular maintenance was put on a surer footing, little permanent improvement was made. The general rate could not contribute enough towards the cost of remedying the major defects of the main channels, and the cost of improvement fell on the Internal Drainage Boards which did not have the necessary resources. The falling price of agricultural products, especially of hay[1] and meat, the main products of the moors, did not warrant the extra capital charge of £5 or £6 per acre which would have been necessary if the districts were to finance their own improvements.[2]

Not only were there these very real economic drawbacks but the Catchment Board began to adopt the timid and apprehensive approach that had characterized other administrative bodies in previous years. It resorted to plans and reports, and to routine maintenance of no lasting consequence, although at this time the extent and severity of flooding were considered to be increasing (Fig. 42).[3] The board became preoccupied with the old problem of excluding tide and silt from the Parrett outfall and explored many grandiose possibilities without agreeing on any one course of action.[4] The object of these schemes was to attempt either to exclude the tide by a clyse or reduce the marine siltation of the outfall by increasing the scouring power of the river with the aid of training walls, groynes, and new cuts, and thereby ensure a quicker evacuation of flood waters. The effect of either might have been to reduce the level of water in the main river, though many people seriously doubted this.

Finally, the various proposals were subjected to extensive

[1] Stuart-Menteath, *Somerset*, 27: 'Hay on West Sedgemoor was offered recently at £1 to £1. 5*s*. a ton. Twenty acres was offered and refused for £2 the lot, if the buyer would cut it.'

[2] If Lunn's plan of 1918 had been put into operation, the cost per acre over the area of the District Boards would have been as follows:

Southern Levels, £6. 4*s*. 8*d*. per acre
King's Sedgemoor, £6. 18*s*. 9*d*. per acre
Brue valley, 14*s*. 10*d*. per acre
Axe valley, 7*s*. per acre

[3] Stuart-Menteath, *Somerset* 29. See also Figure 42A for the extent of 'average annua flooding' in 1936.

[4] The reports included the following: (i) J. Wolfe-Barry, *Report...upon the Question of Tidal Gates or Sluices in the River Parrett* (1920); (ii) H. Van der Veen, *Report on the Improvement of the Lower Course of the Parrett* (1935); (iii) F. B. Goodman, *Résumé of Improvement Schemes* (1936); (iv) R. G. Clark, *Report on the Somerset Rivers Flood Alleviation Scheme* (1937). In the last report it was stated that 44,000 acres were found flooded in 1936, and this was considered to be a 'mild flood'.

experiments and tests by a tidal model.[1] But the first report on these experiments was not available until 1941, eighteen months after the war had begun, by which time startling changes had occurred which were to usher in a new era in the draining of the Levels.

[1] The original reports of the tests by Professor Gibson of Manchester University are in S.R.B./C.B. 6. Useful summaries are given in Du Cane, *River Parrett Estuary Scheme*, and J. Allen, 'Schemes of improvement for the River Parrett in Somerset; an investigation with the aid of a tidal model', *Journ. Inst. Civil Engineers*, XIX (1942), 85.

8

THE NEW SPIRIT

GOOD FORTUNE AND FORESIGHT

In 1939, the flood of reports and recommendations ended, and debate gave way to action. Work was begun on the improvement of the King's Sedgemoor Drain, this being the first significant contribution made to the draining of the Levels for over a hundred years, with the exception of the introduction of the steam pump.[1] The reason for the new attack on the problem is not difficult to understand when one recalls the static and even deteriorating state of the entire drainage system during the previous years. Something had to be done, even at the cost of incurring large debts by borrowing the necessary money. There was, in addition, a marked change of attitude towards the problem of draining with the appointment to the board of a new engineer. A person of great energy and foresight, he raised the post of engineer from what in the past had been a subordinate position in the control and conduct of drainage matters, to a paramount one. All the information regarding drainage problems was at the disposal of the board, and there was no need for recourse to outside authorities for more advice. It merely needed someone with a thorough knowledge of local and general drainage conditions to mould these recommendations into a feasible plan and act upon it.

Another factor which contributed towards the new attitude was the realization that attention had been concentrated on the River Parrett for too long. It was the one river for which there was no agreed solution, and this became more evident as time went on; therefore, with no hope of achieving an immediate solution there, it was better that attention be turned elsewhere. The immediate need was for less ambitious but better conceived schemes that were well constructed and well maintained. They should be planned as part of a larger whole if need be, yet capable of operating independently, bringing some measure of relief to flooded areas, until such time as they were integrated with other schemes. Much of the difficulty

[1] M. Williams, 'Draining activity in the Somerset Levels since 1939', *Geography*, XLIX (1964), 387–99.

during the nineteenth century had arisen from repeated complaints about the parochial attitude towards draining, which had eventually led to much theoretical planning on too large a scale for the resources at the disposal of the authorities; consequently little was done. This was particularly true of the 1930s and arose from the insistence of the Ministry of Agriculture, which was contributing money for prospective schemes of improvement.

Now that there was agreement about the type of scheme needed, work started in King's Sedgemoor, which had the best drainage system in the Levels and was most likely to produce a satisfactory result. The works aimed at rectifying the three defects which had bedevilled the efficient working of the drain since its construction;[1] these were, the narrowness of the drain, especially through the clay belt, the narrow bridges, and the insufficiently large opening and high sill of Dunball Clyse. Barely had work begun when the Second World War started, and, with the threatened diversion of men and materials to more urgent work, the great advance that had at last been made seemed in danger of being lost.

Fortunately this was not so. A new Royal Ordnance Factory was planned near Puriton in the Brue valley for which the Catchment Board was asked to provide a guaranteed daily supply of 4½ million gallons of water; this was not difficult to do during the autumn and winter months but could be difficult during the extremely dry summer conditions which sometimes prevail in central Somerset. With considerable foresight, the Chief Engineer of the board combined the plan for water supply to the factory with a drainage scheme. It was based on a suggestion of Clark's in 1853, and a straight channel, 5 miles long, was cut across the coastal clay belt from the southern portion of the Brue valley at Gold Corner, to the Parrett estuary (Fig. 39). The eastern end of the Huntspill river, as the new channel became known, was then connected to the South Drain. By adding retention sluices at both ends of the new channel, the Huntspill river formed an elongated reservoir, which could be drained gravitationally in the winter to clear flood water, and could be replenished in summer by pumping from the peat moors.

Work began on the new channel in January 1940, less than three months after the scheme was first proposed. The depth of the new channel through the clay belt was planned to be 25 ft in order to

[1] See pp. 149–50 above.

allow gravitational discharge from the moors, but, when excavations started, the weight of the excavated earth proved too great for the peat subsoil, and within a few hours the spoil banks depressed the subsoil and rotated in an arc-like fashion causing welling-up in the bottom of the channel. Extensive investigations showed that the spoil banks

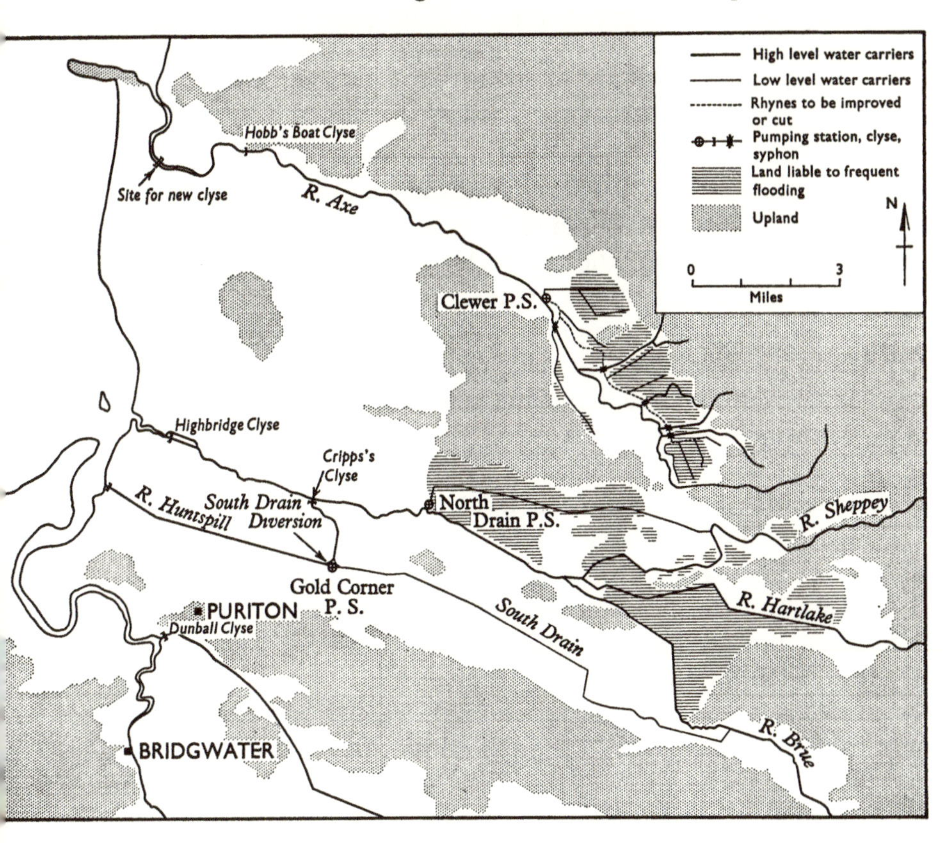

Figure 39. The Brue and Axe valleys in the mid 1960s.

could not be more than 15 ft high and not closer than 30 ft from the edge of the channel, while the channel depth should not exceed 16 ft.[1] This restriction of the depth of the excavation meant that gravitational drainage was impossible and that the small pumping station near Gold Corner had to be enlarged to cope with the whole flow of the South Drain and any excess flood water in the moors. The

[1] L. F. Cooling, 'Soil mechanics and site exploration', *Journ. Inst. Civil Engineers*, XVIII (1941), 59–61.

pumping station was finally situated on the edge of the coastal clay belt which is about 8 ft higher than the inland moors, and it lifted water over the clay belt barrier, into the Huntspill river.[1] By joining the Huntspill river to the South Drain, one of the defects of the Brue Comprehensive Drainage Scheme of 1801 was rectified; the South Drain now had an outlet which was independent of the Highbridge outfall. With the minute control of water-levels afforded by the Gold Corner pumping station, annual flooding was all but eliminated in the southern part of the Brue valley.

After the excavation of the Huntspill river, the length of the South Drain between Gold Corner pumping station and the River Brue was no longer used, but in order to increase the overall effectiveness of the scheme this portion of the South Drain was enlarged, Cripps's sluice built at its northern end and the drain diverted to the west side of the Gold Corner pumping station and into the Huntspill river. The South Diversion, as this channel became known, allowed surplus water from the Brue, impounded by tide-lock, to be drawn into the Huntspill river by the regulation of Cripps's sluice. This has decreased markedly the flooding alongside the Brue.[2] Only the north side of the Brue valley was still subject to regular inundation caused by conditions similar to those that made the South Drain ineffective. In the meantime, work on the King's Sedgemoor Drain was resumed and all intended improvements completed because the drain was made into a secondary source of water supply for the factory, in case the Huntspill source failed.

Thus, in five years, both King's Sedgemoor and the south side of the Brue valley were freed from all but the most severe of autumn and winter floods. Together they amounted to nearly a third of the land in the Levels, which had previously been waterlogged or inundated for the greater part of the year.

PRESENT TRENDS AND SCHEMES

The years since 1945, particularly those since 1950, have seen an unparalleled amount of activity which is fast bringing the flood problem of the Levels under control. Freedom to plan ahead and ability to construct new works have been possible because of the provision of

[1] The pumping station contains four Diesel pumps each of 240 h.p., capable of pumping a total of 1,100 tons of water per minute.

[2] This part of the scheme was not completed until 1955.

adequate funds,[1] the presence of an efficient administration, and the execution of regular maintenance of the existing drainage system. Because of the great mass of technical detail that has been accumulated during recent years[2] concerning the improvement of existing works and the construction of new works, it is impossible, and probably not desirable, to do more than indicate the main trends that have become apparent, and to comment on the most important of the new schemes.

Planning and activity have been concentrated on the following four aspects of the flood problem:

1 The improvement, wherever possible, of existing banks, sluices, drains and sea walls;

2 The provision of new outlets through the coastal clay belt;

3 The improvement of the existing pumping stations and the building of new diesel or electric pumping stations to raise excess water from the moor basins into the main river and drain outlets;

4 The provision, upstream, of flood-relief channels to divert the water from the upper portions of the rivers direct into the relief channels and thus by-pass the moors.

These schemes are aimed at extending the grazing season by eliminating all the intermittent floods of spring and summer, and, if possible, those of early autumn. The severe floods that occur fairly regularly during the late autumn and winter, i.e. November to January, are to some extent uncontrollable, and to construct works capable of completely eliminating them would not be economic. In any case, the Levels are almost wholly under grass, and a temporary winter flood does relatively little harm to the vegetation provided that the flood water can be cleared within about two weeks. This approach to the flood problem, which attempts to provide seasonal relief, is more realistic than the earlier attempts at complete relief, and has brought greater success.

[1] Direct government grants account for approximately 50% of the board's income. Up to 85% of the cost of a particular drainage scheme may be provided for by a government grant.

[2] Perhaps the best and most accessible account of recent drainage works is in the *Annual Reports of the Somerset River Board.* See also E. L. Kelting, 'An outline of the Development of Land Drainage in Somerset', a paper read to the Annual Conference of the River Boards' Association, Weston-super-Mare, 1958.

The improvement and maintenance of existing works

No drainage system can function efficiently without the indispensable requirement of routine and regular attention; therefore, the maintenance and improvement of existing works have been the greatest, though the least spectacular, part of work done in the Levels in recent years. Two technological innovations have altered the whole conception of maintenance; first, the mechanical drag-line is mobile and versatile in its application and can clean and improve rhynes without damaging the banks; secondly, hormonal sprays have been used to kill off vegetation-growth in the rhynes.

The amount of work done has been so great that it is impossible to single out specific schemes. It would be true to say that there is hardly a major rhyne, sluice or embankment in the Levels which has escaped attention in recent years. Over 4 miles of sea walls, particularly in the Northern Levels, have been strengthened or rebuilt; 14·8 miles of tidal embankments, mostly around the Parrett estuary, have been raised, widened and strengthened; over 81 miles of flood banks alongside the Brue, Parrett, Tone and other major channels have been rebuilt and strengthened, and 88·7 miles of major channels improved and maintained by dredging. At present, further work is being done on the South Drain, which is being enlarged and generally renovated. Viewed against the inactivity of previous decades and centuries, this is an impressive catalogue of improvement and maintenance and it could be augmented if one also considered the work done by the Internal Boards.

New outlets

The Huntspill river was the first of the new outlets to be cut across the barrier of the clay belt. Another was in the low peat moors of the Northern Levels, lying mainly between Clevedon, Yatton, and Tickenham. This land was subject to frequent and prolonged flooding mainly because the outfall of the Kenn river was open to tidal influence. In 1949, work started on a new channel leading from the Kenn river across the clay belt for 2½ miles, ending with an outfall sluice about a mile north of the former Kenn estuary (Fig. 25). The effectiveness of the Blind Yeo, as the new channel has become known, was demonstrated immediately after its completion in the winter of 1953–4, when rainfall and tidal conditions, which normally would have produced flood conditions in the moors, resulted in only local waterlogging.

The dependence on pumping

The dependence on pumping is a reversion to the mid nineteenth-century practice, and indicates, to some extent, the abandonment of the principle of gravitational drainage, which had been forcefully advocated for so many years. The conception of one level of drainage in the Levels is not realistic. Under normal conditions of rainfall and tide, the level of the main rivers that flow from the uplands to the sea is barely lower than the peat moors through which they flow. When rainfall and tidal conditions are not favourable, however, the level of the main rivers is well above that of the surrounding land. Under these conditions of high tides, tide-lock, and heavy rainfall, gravitational discharge from the moors is totally impossible. Thus, there are two levels of drainage; one is the high level drainage of the large rivers which are fed from the uplands; the other is the low-level, internal drainage of the moors which requires pumping. The acceptance of this fact of two levels of drainage has played an important part in the present improvement in drainage.

Pumping has been encouraged by the advantages of the diesel engine which is efficient, economical,[1] quick to start in an emergency, and well suited for intermittent work. Thus the old steam engines in Saltmoor, Northmoor, Stanmoor, Southlake Moor, and the Westonzoyland Level were replaced by more efficient and powerful diesel engines in 1941, 1942, 1948, 1948 and 1950 respectively (Fig. 40). In 1944, a new pumping station was constructed for West Sedgemoor, and, in 1955, the Athelney steam station was abandoned and replaced by the new Curry Moor diesel station. In an attempt to obviate flooding on the northern side of the Brue valley, the North Drain pumping station was constructed in 1959 and the old North Drain widened. Flood waters now pass quickly to the pump and are lifted into the Brue, irrespective of the height of the flood (Fig. 39).[2]

[1] Tests were made by the S.R.B. on the old steam-powered pump at Athelney and the new Curry Moor diesel-powered pump built to replace it. While the diesel engine could pump against a static head of 11·5 ft, the steam pump had to be stopped at 5 ft. Moreover, the cost of coal for the steam pump per water horse-power hour was about 1*s.* while the corresponding cost of fuel oil for the diesel pump was 1½*d.* See *Fifth Annual Report of Somerset River Board,* (1954–5), 14.

[2] The flood level in the River Brue is approximately 13 ft O.D. while the level of the peat moors varies from between 7·5 and 10 ft O.D. See E. L. Kelting, *Municipal Journal*, 12 April

In the same way as the diesel proved more efficient and economical than the steam engine in the years immediately after the Second World War, so the electric engine is now replacing the diesel engine,

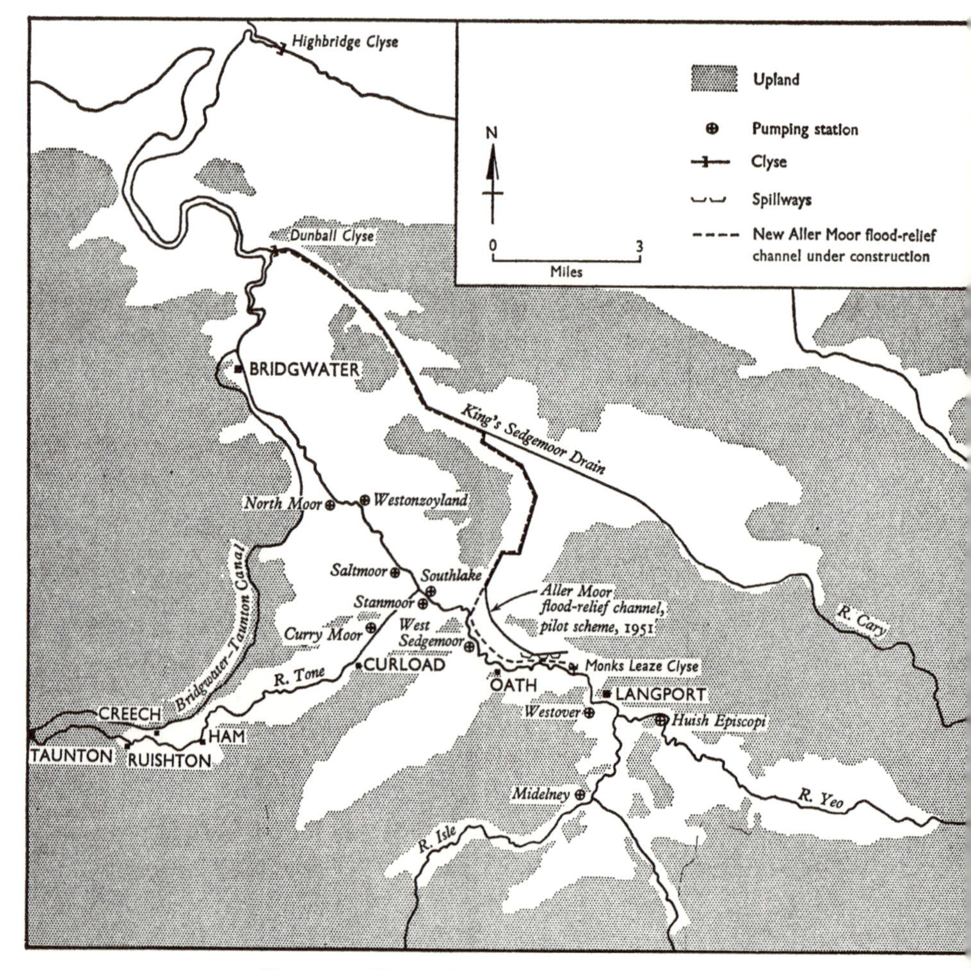

Figure 40. The Southern Levels in the mid 1960s.

and has been installed in three new pumping stations which have been constructed in the moors above Langport between 1960 and 1966 (see Fig. 40). These are the Midelney station, which drains West Moor; the Huish Episcopi station, which drains Wet, Hay and

1957, 792–3. See map of flooding in Northern Brue valley accompanying the Agricultural Land Commission Report, *Land in the Brue Valley and North Moor, Somerset* (1955).

Ablake moors, and the Muchelney Level; and the Westover station, which drains the Huish Level and Perry Moor. Because the electrically-driven pump is capable of automatic operation, depending upon the flood level in the main rivers, it makes more economical use of the pumping time, and also saves in the use and cost of skilled labour which is an important factor in the operation and maintenance of diesel engines. Another advantage of electric motors is that they can be more easily erected above, rather than alongside, the pumps, thus saving floor space and consequently some of the very great expense involved in piling for the station's foundations in the peat and alluvium.[1] So economical and efficient are the electrically-driven, off-peak supplied pumps, that they will be used in all future stations. Already, one of the four diesel pumps at Gold Corner is to be converted to electric motive power, and the new pumping station planned for the Axe valley will have electrically-driven engines.

The new station in the predominantly flood-free Axe valley will drain a small pocket of persistently flooded and waterlogged land in Stoke, Draycott, Cheddar, Westbury and Wedmore Moors. This flooding is caused by the rapid run-off of water via upland carrier drains into the Axe, which causes a quickly rising level in the river and therefore cessation of gravitational drainage from the moors. The new scheme envisages the construction of a new pumping station[2] near Clewer and the reorientation and improvement of the low-level moorland rhynes either side of the Axe so that they will all lead towards the station and their water be pumped up into the Axe (Fig. 39).

Wherever new pumps have been installed, a vast improvement has resulted, especially in those areas where completely new stations have been erected, as at West Sedgemoor and the North Drain. A strict control of water-levels is now possible and surface water can be cleared within hours of appearing. Very few new rhynes were excavated for the pumping stations because of the cost of purchasing new land, the possibility of severing property, and the need to provide access bridges; instead, old rhynes were enlarged, deepened and generally reconditioned. Consequently, modern drainage works have

[1] The expense of piling can be appreciated when it is realized that the average depth of 70 piles for the North Drain station was 96·5 ft.

[2] A small centrifugal pump had been erected by the Upper Axe Drainage Board as early as 1915 to deal with this flooding, but it was too small for the work it needed to do, and its use was controversial; it was felt that the pump increased the level of water in the Axe and affected gravitational discharge from the moors; and it consequently fell into disuse.

brought about very little change in the early nineteenth-century pattern of drainage channels in the Levels (Fig. 26).

As a result of the recent increase in pumping and the general improvement in drainage, two difficulties may be expected. The first is that of peat shrinkage and wastage through improved drainage. Experience in the Fenland of eastern England has shown what enormous problems these could become. Nevertheless, it appears unlikely that the same changes would take place in the Levels where all conditions tend to keep the peat saturated. Foremost amongst these conditions is the regional pattern of farming, which is overwhelmingly pastoral in nature, being based on relatively high rainfall and therefore tolerating a high degree of soil moisture. Over 90 % of the farms are devoted to dairying and stock-rearing, and over 90 %, on average, of the area of a Levels parish is given over to permanent grass or rough grazing, cultivation being confined almost entirely to flood-free 'islands' and upland edges. A low water-table is therefore not essential, and the peat is not allowed to dry out. The saturation of the peat is also consciously aided by the practice of 'wet draining', whereby water is kept in the rhynes for fencing and irrigation purposes, and, it is averred by some farmers, to prevent the welling-up of the rhyne bottoms, which would occur if the weight of the water was absent. In addition to these factors there is still the controlled flooding of the moors for warping and pest control (although these practices are of declining importance), and there is the possibility of uncontrolled floods, both of which cause the saturation of the peat. In the experience of the River Board, peat shrinkage 'has not been measurable'; however, the possibility of the danger arising with improved farming techniques is recognized, and all pump inflows have been constructed so that a possible 4 ft lowering of the surface could be accommodated in the future.

The second difficulty arising from the trend towards pumping is that it will increase the already high level of water in the main channels. This danger has been made more real by the continual construction of stronger and higher embankments, which make it more difficult for the swollen rivers to break through and dissipate their volume in the surrounding moors. This consideration, then, leads one to the fourth characteristic of modern drainage activity in the Levels, the provision of relief channels.

The provision of relief channels

Within the main channels of the Parrett and its tributaries, the problem of the high freshwater flood levels remains to be solved. It is severely aggravated by the unrestricted entry of the tide at the outfall and by the narrow and winding parts of the river course, which restrict the discharge of the flood water. The solution to the problem of the high freshwater flood levels is no longer sought at the outfall as it was before 1939, for no further deepening or regrading would either prevent the extension of silting or alter the tidal levels of the sea in the river channel. Even if silting could be prevented and the tidal level altered, it is by no means certain that this would result in a lowering of the surface gradient of the water upstream.[1] An alternative solution would be to construct a sluice near Bridgwater, but the benefit that would result is not certain and probably not commensurate with the immense expense involved in its construction. To widen and straighten the narrow and winding parts of the channel is difficult, particularly where roads and houses occupy the tops of the flood-free banks. In the face of these difficulties and uncertainties, present opinion seems to favour the construction of relief channels to by-pass difficult sections of the river. The upland water would be tapped in the upper reaches and the lower portion of the main river would be left to cope with the local pumped drainage of the moorland basins.

The problem of the Parrett, Tone and Cary rivers and its possible solutions are best introduced by a diagrammatic representation of the flood situation during a period of fairly severe flooding. Figure 41 shows the average position during a ten-day peak period in the middle of flooding from 6–20 November 1951. This November was the wettest in the Levels since 1929, when serious flooding occurred;[2] it was also one of the wettest Novembers during the last 75 years, the rainfall over the Parrett catchment area from 6–20 November being 6·73 in. From this information the discharges of the rivers have been calculated, assuming a 100 % run-off, which in the west of the British Isles at this time of the year should not be too far removed from actual conditions. The observed discharges are the mean of the readings over the ten peak days of flooding.[3]

[1] See the reports on the investigations with the Parrett estuary tidal model. For references, see p. 236, n. 1, above. [2] See p. 234 above.

[3] Information on the observed discharges was kindly supplied by the Somerset River Board.

The diagram shows the constriction of the outfall of the River Tone and the need for widening the channel. But widening would have been difficult because so much building had taken place along

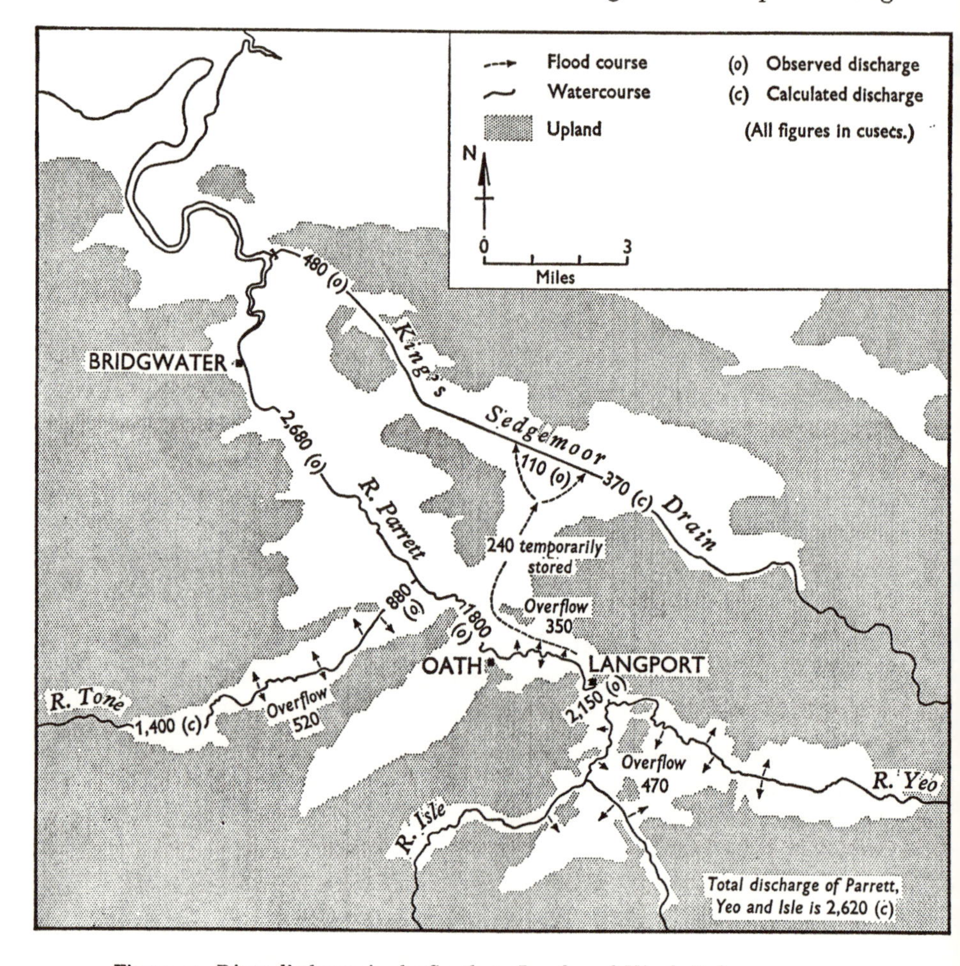

Figure 41. River discharge in the Southern Levels and King's Sedgemoor, 1951.

the relatively flood-free banks; houses stretched for about two miles on the top of the Stanmoor or south-eastern bank, between Curload and Burrow bridge.[1] The cost of purchasing these properties would have been prohibitive and therefore, with no prospect of widening the river, a concrete core was placed in the centre of the bank, during

[1] For details of this characteristic form of Levels settlement, see B. M. Swainson, 'Rural settlement in Somerset', *Geography*, xx (1935), 116.

the years from 1956 to 1964, to prevent seepage and breaks which revealed themselves in time of floods. But this was only a temporary solution, because the channel was still incapable of carrying off all the water that descended from the western hills, and flooding occurred regularly in the surrounding moors. The true magnitude of the

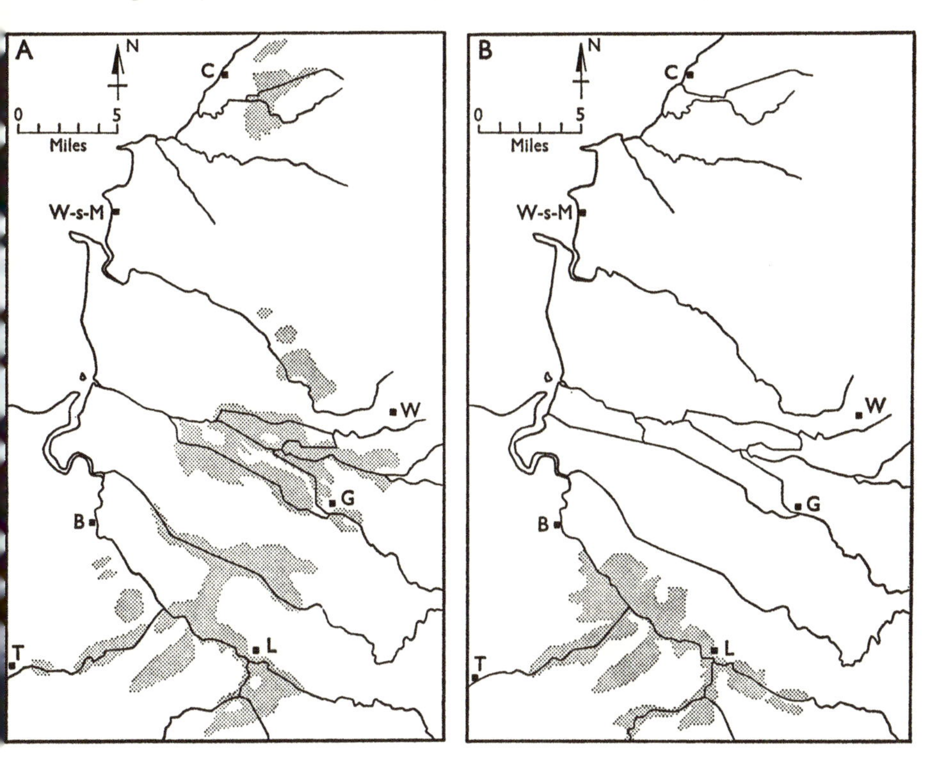

Figure 42. Flooding in the Levels: A—in 1936; B—in 1960. B, Bridgwater; C, Clevedon; G, Glastonbury; L, Langport; T, Taunton; W, Wells; W-s-M, Weston-super-Mare. (Sources: 1936—T. Stuart-Menteath, *Somerset*, Part 86 of the *Land of Britain*, ed. L. Dudley Stamp (1938), 21. It shows 'approximate areas flooded annually'; 1960—C. H. Dobbie and Partners, *Somerset River Board, River Parrett Flood Relief Channels; Technical Report* (1961) (mimeographed). This shows the extent of flooding in October and November 1960.)

problem was starkly revealed by the unprecedented floods of October 1960, when rainfall over the entire catchment area of the Levels averaged 9·72 in. or 249 % of the normal. It was the wettest month of the century. Flooding was widespread throughout the Tone moors, North Moor, West Sedgemoor, Aller Moor and the upper Parrett valley, and it was particularly severe in Taunton itself where a large part of the commercial section of the town was under 3 ft of water (Fig. 42).

A scheme was prepared immediately to alleviate flooding by utilizing the abandoned Bridgwater–Taunton Canal as a flood-relief channel (see Fig. 40 for location). The canal has considerable advantages for, although it is circuitous, it is located mainly within the upland areas where there is little danger of flood water escaping into the lowland basins. Also the need for high embanked channels, with unstable peat foundations, on moors would have been avoided. Yet the cost of nearly £1,700,000 was not considered commensurate with the probable benefits that would result, and so a modified scheme at about one-third of the cost was prepared and this is being implemented now.[1]

The modified scheme provides for a discharge of about 4,500 cusecs through Taunton, this discharge having a calculated frequency of once in 30 years; but there is some freeboard with it, so that overflow will not take place until the discharge reaches 6,000 cusecs, which has a frequency of about once in 100 years. The villages of Creech St. Michael, Ham, and Ruishton, below Taunton, will be provided with the same standard of protection. In order to do this the improvement of the channel is being undertaken; banks are being strengthened and raised, cut-offs made of narrow and winding portions of the river immediately below Taunton, old navigation locks and mill weirs demolished, and bridges widened. Below Taunton and the villages, the river channel is being widened to cope with a flow of between 2,000 and 2,500 cusecs, and any excess will be stored in Curry Moor and pumped out later. But the problem of the narrow final stretch remains, and a long-term programme of purchasing properties on the North Moor or north-west bank when they come on the market is being undertaken, so that eventually the entire bank can be moved back and the outfall into the Parrett widened.

The diagram in Figure 41 shows that severe flooding can also take place in three other areas of the Southern Levels, in the moors above Langport, in Aller Moor and in King's Sedgemoor. In the Parrett river just above Langport the observed flow of 2,150 cusecs was 470 cusecs less than the incoming calculated flow above this point, therefore 470 cusecs must have been disposed of by the overflowing of the

[1] C. H. Dobbie and Partners, *Somerset River Board, River Parrett Flood Relief, Technical Report* (1961; mimeographed), and *idem, Somerset River Board, Technical Report on the Tone Valley Scheme* (1963; mimeographed).

banks. The exact amount of this overflow may be questioned because of the possibility of exaggeration in the calculated inflow figure; nevertheless, experience has shown that the moors in the vicinity of the junction of the rivers Parrett, Yeo and Isle are a centre of habitual and prolonged flooding, due mainly to the constriction of the Parrett channel below this point. The three new pumping stations at Midelney, Huish Episcopi and Westover have been constructed to combat the worst effects of this constant threat.[1]

Whatever reliance is placed upon the calculated discharges, the differences between the observed discharges at Langport and Oath cannot be gainsaid. The overflow of a substantial proportion of the Parrett water from this repeatedly troublesome section alongside Aller Moor is an established fact. From the difference between the calculated inflow of the Cary and the discharge at Dunball Clyse it is quite obvious that most of this water flows into King's Sedgemoor, two-thirds of the water stagnating on the moor before it gradually drains away into the Main Drain. Therefore, there is an obvious need for a flood-relief channel through Aller Moor to prevent flooding in Aller Moor, King's Sedgemoor, and in the moors above Langport. Clark had suggested such a relief channel in 1853, and an experimental one, the Langacre rhyne, was excavated in 1951 in almost the same position; it is fed by a spillway in the Aller Moor bank (compare Fig. 40 with Fig. 24). The feasibility of passing Parrett flood water into the King's Sedgemoor Drain has been proved beyond doubt and the pressure of flood water near Langport and in Aller Moor has been somewhat relieved.

However, the floods of 1960 proved disastrous for the area, and, as in the case of the River Tone, proposals for improvement were put forward immediately, only to be rejected by the Ministry of Agriculture, Fisheries and Food as being too costly in view of the likely benefit. A new, modified scheme was prepared in 1963, which is now being carried out.[2] The original scheme was to consist of a new channel, comparable to the Langacre rhyne, which would have taken a very large proportion of the Parrett water through King's Sedgemoor, leaving the Parrett channel below the relief-channel spillways

[1] For further details of flooding in the moors above Langport, see 'Further Food Production from the Langport Area by Better Control of Flooding', a paper read before the Somerset National Farmers' Union at Langport, 11 July 1952. Copy in S.R.B.

[2] See Dobbie and Partners, *River Parrett Flood Relief*, and *idem*, *Technical Report on the River Parrett Relief Channel* (1964; mimeographed).

to cope with water pumped from the moorland basins and with the flow from an improved River Tone. The new scheme, on the other hand, has a similar relief channel, but with a capacity reduced by 40%; nevertheless, all structures will be made large enough to accommodate a channel capable of taking as much of the Parrett flow as was envisaged in the original scheme. A new urgency has been given to the work by the operation of the three new pumping stations above Langport. The King's Sedgemoor Drain will have to be widened considerably, and the Dunball Clyse, which can now deal with a discharge of only about 610 cusecs, will have to be demolished in the near future and replaced by one capable of passing 1,210 cusecs, and, when the whole scheme is completed, a total of 1,670 cusecs.

PROGRESS AND PROSPECTS

The years since 1939 constitute a period of intensive, rapid and spectacular activity in the draining of the Levels which is in complete contrast to the stagnation and apathy during the earlier part of this century and the bulk of the previous century. The accelerated activity of recent years shares with the period from 1770 to about 1830 and the less easily definable years between about 1230 and 1330 the distinction of standing out as an era of marked geographical alteration, indicated by changes alike in the drainage pattern, in the economy, and in the very landscape itself.

The major changes in the drainage pattern have been described already, and the continuance of the success they are bringing, and the future prospect of a solution to the problem of flooding, seem to be assured. The policy of pumping and of making relief channels and new outlets is realistic, correct, and likely to produce as good a result in the future as it is doing now. In particular, the improvement promised by the widening of the River Tone and the construction of the Aller Moor flood-relief channel holds out the prospect of revolutionary changes in the Southern Levels, with the almost permanent exclusion of normal autumn and winter floods.

Financially, and administratively, draining seems unlikely to languish again either from the want of adequate funds or from the frustration of antiquated laws. The problems of land drainage, especially on the scale encountered in the Levels, are slowly coming to be regarded as a national responsibility, and, as such, are being kept under constant review. A good example of this is the new Somerset River Authority, which was constituted in October 1964, and which replaced the old Somerset River Board. Amongst other things, the new authority gives greater representation to agricultural and water-supply interests and less to the Internal Drainage Boards, thus reflecting the widening interests and responsibilities of the authority in order to control the total hydrological features of the Somerset catchments. Of course, the Internal Boards still exist, and there is always the possibility that their interests will conflict with those of the authority; yet their demands and differences are gradually diminished by the comprehensive planning and central control of the River Authority in Bridgwater.

Future activity will continue to be based loosely on the five main drainage regions, which the physique of the Levels and the historical development of draining have suggested. Nevertheless, the construction of the Aller Moor flood-relief channel will destroy a division that previously existed between the Southern Levels and King's Sedgemoor, and the drainage of the two areas will become increasingly integrated. This is a reversal of the position after 1791 when the Cary was diverted away from the Parrett and into the King's Sedgemoor Main Drain. The possibility of integrating the Axe and Brue drainage systems by diverting some of the upper Brue water via the line of the old course of the Brue to the better drained Axe valley, is now remote, and indeed, the drainage of the southern Axe valley has been incorporated in the Brue system. The construction of the Huntspill river and of the pumping stations at Gold Corner and North Drain has greatly improved conditions in the peat moors in the Brue valley and therefore obviated the need for an alternative scheme.

From an agricultural point of view, the recent improvement in drainage conditions has resulted in an intensification of the pastoral economy, rather than in its replacement by arable farming. The reduced likelihood of late spring and summer floods has led to a decrease in mowing and an increase in the stock-numbers kept on the moors. It has also led to an increased use of fertilizer, which was formerly leached out of the soil, and to the establishment of leys on the better peat soils. Although some rearing and fattening of stores is still undertaken, and some pigs, poultry and sheep are kept, the region is now overwhelmingly a centre of summer milk production, and shares with Cheshire the greatest density of dairy cattle in Britain, outside the milk-producing zones on the periphery of the main urban areas.[1] With this change has come a moderate degree of prosperity to the Levels farmers. Cattle are turned out onto the moors in early spring and milked either in the open by hand, or more commonly by mechanical bail, and not until the moors become waterlogged in the late autumn or winter are the cattle removed to higher and drier ground. This emphasis on milking in the moors has given rise to a new problem; assured access to the moors is now a necessity, for milk must be collected daily and the bails removed if the weather deteriorates temporarily. The condition of the droveways is now almost as great an obstacle to the successful utilization of the Levels as flooding used to be; they are little more than unmetalled dirt-tracks which are impassable after rain, their deterioration being hastened by the increased weight of tractors, implements and lorries.[2]

A contributory factor, of undoubted importance, in the increase and predominance of stock has been the widespread development of irrigation. The requirements of drainage and those of irrigation are conflicting and hard to reconcile as the one demands the maintenance of a low water-level and the other a high water-level; the problem is given added complication by the different elevations of the coastal clay belt and the lowland peat moors further inland. In the past, the River Board has had to compromise between the optimum requirements for each purpose, with the consequent result of satisfying nobody. Recently, however, a system of irrigation has been completed, and the water has been penned, in the summer, at levels acceptable to both parties. The higher ground of the coastal clay belt and Bleadon Level is supplied by half-a-dozen small pumping engines which use water from the Axe and Huntspill rivers; the lower ground of the inland moors, such as Salt Moor, Stan Moor, Aller Moor, and King's Sedgemoor and the River Kenn moors, is supplied by gravity, the volume being controlled by penning sluices. This impressive pro-

[1] See J. T. Coppock, *Agricultural Atlas of England and Wales* (1964), Figure 159, which shows the number of dairy cows per 100 acres of agricultural land.

[2] For accounts of the problems and trends in the modern farming of the Levels see: (i) *Land in the Brue Valley and North Moor, Somerset,* Report of the Agricultural Land Commission (H.M.S.O. 1955); (ii) J. W. Dallas, 'A Symposium on some problem areas in the South-West. (1) The Somerset Moors', *Journ. Roy. Agricult. Soc.* CXII (1951), 9–14; (iii) *Idem,* 'Some aspects of West Country farming. (2) Somerset', *Journ. Roy. Agricult. Soc.* CXVIII (1957), 9–13; (iv) *Idem,* 'Farming in Somerset', *Agriculture, Journ. Ministry of Agriculture,* LXV (1958), 172–6; (v) C. M. MacInnes and W. F. Whittard (eds.), *Bristol and its adjoining Counties* (1955); see Chapter 7, 111–21, 'Agriculture and Horticulture', for a section on Somerset by J. W. Dallas.

gramme of providing water for irrigation and for stock in summer, has had a marked effect, and has resulted in an extended and more luxuriant growth of grass; it has also meant a vast extension of the responsibility of the River Authority.

Dairying is undoubtedly the most profitable farming enterprise in the Levels and perhaps the one most suited to the peculiarities of the environment. The future prosperity of the Levels farmer probably lies in even greater specialization in the product he is best able to produce. Any augmentation of income in the future will come with greater intensification of the dairy enterprise aided by improved techniques of draining, pasture management, feeding, and fertilizing. But the dependence of the prosperity of the region on a single commodity raises problems, the main one being the over-production of milk with no compensatory increase in liquid consumption and a resultant falling in price. If this happened on a national scale the hardship for the Levels farmer would be very real. Some indication of this tendency was given in a White Paper accompanying the annual review and determination of guarantees in 1958, in which the government objectives were stated to be 'the production of less milk, pig-meat and eggs' and the expansion of beef and lamb production. Whatever the desirability and applicability of this objective on a national scale, the result in the Levels would be very unfortunate as the choices of alternative land use are limited. The winter grazing of sheep, if it is undertaken with care, would be increased and would improve the quality of pastures without diminishing the supply of grass available to dairy cows from April onwards. But other than this it is hard to see any alternative form of land use being entertained seriously. The extension of arable cropping is severely restricted by the drainage condition of the moors, and the extension of mixed farming is difficult because of the varied water-levels that would have to be achieved for arable crops and grassland in a mixed system.[1] In either case, arable or mixed farming, the ideal water-table of 18 in. for pasture land (which is not functioning in most places at the present time) would have to be lowered by at least another 12 in. for the successful cultivation of root crops, and by up to 24 in. for grain, and, what is more, it would have to be maintained at these levels. The

[1] G. F. C. Mitchell, 'The Central Somerset Lowlands: The Importance and Availability of Alternative Enterprises in a Predominantly Dairying District'. *Selected Papers in Agricultural Economics*, VII, No. 5. University of Bristol, 1962.

possibility of peat shrinkage and wastage then looms spectre-like, with consequences that have been touched upon before.[1] But the consequences are even more far-reaching than this, for the whole policy of draining which has been evolved would have to be abandoned; controlled flooding could not be practised, and the accepted possibility of flooding from November to February could not be tolerated; consequently larger, more extensive and much more costly drainage works would have to be built. Indeed it does seem that it is to the advantage of all concerned to see that there is no appreciable change in the agricultural economy of the Levels, a state of affairs which, from the point of view of the geographer, implies a more or less static landscape.

Balanced against this more theoretical possibility of peat shrinkage which might result from a change in the economy of the Levels is the ever-present prospect of the occurrence of a particularly unfavourable combination of high tides, adverse wind and barometric pressure conditions, and heavy and prolonged rainfall. This combination would make short work of man's effort to control the adverse physical environment of the Levels.

THE LANDSCAPE

For one who has crossed and re-crossed the Levels, and has followed on foot at least part of the course of nearly every major rhyne and embankment, there remain some abiding impressions of its landscape.

Foremost is the feeling that the answers to many of the problems raised in this study lie in the very landscape itself, in the abandoned channels and embankments, as well as in the pattern of fields and enclosures. Each rise or depression tells either of some past work of flood control or natural deposition, now long since smoothed out by weathering and flood water, clothed in a rich, green sward of grass, but still important in the history of the region. Maitland likened the map to a palimpsest on which the vestiges of previous occupation were never erased;[2] yet the map, after all, is but a two-dimensional representation of the land, and it is the land itself that poses the problems and gives some of the answers. In the lowland landscapes of Britain, besides the tangible feature of past draining activity, there are those channels and ditches which now reveal themselves only

[1] See p. 246 above.

[2] Maitland, *Domesday Book and Beyond*, 15.

through the medium of air photographs. Much more needs to be done to unravel these facets of the drainage pattern through trial borings and analysis of soil deposits, a task which, if taken to its logical conclusion, would surely be beyond the immediate objective and the competence of this writer. This is all to say one thing, that the landscape is a document in itself, and although this study has relied on the evidence of the library and archive, these materials have been related to the local scene to aid its interpretation.

But what of the landscape of today; what scene confronts the traveller to this part of Britain? Most people travel through the region by the main road or railway that run nearly parallel to each other along the coastal clay belt joining the main centres of population in Bristol, Bridgwater and Taunton. For them there is a brief glimpse of a flat, lush, green land, of fields corrugated into gentle folds by underdraining, and divided by many small open rhynes, full of water. But it is a monotonous landscape, and the view is obstructed and enclosed, in many places, by hedges and trees; the scene is barely noted and quickly forgotten, and the distant prospect of either the Mendip Hills to the north or the Quantock Hills to the south beckons the traveller with their promise of more exciting landscapes and greater variety. In this way it is quite easy to cross the major outfalls of the Axe, Brue and Huntspill rivers and the King's Sedgemoor Drain, which are sunk deep into the clay belt, without seeing them and so without asking either what is the reason for their existence or what lies behind these massive and sometimes straight and obviously artificial channels. There are few settlements of any consequence, except for Huntspill, and even if the traveller pauses in the larger coastal resorts of Burnham and Weston-super-Mare, his gaze is orientated out to sea and not inland.

It is inland, on the less frequented roads, that one must go to see the peculiar impress that has been made on the land by draining. The hedges and trees become fewer and the view lengthens, never infinite, but always bounded in the distance by a ridge or rise that marks the rim of the Levels or an island in it. In the Brue valley, the peat grasses are browner than those on the clay belt, particularly in the summer; dirt droveways and open rhynes intersect the landscape with regularity and straightness, all the droveways built up slightly above the general level of the moors on the refuse cleaned out of the rhynes that bound them on either side. Further east, on the raised bogs, the open moors give way to scrubby heath land, sometimes wildly corrugated and uneven, the result of centuries of peat digging, sometimes neatly cut by deep, straight, steep-sided, black-looking trenches of current diggings, which are surrounded by the distinctive beehive-shaped mounds of drying blocks of peat. In the Southern Levels, droves and rhynes are lined by rows of short, thick-trunked willow trees, heavy and bushy with trailing foliage in the spring and summer, but later on, when cut and pollarded, looking naked and gaunt, like so many giant, bulbous-topped, knobkerries stuck into the ground. The rhynes are then full of withy shoots, retting in the water, ready to be taken away to one of the multitude of small drying-sheds with their characteristic tall chimneys, which are dotted around the moors. The finished products of furniture and basket-work can be found displayed for sale in many a roadside spot. Quietness hangs heavily over all the moors and is punctuated only by the occasional passing car, the chugging of a distant milking bail, or the lowing of the ubiquitous black and white Friesian cows, grazing in a scene more reminiscent of their ancestral home bordering the North Sea than of the west of England.

The moors are not unnaturally, bereft of settlements, and as one walks towards the main rivers and rhynes and climbs up the banks, there are the white-washed cottages and houses stretched out in long lines beside the roads on the tops of the embankments, looking like so many polder villages. These straggling, loosely agglomerated settlements have no recognizable core and do not compare with the older, tightly-knit communities that top the islands and hug the up-land edges. These all have the random morphology of nucleated settlements of great antiquity, each with its pub, garage, post office, half-a-dozen or so general stores and a magnificent church, far too large for the community today, its characteristically Somersetshire tower—tall, square and slender—a symbol of modest prosperity and of a society and an age which were probably more pious and certainly more parochial than those of today. These truly island villages still possess an air of isolation which is sometimes dramatically emphasized by an extensive flood, and the time has not yet gone when one can catch a glimpse of a boat tied up at the bottom of a sloping garden or orchard, ready to be used when the water rises.

At the drier margins of the Levels is a ring of large thriving urban centres with populations varying between 3,000 to over 30,000 and with a corresponding complexity of urban services and functions, concentrating increasingly the life of the surrounding smaller villages upon them. Such are Burnham, Bridgwater, Taunton, Street, Glastonbury, Wells, Weston-super-Mare and Clevedon, all growing, some more rapidly than others, as new industries augment their purely central functions of more ancient times. In between and around about the towns on the edges of the Levels is an equal number of smaller, but locally important settlements such as North Petherton, North Curry, Ilminster, Langport, Wedmore, Cheddar, Winscombe, Axbridge, Nailsea, Yatton and Wrington. This peripheral distribution of towns and large villages reflects the drainage condition of the Levels.[1]

Here on the margins of the Levels, the landscapes of the surrounding upland country impinge upon the moors; in the north, the cold, off-white Carboniferous Limestone farmhouses of the Mendip Hills and the dry stone walls that separate the fields, touch upon the Axe valley; but, in contrast, in the south, the deep, red, warm loams of the Old Red Sandstone in the Vale of Taunton Deane spill into the Levels; the houses are red and seem part of the earth itself, and a pink dust powders the leaves of the overhanging hedges of the country lanes. But this is more than a boundary between dry and wet land, or between different-coloured soils, for if one goes a little further west one is conscious of crossing a deep cultural divide, the passing from lowland Britain into highland Britain, from an extension of the English plain to the bocage country of the west. Deeply-incised lanes climb up the steep hill edges and out of the Levels; massive earthen and stone walls, surmounted by hedges, divide the fields; and hamlets increasingly become more frequent than nucleated villages.[2]

These are scenes for all to see at any time, yet some of the most enduring impressions are more momentary and more rarely witnessed: the rush and noise of the bore moving up the Parrett ahead of a spring tide; water lapping the tops of the river banks, ten feet or more above the surrounding land, and then spilling into the moors; or

[1] For an interesting study of the distribution of centres of varying importance in Somerset, see H. E. Bracey, 'English central villages: identification, distribution and function', *Proc. Int. Geog. Union Symposium on Urban Geography*, ed. Knut Norborg (1960), 169–90, and particularly the map on p. 176. Other information on the influence of the large towns on the country areas is in *idem*, 'Towns as rural service centres. An index of centrality with special reference to Somerset', *Trans. Instit. Brit. Geog.* XIX (1953), 95–105.

[2] B. M. Swainson, 'Dispersion and agglomeration of rural settlement in Somerset', *Geography*, XXIX (1944), 1–8.

stranded stock being taken off an isolated hillock or bank in a rowing boat. But there is one scene I think of more than most; as one looks south from the high edge of the Mendips, just after sunrise, the Levels are often covered with a thick mist, like a sea. Nothing rises above the flat white surface except the smooth Tor of Glastonbury surmounted by the slender shape of St. Michael's Tower, looking for all the world like the Isle of Avalon it is claimed to be. This relic of a long and distant past, seemingly surrounded by a vast sheet of water, yet untouched, hints of the patient struggle to rid the land of the recurring threat of water. Here past and present are intimately entwined and symbolize in the imagination the story of the draining of the Levels.

SOURCES AND BIBLIOGRAPHY

UNPUBLISHED

AT THE PUBLIC RECORD OFFICE

Early Chancery Proceedings (miscellaneous bundles).
Exchequer Depositions by Commission.
Exchequer Special Commissions.
Home Office Documents.
Patent Rolls.
Rentals and Surveys (misc. books: Land Revenue).
State Papers Domestic.
Subsidy Rolls.

AT THE BRITISH MUSEUM

Additional MSS.
Additional Rolls.
Egerton MSS.
Harleian MSS.
Royal and Kings MSS.
Stowe MSS.

AT THE BODLEIAN LIBRARY

Ballard MSS.
Gough MSS.
North MSS.
Rawlinson MSS.

AT THE SOMERSET RECORD OFFICE

Carver MSS.
Chilton-upon-Polden Documents (Civil Records).
Combe MSS.
Dickinson MSS.
Dodson and Pulman MSS.
Enclosure Awards.
Glebe Terriers.
Kemys-Tynte MSS.
Meade King MSS.
Sexey MSS.
Somerset Archaeological Society, Parochial MSS.
Somerset Archaeological Society, Serel MSS. (which include Locke's 'Survey').
Strangeways MSS.
Wharton MSS.

Sources and bibliography

AT THE SOMERSET RIVER BOARD OFFICES

In the Committee Room

The Records of the Commissioners of Sewers for Somerset:
- Axbridge Division, 1759–1783, 1784–1826 and 1827–81.
- Bridgwater (Western) Division, 1759–1826 and 1827–81.
- Brue Division, 1790–1833.
- Glastonbury (Middle) Division, 1827–81.
- Langport Division, 1790–1826 and 1827–81.
- Wells (Middle) Division, 1826–81.
- Wrington Division, 1834–1934 (6 vols).

The Records of some Juries of Sewers:
- Banwell Jury of Sewers, 1871–1934.
- North Weston Jury of Sewers, 1869–92.
- Worle and Kewstoke Jury of Sewers, 1881–1931.
- Yatton Jury of Sewers, 1774–1856, 1857–92, 1892–1908 and 1908–37.

Sedgemoor Drainage Expenditor Accounts, 1825–36.

Proceedings of the Commissioners for the Brue Drainage, 1801–6 and 1812–81.

In the Record Room

Axe Drainage Papers (A.D. 1–4):
- Correspondence, 1801–14; plans; petitions; legal opinions; Sutcliffe's reports; Easton's report; etc.

Brue Drainage Papers (B.D. 1–4):
- Correspondence, 1801–6; plans; petitions; etc.

Maps and plans.

Newspapers:
- Portions of the *Bridgwater Mercury; Bridgwater Times; Langport and Somerton Herald; Somerset County Gazette; Taunton Courier.*

Parrett Navigation Papers (P.N. 8):
- Acts; awards; petitions; etc.

Records of the Somerset Drainage Commissioners (1881–1930):
- D.C. 1–12. Correspondence.
- D.C. 13–15. Accounts, 1881–1930.
- D.C. 27–9. Aller Gap Appeals.
- D.C. 30–2. Engineer's reports, 1881–1929.
- D.C. 39. Slime-batches, Langport Town Embankment.
- D.C. 48–9. Highbridge Clyse.

Records of the Somerset Rivers Catchment Board (1931–50).

Records of Internal Drainage Boards:
- Curry Moor (C.M. 1–2); King's Sedgemoor (K.S. 1); West Sedgemoor (W.S. 1).

Reports and drawings by Professor Gibson on the working and results of the Parrett estuary tidal model (C.B. 6).

River flow and rainfall statistics.
Schedules of Sewerable Lands (South Case 4).
Wrington Sewers Papers; Congresbury and Weston Drainage:
S.W. 4. Correspondence, 1809–33; Proceedings of Commissioners of Sewers, 1809–12; Proceedings of Commissioners of Weston Drainage, 1809–11; Proceedings of Commissioners of Congresbury Drainage, 1819–33; plans; and petitions.
S.W. 1. Minutes of Wrington Sewers, 1800–19 and 1862–96.
S.W. 3. Returns and fines, 1761–99.
S.W. 5. Returns and fines, 1820–82.

AT THE MINISTRY OF HOUSING AND LOCAL GOVERNMENT, WHITEHALL, LONDON

Air photograph coverage of the Levels, 1947.

THESES

Gramolt, D. W. 'The Coastal Marshland of East Essex between the seventeenth and mid-nineteenth Centuries'. M.A., University of London, 1961.
Hardman, D. B. 'The Reclamation and Agricultural Development of North Cheshire and South Lancashire Mossland Areas'. M.A., University of Manchester, 1961.
Keil, I. J. E. 'The Estates of the Abbey of Glastonbury in the Later Middle Ages'. Ph.D., University of Bristol, 1964.
Lawrence, J. F. 'Somerset, 1800–1835'. M.A., University of Durham, 1938.

PUBLISHED

Acland, T. D. 'On the farming of Somersetshire'. *Journal of the Royal Agricultural Society of England*, XI (London, 1850), 666. Reprinted the following year with William Sturge's essay (q.v.) and entitled *The Farming of Somersetshire*. London, 1851.
Acres, W. M. *A Brief History of Wedmore*. Wedmore, 1954.
Admiralty Tidal Handbook (*European Waters*). London, 1959.
Agricultural Land Commission. *Land in the Brue Valley and North Moor, Somerset*. H.M.S.O. London, 1955.
Allen, D. G. C. 'The Rising in the West, 1628–31', *Economic History Review*, V (London, 1952), 76.
Allen, J. 'Schemes of improvement for the River Parrett in Somerset; an investigation with the aid of a tidal model'. *Journal of the Institution of Civil Engineers*, XIX (London, 1942), 85.
Anonymous. *A true report of certain wonderful over-flowings of waters, now lately in Summerset-shire, Norfolke and other places of England*. London, 1607.

A second edition was issued in the same year. A copy of the Somerset section of the second edition was printed in Weston-super-Mare, 1884, ed. E. E. Baker.

Anonymous. *Observations made during a Tour through part of England, Scotland and Wales*. London, 1780.

Anonymous. 'The 1607 Flood'. *Gentleman's Magazine*, XXXII (London, 1762), 306.

Armstrong, W. *Report on the Navigation of the River Tone from Taunton to Bridgwater*. Bristol, 1824.

Avery, B. W. *The Soils of the Glastonbury District of Somerset*. Memoirs of the Soil Survey of Great Britain. H.M.S.O. London, 1955.

Barnes, T. G. *Somerset 1625–1640: A County's Government during the "Personal Rule"*. London, 1961.

Bassindale, R. 'Studies on the biology of the Bristol Channel, XI'. *Journal of Ecology*, XXXI (Cambridge, 1943), 21.

Bates, E. H. (ed.). 'The Cartularies of Athelney and Muchelney Abbeys'. *Somerset Record Society*, XIV (Taunton, 1899).

(ed.). 'A Particular Description of the County of Somerset, Drawn up by Thomas Gerard of Trent, 1633'. *Somerset Record Society*, XV (Taunton, 1900).

(ed.). 'Text of the Somerset Domesday', *Victoria County History of Somerset*, I (London, 1906), 434.

Bazalgette, J. W., and Whitehead, A. *Report on the Yeo, Parrett and Isle Drainage*. Langport, 1869.

Beaumont, J., and Disney, H. *A New Tour thro' England performed in 1765, 6 and 7*. London, 1768.

Billingsley, J. *A General View of the Agriculture of the County of Somerset*. London, 1794. Reprinted with amendments, Bath, 1798.

'On the uselessness of commons to the poor'. *Annals of Agriculture*, XXXI (London, 1798), 27.

'On the best method of inclosing, dividing and cultivating waste lands'. *Letters and Papers of the Bath and West of England Society*, XI (Bath, 1809), 1.

Birch, T. (ed.). *The Diary of Thomas Burton, Esq., for 1656–1659*. 4 vols. London, 1828.

Birch, W. de Gray. *Cartularium Saxonicum*. 3 vols. London, 1885–93.

Bird, W. H. B. and Baildon, W. P. (eds). *The Calendar of the Manuscripts of the Dean and Chapter of Wells*. I, 'Books' ed. W. H. B. Bird; and II, 'Accounts, Books and Deeds', ed. W. P. Baildon. *Historical Manuscripts Commission*, ser. XII, London, 1907 and 1914. For an earlier report on these manuscripts, see J. A. Bennet, in *Tenth Report of the Historical Manuscripts Commission*. Pt. III. (London, 1885).

Bloome, T. *Britannia*. London, 1673.

Bracey, H. E. 'Towns as rural service centres. An index of centrality with special reference to Somerset'. *Transactions of the Institute of British Geographers*, XIX (London, 1953), 95.

'English central villages: identification, distribution and function'.

Proceedings of the International Geographical Union Symposium on Urban Geography, ed. Knut Norborg (Lund, 1960), (Lund Studies in Geography, Series B, 1962, No. 24).

British Parliamentary Papers.

British Rainfall 1956. H.M.S.O. London, 1957.

Buckland, E. and Conybeare, W. D. 'Observations on the south-western coal district of England'. *Transactions of the Geological Society of London*, I (London, 1824), 210. Afterwards continued as the *Quarterly Journal of the Geological Society.*

Bulleid, A. and Gray, H. St. George. *The Glastonbury Lake Village.* 2 vols. Taunton, 1911.

The Meare Lake Village. Vol. I: Taunton, 1948; vol. II: Taunton, 1953.

Calender of Inquisitions Post-Mortem.

Calendar of Patent Rolls.

Calendar of State Papers, Domestic.

Carr, A. P. 'Cartographic record and historical accuracy'. *Geography*, XLVII (Sheffield, 1962), 135.

Chadwick, E. (comp.) 'The sanitary effects of land drainage'. Report of the Poor Law Commissioners on an Inquiry into the Sanitary Condition of the Labouring Population of Great Britain. 1842. Extracts printed in *Journal of the Royal Agricultural Society of England*, IV (London, 1843), 151.

Chadwyck-Healey, C. E. H. (ed.). 'Somersetshire Pleas from the Rolls of the Itinerant Justices. Close of Twelfth Century to 41 Henry III'. *Somerset Record Society*, XI. Taunton, 1897.

Chambers, J. D. 'Enclosure and labour supply in the Industrial Revolution'. *Economic History Review*, V (London, 1953), 319.

Chubb, T. *A Descriptive List of the Printed Maps of Somersetshire, 1575–1914.* Taunton, 1914.

Clapham, Sir John. *An Economic History of Modern Britain*, I. Cambridge, 1926.

Clark, J. Aubrey. 'On the Bridgwater and other Levels of Somersetshire'. *Journal of the Bath and West of England Society*, II (Bath, 1854), 99. Reprinted as a separate pamphlet in the same year and entitled, *Prize Essay on the Somerset Levels, their Present State and Capabilities of Improvement.*

'On the scouring lands of Central Somerset'. *Journal of the Bath and West of England Society*, III (Bath, 1855), 52.

Clark, R. G. *Report on the Somerset Rivers Flood Alleviation Scheme.* March, Cambs. 1937.

Clarke, J. Algernon. 'Trunk Drainage'. *Journal of the Royal Agricultural Society of England*, XV (London, 1854), 1.

Clayton, C. H. J. *Flooding in the Drainage Areas of the Rivers Parrett, Tone, Cary, Brue and Axe.* London, 1918.

Climatological Atlas of the British Isles. H.M.S.O. London, 1952.

Cole, S. D. *The Sea Walls of the Severn.* Bristol, 1912.

Collinson, J. *The History and Antiquities of the County of Somerset: Collected from Authentick Records and an Actual Survey made by the late Mr. Edmund Rack.* 3 vols. Bath, 1791.

Conybeare, J. W. E. *Alfred in the Chroniclers.* London, 1900. 2nd ed., Cambridge, 1914.

Cooling, L. F. 'Soil mechanics and site exploration'. *Journal of the Institution of Civil Engineers,* XVIII (London, 1941), 37.

Coppock, J. T. *Agricultural Atlas of England and Wales.* London, 1964.

Cressy, E. *An Outline of Industrial History.* London, 1915.

Dallas, J. W. 'A Symposium on some problem areas in the South-West. (1) The Somerset Moors'. *Journal of the Royal Agricultural Society of England,* CXII (London, 1951), 9.

'Some aspects of West Country farming. (2) Somerset'. *Journal of the Royal Agricultural Society of England,* CXVIII (London, 1957), 9.

'Farming in Somerset'. *Agriculture, the Journal of the Ministry of Agriculture,* LXV (London, 1958), 172.

Darby, H. C. *The Medieval Fenland.* Cambridge, 1940.

'The changing English landscape'. *Geographical Journal,* CXVII (London, 1951), 377.

The Domesday Geography of Eastern England. Cambridge, 1952.

'On the relations of geography and history'. *Transactions of the Institute of British Geographers,* XIX (London, 1953), 8.

The Draining of the Fens. Cambridge, 1940. New ed., Cambridge, 1956.

An Historical Geography of England, before A.D. 1800. Cambridge, 1936.

and Terrett, I. B. *The Domesday Geography of Midland England.* Cambridge, 1954.

and Campbell, Ella M. J. *The Domesday Geography of South-East England.* Cambridge, 1962.

and Finn, R. Welldon (eds). *The Domesday Geography of South-West England.* Cambridge, 1967.

Darby, J. 'An account of the reclamation and improvement of land in the Vale of Taunton Deane'. *Journal of the Bath and West of England Society,* x (Bath, 1862), 77.

'The farming of Somerset'. *Journal of the Bath and West of England Society,* v (Bath, 1873), 96.

'Stream and flood in relation to agriculture'. *Journal of the Bath and West of England Society,* XII (Bath, 1880), 149.

Davis, T. 'On the management of marsh lands...'. *Letters and Papers of the Bath and West of England Society,* x (Bath, 1805), 324.

Dawe, C. V., and Blundel, J. E. *An Economic Survey of the Somerset Willow Growing Industry.* University of Bristol Bulletin No. 9, Bristol, 1932.

Defoe, D. *The Storm: or, a Collection of the most remarkable casualties and disasters which happen'd in the late dreadful tempest both by sea and land.* London, 1704.

A tour thro' the whole island of Great Britain, 1724, 1725 and 1727. 2 vols. Ed. G. D. H. Cole, London, 1927.

Denton, J. B. 'On the discharge from under-ground drainage and its effect on the arterial channels and outfalls of the country'. *Minutes of the Proceedings of the Institution of Civil Engineers*, XXI (London, 1861–2), 48–81; and discussion, 82–130.

Dickinson, F. A. (ed.). 'Kirby's Quest for Somerset'. *Somerset Record Society*, III. Taunton, 1889.

Dobbie, C. H. and Partners. *Somerset River Board, River Parrett Flood Relief, Technical Report*. London, 1961.

Somerset River Board, Technical Report on the Tone Valley Scheme. London, October 1963.

Technical Report on the River Parrett Relief Channel. London, March 1964.

Drayton, Michael. *Polyolbion, or A chorographical description of Great Britain*. London, 1613.

Du Cane, C. G. *River Parrett Estuary Scheme*. London, 1941.

Dugdale, W. *History of Imbanking and Drayning*. London, 1662. 2nd ed. by C. N. Cole. London, 1772.

Monasticon Anglicanum. 3 vols. London, 1655–73. 2 additional vols. by John Stevens, London, 1722–3. New ed. in 6 vols., London, 1817–30; reprinted in 1846.

East, G. 'Land utilization in England at the end of the eighteenth century'. *Geographical Journal*, LXXXIX (London, 1937), 156.

Easton, J. *Prospectus of a Plan for rendering Navigable the King's Sedgemoor and other Principal Drains between Dunball, Somerton, Ashington and Yeovil*. Taunton, 1829. To which is attached a report on the same by G. Rennie. London, 1829.

Ekwall, E. *The Concise Oxford Dictionary of English Place-Names*. Oxford, 1936.

Elton, C. J. (ed.). 'Rentalia et Custumaria: Michaelis de Amesbury, 1235–52, et Rogeri de Ford, 1252–61'. *Somerset Record Society*, V. Taunton, 1891.

Ernle, Lord. *English Farming, Past and Present*. London, 1912, 1917, 1922, 1927 and 1936.

Eyton, R. W. *Analysis and Digest of the Somerset Survey*. London, 1880.

Fiennes, C. *The Journeys of Celia Fiennes*, (*c*. 1702). 1st ed. by E. W. Griffiths, London, 1888. Another by C. Morris, London, 1946; and revised in 1949.

Findlay, D. C. *The Soils of the Mendip District of Somerset*. Memoirs of the Soil Survey of Great Britain. H.M.S.O. London, 1965.

Flower, C. T. (ed.). *Public Works in Medieval Law*. Selden Society, XXXII (London, 1915), and XL (London, 1923).

Fussell, G. E. 'Cornish farming, A.D. 1500–1910'. *Amateur Historian*, IV (1960), 338–45.

Galton, E. 'An account of improvement of a shaking bog at Meare in Somersetshire'. *Journal of the Royal Agricultural Society of England*, VI (London, 1845), 182.

Gibson, A. H. (chairman). *Severn Barrage Committee Report*. H.M.S.O. London, 1933.

Glyn, J. 'Draining by steam power'. *Transactions of the Royal Society of Arts*, LI (London, 1837), 3–24.

Godwin, H. 'The origin of Roddons'. *Geography Journal*, XCI (London, 1938), 241.

'Studies in the post-glacial history of British vegetation: VI, Correlations in the Somerset Levels'. *New Phytologist*, XI (Cambridge, 1941), 108.

'The coastal peat beds of the British Isles and the North Sea'. *Journal of Ecology*, XXXI (Cambridge, 1943), 199.

'Studies in the post-glacial history of British vegetation: X, Correlations between climate, forest composition, prehistoric agriculture and peat stratigraphy in the Sub-Boreal and Sub-Atlantic peats of the Somerset Levels'. *Philosophical Transactions of the Royal Society*, B CCXXXIII (London, 1948), 275.

'The botanical and geological history of the Somerset Levels'. *Proceedings of the British Association for the Advancement of Science*, XII, No. 47 (London, 1955), 319.

'Studies in the post-glacial history of British vegetation: XIII, The Meare Pool region of the Somerset Levels'. *Philosophical Transactions of the Royal Society*, B CCXXXIX (London, 1955/6), 169.

and Clapham, A. R. 'Studies in the post-glacial history of British vegetation: IX, Prehistoric trackways in the Somerset Levels'. *Philosophical Transactions of the Royal Society*, B CCXXXIII (London, 1948), 249.

Goodman, F. B. *Résumé of Improvement Schemes*. Bridgwater, 1936.

Grantham, R. B. 'The Land Drainage Act, 1861: its provisions, its working, and its results'. *Journal of the Bath and West of England Society*, XIII (Bath, 1865), 118.

'Arterial drainage and outfalls'. *Journal of the Bath and West of England Society*, XV (Bath, 1867), 251.

Report on the Floods in Somersetshire in 1872–73. London, 1873.

Great Britain. Parliament. Royal Commission on Agriculture. Report on Devon, Cornwall, Dorset and Somerset. *British Parliamentary Papers*, XIV (1882), 423.

Joint Committee on the Ouse Drainage Bill. Report. *British Parliamentary Papers*, VI (1927), 145 (Sessional Paper, 113).

Joint Committee on the Doncaster Area Drainage Bills. Report. *British Parliamentary Papers*, IV (1928–9), 723 (Sessional Paper, 88).

Royal Commission on Land Drainage. Report. *Parliamentary Papers*, X (1927), 1057 (Cmd. 2993).

Green, E. 'On Somerset Chap Books'. *The Proceedings of the Somerset Archaeological and Natural History Society*, XXIV (Taunton, 1872), 50.

Bibliotheca Somersetensis: A catalogue of books, pamphlets, single sheets, and broadsheets in some way connected with the county of Somerset. 3 vols. Taunton, 1902.

(ed.). 'Pedes Finium. Feet of Fines, 1196–1307'. *Somerset Record Society*, VI. Taunton, 1892.

(ed.). 'Pedes Finium. Feet of Fines, 1307–1346'. *Somerset Record Society*, XII. Taunton, 1898.

Grenville, R. N. 'Somerset Drainage'. *The Proceedings of the Somerset Archaeological and Natural History Society*, LXXII (Taunton, 1926), 1.

Grigg, D. B. 'Changing regional values during the Agricultural Revolution in South Lincolnshire'. *Transactions of the Institute of British Geographers*, XXX (London, 1962), 91.

Grundy, G. B. 'The ancient highways of Somerset'. *Archaeological Journal*, XCVI (London, 1930), 226.

'The Saxon charters and field names of Somerset'. Issued as a supplement to the *Proceedings of the Somerset Archaeological and Natural History Society*, LXXIII (Taunton, 1927), and LXXX (Taunton, 1934).

Hadfield, E. C. R. *The Canals of Southern England*. London, 1955.

Hallam, H. E. *Settlement and Society: A Study of the Early Agrarian History of South Lincolnshire*. Cambridge, 1965.

Hammett, R. *Report on the Glastonbury Canal*. Taunton (?), 1826.

Hammond, J. L. and B. *The Village Labourer, 1760–1832*. 3 vols. London, 1911. 2nd ed. 1917–20.

Harris, L. E. *Vermuyden and the Fens. A Study of Sir Cornelius Vermuyden and the Great Level*. London, 1953.

Harting, E. 'Wild fowl decoys in Somerset'. *Downside Review*, V (London, 1886), 218.

Hartlib, S. *His Legacie: or An Enlargement of the Discourse of Husbandry...* London, 1651. Reprinted with additions in the next year.

Harvey, Barbara F. 'The Population Trend in England Between 1300 and 1348'. *Transactions of the Royal Historical Society*, 5th ser., XVI (London, 1966), 23–42.

Haverfield, H. J. 'Romano-British Somerset'. *Victoria County History of Somerset*, I (London, 1906), 207.

Hearne, T. (ed.). *Peter Langtoft's Chronicle*. 2 vols. Oxford, 1725.

Joannis confratis et monachi Glastoniensis chronica, sive historia de rebus Glastoniensibus. 2 vols. Oxford, 1726.

Adami de Domerham. Historia de rebus gestis Glastoniensibus. Oxford, 1727.

Helm, P. J. 'The Somerset Levels in the Middle Ages'. *Journal of the British Archaeological Association*, XII (London, 1949), 37.

Hilton, R. H. 'A study of the pre-history of English enclosure in the fifteenth century'. *Studi in Onore de Armando Sapori*. 2 vols. (Milan, 1957).

Historical Manuscripts Commission. *Seventh Report, Part I, Report and Appendix*. London, 1879; *Ninth Report*. London, 1884; and *Tenth Report, Parts v and vi*. London, 1885–7.

Hobhouse, Bishop (ed.). 'Calendar of the Register of John de Drokensford, Bishop of Bath and Wells, 1309–1327'. *Somerset Record Society*, I. Taunton, 1887.

'Remarks on Domesday Map'. *Proceedings of the Somerset Archaeological and Natural History Society*, xxxv (Taunton, 1889), ix.

(ed.). 'Pre-Reformation church-wardens' accounts of Croscombe, Pilton, Yatton, Tintinhull, Morebath, and St. Michael's Wells, ranging from A.D. 1349 to 1560'. *Somerset Record Society*, IV. London, 1890.

Hobhouse, E. (ed.). *The Diary of a West Country Physician, A.D. 1684–1726.* London, 1934.

Holsworthy, R. (ed.). *The Hearth Tax for Somerset.* E. Dwelly's National Records, I. Fleet, Hants., 1916.

Hoskins, W. G. *The Making of the English Landscape.* London, 1955.

Humphreys, A. L. *Somersetshire Parishes: A Handbook of Historical Reference to all Places in the County.* 2 vols. London, 1906.

Hydraulic Engineer (Pseudonym). *Report on the Best Scheme for the Prevention of the Floods at Taunton.* Taunton, 1889.

Jackson, J. E. (ed.). *Liber Henrici de Soliaco Abbatis Glaston: An Inquisition of the Manor of Glastonbury, 1189.* Roxburghe Club, London, 1882.

Jeboult, E. *A General Account of West Somerset. Description of the Valley of the Tone and the history of the Town of Taunton.* Taunton, 1873. Re-issued and revised as *A Popular History of West Somerset.* Taunton, 1893.

Jessop, J. *Prospectus of the Proposed Improvements of the Port and Harbour of Bridgwater.* Bridgwater, 1829.

Jones, J. L. 'Draining and irrigating the Somerset peat moors'. *Farmer's Weekly*, XLI, No. 8 (London, 1954), 72.

Journals of the House of Commons.

Keen, B. A. 'Land drainage: The area of benefit'. *Agriculture, Journal of the Ministry of Agriculture*, XLIII (London, 1936), 521.

Kelting, E. L. 'The dredging plant of the Somersetshire Drainage Commissioners and the Somerset Rivers Catchment Board'. *Fourth Annual Report of the Somerset River Board.* Bridgwater, 1954.

'An Outline of the Development of Land Drainage in Somerset'. A paper read before the Annual Conference of the River Boards' Association. Weston-super-Mare, 1958.

King, J. E. *Inventory of Parochial Documents in the Diocese of Bath and Wells and the County of Somerset.* Taunton, 1938.

Kirk, R. E. G. and Bennett, J. A. (eds.). 'The Manuscripts of the Earl of Carlisle'. *Historical Manuscripts Commission*, 15th report, ser. xlii. London, 1897.

Kitchen, G. W. (ed.). 'The Compotus Rolls of the Obedientiaries of St. Swithin's Priory, Winchester'. *Hampshire Record Society.* Winchester, 1892.

Knight, F. A. *The Heart of Mendip: an Account of the History, Archaeology and Natural History.* London, 1915.

Lambert, J. M., Jennings, J. N., Smith, C. T., Green, Charles and Hutchings, J. N. *The Making of the Broads.* Royal Geog. Soc. Research Series No. 3. London, 1960.

Landon, L. (ed.). 'Somerset Pleas from the Rolls of the Itinerant Justices 1255–1272'. *Somerset Record Society*, XXXVI. Taunton, 1921.

(ed.). 'Somerset Pleas from the Rolls of the Itinerant Justices, 1280'. *Somerset Record Society*, XLIV. Taunton, 1929.

Legg, L. G. W. (ed.). 'A relation of a short survey of the Westerne Counties, August, 1635'. *Camden Miscellany XVI*, Camden Society, 3rd ser., vol. LII. London, 1936.

Lennard, R. 'The demesnes of Glastonbury Abbey in the eleventh and twelfth centuries'. *Economic History Review*, VIII (London, 1955), 355.

Lewis, E. A. 'Welsh Port Books, 1550–1603: with an analysis of the custom revenue accounts of Wales'. *Cymmrodorion Record Series*, No. 12. London, 1927.

Locke, J. G. *Book of the Lockes*. Boston and Cambridge, Mass., 1853.

Locke, R. 'On the improvement of meadow land with a short history of a part of Somersetshire'. *Letters and Papers of the Bath and West of England Society*, V (Bath, 1793), 179.

'An historical account of the marsh-lands of the County of Somerset'. *Letters and Papers of the Bath and West of England Society*, VIII (Bath, 1796), 259.

An Essay on the Subject of Draining the Flat Part of Somersetshire by means of a new Sluice at Highbridge, being a Letter to Arthur Phippen. Taunton, 1800.

A Letter to George Templer, Esq., On the Propriety of imposing a new Tax on Eighteen Parishes in the County of Somerset. Taunton, *c.* 1800.

The Customs of the Manor of Taunton and Taunton Deane. Taunton, 1816.

Lowrey, F. *Report on the River Parrett Floods Prevention*. Bridgwater, 1893.

Lunn, W. *Report on the River Parrett Floods Prevention*. Bridgwater, 1898.

Flood Prevention and Obstructions in the River. Bridgwater, 1912.

Flooding in Districts: Report and Estimates by Mr. Lunn for Improvements in the Rivers Parrett, Tone, Brue, Axe, King's Sedgemoor Drain and Cary River. Bridgwater, 1919.

Luttrell, Narcissus. *A Brief Historical Relation of State Affairs from September 1678 to April 1714*. 6 vols. Oxford, 1857.

MacDermot, E. T. *The History of the Great Western Railway*. London, vol. I, 1927; vol. II, 1931. New ed. by R. Clinker, London, 1964.

MacInnes, C. M. and Whittard, W. F. (eds.). *Bristol and its adjoining Counties. A Handbook for the British Association for the Advancement of Science*. Bristol, 1955.

Maitland, F. W. *The Domesday Book and Beyond*. Cambridge, 1897. Re-issued, 1907 and 1960.

Margary, I. D. *Roman Roads in Britain*. Vol. I: *South of Fosse Way—Bristol Channel*. London, 1955.

Marshall, W. *The Rural Economy of the West of England*. London, 1796. 2nd ed. with amendments, 1805.

Mingay, G. E. 'The Agricultural Revolution in English history: a re-consideration', *Agricultural History*, XXXVII, No. 3 (Urbana, Illinois, 1963), 123.

Mitchell, G. F. C. 'The Central Somerset Lowlands: The Importance and Availability of Alternative Enterprises in a Predominantly Dairying District'. *Selected Papers in Agricultural Economics*, VII, No. 5. University of Bristol, Bristol, 1962.

Moore, Adam. *Bread for the Poor...Promised by the Enclosure of the Wastes and Common Grounds of England*. London, 1653.

Morgan, F. W. 'The Domesday geography of Somerset'. *Proceedings of the Somerset Archaeological and Natural History Society*, LXXXIV (Taunton, 1938), 139.

Morland, J. 'The Brue at Glastonbury'. *Proceedings of the Somerset Archaeological and Natural History Society*, LXVIII (Taunton, 1922), 64.

Moss, C. E. *The Geographical Distribution of Vegetation in Somerset: Bath and Bridgwater District*. Royal Geographical Society Publication, London, 1907.

National Farmer's Union, Somerset. 'Further Food Production from the Langport Area by better Control of Flooding'. A paper read to the above Union, at Langport, 11 July 1952.

Neilson, N. 'Customary Rents', in *Oxford Studies in Social and Legal History*, II. Oxford, 1910.

Nesbitt, A. 'The Manor House, Meare, Somerset'. *Archaeological Journal*, X (London, 1853), 130.

Olivey, H. P. *North Curry: Ancient Manor and Hundred*. Taunton, 1901.

Pelham, R. A. 'Fourteenth Century England', being Chapter VI of *An Historical Geography of England before A.D. 1800*, ed. H. C. Darby. Cambridge, 1936.

Petrie, H. and Sharpe, J. *Monumenta Historica Britannica*. 2 vols. London, 1848.

Phelps, W. *Observations on the Great Marshes and Turbaries of the County of Somerset, with Suggestions for their Improvement*. Dorchester, 1836.

The History and Antiquities of Somersetshire. London, vol. I, 1836; vol. II, 1839.

'On the formation of peat bogs and turbaries which extend from the Bristol Channel in the central part of Somersetshire'. *Proceedings of the Somerset Archaeological and Natural History Society*, IV (Taunton, 1853), 91.

Poole, W. R. *The Somerset Drainage Act, 1877*. Bridgwater, 1881.

Postan, M. M. 'The chronology of labour services'. *Transactions of the Royal Historical Society*. 4th ser., XX (London, 1937), 189.

'The Glastonbury estates in the twelfth century'. *Economic History Review*, V (London, 1953), 358.

'The Glastonbury estates in the twelfth century: A reply'. *Economic History Review*, IX (London, 1956), 106.

and others. 'L'Economie européenne des deux derniers siècles du Moyen-âge'. *Reports of the 10th International Congress of Historical Science*, Vol. 3: *Storia del Medioevo* (Firenze, 1955), 657.

Preston, S. *Notes...on Reports re Floods in the Districts of the Parrett, Tone, Cary, Brue and Axe*. London, 1919.

Price, H. H. *Report on the Establishment of a Ship Canal and Docks at the Port of Bridgwater*. Bridgwater, 1835.

Rollinson, W. 'Schemes for the reclamation of land from the sea in North Lancashire during the eighteenth century'. *Transactions of the Historical Society of Lancashire and Cheshire*, cxv (Liverpool, 1964), 107.

Roseveare, J. C. A. 'Land Drainage in England and Wales'. A paper read before the Institution of Water Engineers. London, 1932.

Ross, C. D. (ed.). 'Cartulary of St. Mark's, Bristol'. *Bristol Record Society*, xxi. Bristol, 1959.

Round, J. H. 'An introduction to the Somerset Domesday'. *Victoria County History of Somerset*, i (London, 1906), 383.

Savage, J. *The History of Taunton*. Taunton, 1822.

Schofield, R. S. 'The geographical distribution of wealth in England, 1334–1649'. *Economic History Review*, xviii (London, 1965), 483.

Scott-Holmes, T. 'Glastonbury Abbey'. *Victoria County History of Somerset*, ii (London, 1911), 81.

(ed.). 'The Register of Bishop Bubwith, 1407–1424'. *Somerset Record Society*, xxix. Taunton, 1913.

Shaw, S. *A Tour of the West of England in 1788*. London, 1789.

Sheppard, J. A. *The Draining of the Hull Valley*. Publication No. 8 of the East Yorkshire Local History Society. York, 1958.

The Draining of the Marshlands of South Holderness and the Vale of York. Publication No. 20 of the East Yorkshire Local History Society. York, 1966.

Smirke, E. 'Notice of the Custumal of Bleadon Somerset, and of the Agricultural Tenures of the 13th century'. *Proceedings of the Archaeological Society of Great Britain and Ireland, held at Salisbury*, 1849 (London, 1851), 182.

Smith, L. Toulmin (ed.). *Leland's Itinerary in England and Wales in or about the Years 1535 to 1543*. London, 1907. Re-issued, 5 vols., London, 1964.

Smith, R. 'On bringing moorland into cultivation'. *Journal of the Bath and West of England Society*, v (Bath, 1857), 111.

Smith, R. A. L. 'Marsh embankment and sea defences in medieval Kent'. *Economic History Review*, x (London, 1940), 29.

Christchurch Cathedral Priory. Cambridge, 1943.

'The Benedictine contribution to medieval English agriculture'. In *Collected Papers*. London, 1947.

Sollas, W. G. 'The estuaries of the Severn and its tributaries; an inquiry into the nature and origin of the tidal sediment and alluvial flats'. *Quarterly Journal of the Geological Society*, xxxix (London, 1883), 611.

Somerset River Board. *Annual Reports* Nos. 1–15. Bridgwater, 1950–1 to 1964–5.

Spence-Sales, H. and Bland, J. *England's Water Problem*. London, 1939.

Statutes of the Realm (Record Commission, London).

Stewart, J. G. *Report to the Board of Agriculture on the Drainage of the Valleys of the Parrett, Tone, Cary, Brue and Axe*. Bridgwater, 1919.

Stuart-Menteath, T. *Somerset.* Part 86 of the *Land of Britain,* edited by L. Dudley Stamp. London, 1938.

Stubbs, W. (ed.). *Willelmi Malmesbiriensis Monachi: De Gestis Regum Anglorum.* 2 vols. London, 1887 and 1889. (Rolls Series No. 90.)

Sturge, W. 'On the farming of Somersetshire', Published with T. D. Acland's essay in *The Farming of Somersetshire.* London, 1851.

Swainson, B. M. 'Rural settlement in Somerset'. *Geography,* xx (Sheffield, 1935), 112.

'Dispersion and agglomeration of rural settlement in Somerset'. *Geography,* xxix (Sheffield, 1944), 1.

Tate, W. E. *Somerset Enclosure Acts and Awards, also Detailed List of Enclosure Award Maps.* Somerset Archaeological and Natural History Society, Frome, 1948.

Thirsk, Joan. 'The Isle of Axholme before Vermuyden'. *Agricultural History Review,* I (Oxford, 1953), 16.

English Peasant Farming: An Agrarian History of Lincolnshire from Tudor to recent times. London, 1957.

Tudor Enclosure. Historical Association, general ser., No. G.41. London, 1959.

(ed.). *The Agrarian History of England and Wales, 1500–1640.* IV. Cambridge, 1967.

Thomas, William L., Jr. (ed. with the collaboration of Carl O. Sauer, Marston Bates, and Lewis Mumford). *Man's Role in Changing the Face of the Earth.* Chicago, 1956.

Titow, J. 'Evidence of weather in the Account Rolls of the Bishopric of Winchester, 1209–1350', *Economic History Review,* XII (London, 1960), 360.

University of Bristol: Reconstruction Research Group. *Land Classification: Gloucestershire, Somerset, Wiltshire.* Bristol, 1947.

Ussher, W. A. E. *The Geology of the Quantock Hills and of Taunton and of Bridgwater.* Memoirs of the Geological Survey. H.M.S.O. London, 1908.

Van der Veen, H. *Report on the Improvement of the Lower Course of the Parrett.* Bridgwater, 1935.

Van Veen, J. *Dredge, Drain, Reclaim: The Art of a Nation.* The Hague, 1948.

Victoria County History of Somerset. Ed. W. Page. London, vol. I, 1906; vol. VII, 1911.

Warner, R. *A Walk through some of the Western Counties of England.* Bath, 1800.

A History of the Abbey of Glaston and of the Town of Glastonbury. Bath, 1826.

Warry, G. D. *Ubi Voluntas Via Fit: The Somerset Drainage Act, its Principles and Practice.* Bridgwater, 1880.

Watkin, Dom. A. (ed.). 'The Great Chartulary of Glastonbury'. *Somerset Record Society,* 3 vols.: LIX (Taunton, 1944); LXII (1948); and LXIV (1949/50).

Webb, S. and B. *English Local Government; Statutory Authorities for Special Purposes.* 8 vols. London, 1906–29. Vol. IV, 1922.

Whale, T. W. 'The principles of the Somerset Domesday'. *Transactions of the Bath Field Club.* Bath, 1902.

Wharton, H. *Anglia Sacra.* 2 vols. London, 1691.

Wheeler, W. H. *Report on the River Parrett proposed Improvements.* Boston, Lincs., 1896.

Whitehead, A. *The Improvement of the Somerset Level.* Weston-super-Mare, 1860.

Willard, J. F. *Parliamentary Taxes on Personal Property, 1290 to 1334.* Monograph of the Medieval Academy of America No. 9. Cambridge, Mass., 1934.

Williams, M. 'River diversions on the Parrett in the 17th and 18th centuries'. *Thirteenth Annual Report of the Somerset River Board* (Bridgwater, 1963), 55.

'The draining and reclamation of Meare Pool, Somerset'. *Thirteenth Annual Report of the Somerset River Board* (Bridgwater, 1963), 51.

'The draining and reclamation of the Somerset Levels, 1770–1833'. *Transactions of the Institute of British Geographers,* XXXIII (London, 1963), 163.

'Draining activity in the Somerset Levels since 1939'. *Geography,* XLIX (Sheffield, 1964), 387.

Wolfe-Barry, J. *Report...upon the Question of Tidal Gates or Sluices in the River Parrett.* London, 1920.

Woodward, H. B. *The Geology of the Somerset and the Bristol Coal-Fields.* Memoirs of the Geological Survey. H.M.S.O. London, 1876.

Woolrych, H. W. *A Treatise of the Law of Sewers including the Drainage Act.* 2nd ed. London, 1849.

Young, A. *A Farmer's Tour through the East of England.* 4 vols. London, 1771.

'A farming tour of the South and West of England'. *Annals of Agriculture,* XXX (London, 1798), 302.

INDEX

Index

Index

Index

Index

www.ingramcontent.com/pod-product-compliance
Ingram Content Group UK Ltd.
Pitfield, Milton Keynes, MK11 3LW, UK
UKHW040703180125
453697UK00010B/380